Lecture Notes in Mathematics

Edited by J.-M. Morel, F. Takens and B. Teissier

Editorial Policy
for the publication of monographs

1. Lecture Notes aim to report new developments in all areas of mathematics – quickly, informally and at a high level. Monograph manuscripts should be reasonably self-contained and rounded off. Thus they may, and often will, present not only results of the author but also related work by other people. They may be based on specialized lecture courses. Furthermore, the manuscripts should provide sufficient motivation, examples and applications. This clearly distinguishes Lecture Notes from journal articles or technical reports which normally are very concise. Articles intended for a journal but too long to be accepted by most journals, usually do not have this "lecture notes" character. For similar reasons it is unusual for doctoral theses to be accepted for the Lecture Notes series.

2. Manuscripts should be submitted (preferably in duplicate) either to one of the series editors or to Springer-Verlag, Heidelberg. In general, manuscripts will be sent out to 2 external referees for evaluation. If a decision cannot yet be reached on the basis of the first 2 reports, further referees may be contacted: the author will be informed of this. A final decision to publish can be made only on the basis of the complete manuscript, however a refereeing process leading to a preliminary decision can be based on a pre-final or incomplete manuscript. The strict minimum amount of material that will be considered should include a detailed outline describing the planned contents of each chapter, a bibliography and several sample chapters.
Authors should be aware that incomplete or insufficiently close to final manuscripts almost always result in longer refereeing times and nevertheless unclear referees' recommendations, making further refereeing of a final draft necessary.
Authors should also be aware that parallel submission of their manuscript to another publisher while under consideration for LNM will in general lead to immediate rejection.

3. Manuscripts should in general be submitted in English.
Final manuscripts should contain at least 100 pages of mathematical text and should include
– a table of contents;
– an informative introduction, with adequate motivation and perhaps some historical remarks: it should be accessible to a reader not intimately familiar with the topic treated;
– a subject index: as a rule this is genuinely helpful for the reader.

Continued on inside back-cover

Lecture Notes in Mathematics

Edited by J.-M. Morel, F. Takens and B. Teissier

Editorial Policy
for the publication of monographs

1. Lecture Notes aim to report new developments in all areas of mathematics – quickly, informally and at a high level. Monograph manuscripts should be reasonably self-contained and rounded off. Thus they may, and often will, present not only results of the author but also related work by other people. They may be based on specialized lecture courses. Furthermore, the manuscripts should provide sufficient motivation, examples and applications. This clearly distinguishes Lecture Notes from journal articles or technical reports which normally are very concise. Articles intended for a journal but too long to be accepted by most journals, usually do not have this "lecture notes" character. For similar reasons it is unusual for doctoral theses to be accepted for the Lecture Notes series.

2. Manuscripts should be submitted (preferably in duplicate) either to one of the series editors or to Springer-Verlag, Heidelberg. In general, manuscripts will be sent out to 2 external referees for evaluation. If a decision cannot yet be reached on the basis of the first 2 reports, further referees may be contacted: the author will be informed of this. A final decision to publish can be made only on the basis of the complete manuscript, however a refereeing process leading to a preliminary decision can be based on a pre-final or incomplete manuscript. The strict minimum amount of material that will be considered should include a detailed outline describing the planned contents of each chapter, a bibliography and several sample chapters.
Authors should be aware that incomplete or insufficiently close to final manuscripts almost always result in longer refereeing times and nevertheless unclear referees' recommendations, making further refereeing of a final draft necessary.
Authors should also be aware that parallel submission of their manuscript to another publisher while under consideration for LNM will in general lead to immediate rejection.

3. Manuscripts should in general be submitted in English.
Final manuscripts should contain at least 100 pages of mathematical text and should include
– a table of contents;
– an informative introduction, with adequate motivation and perhaps some
   historical remarks: it should be accessible to a reader not intimately familiar
   with the topic treated;
– a subject index: as a rule this is genuinely helpful for the reader.

Continued on inside back-cover

Lecture Notes in Mathematics　　　1791

Editors:
J.-M. Morel, Cachan
F. Takens, Groningen
B. Teissier, Paris

Springer
Berlin
Heidelberg
New York
Barcelona
Hong Kong
London
Milan
Paris
Tokyo

Manfred Knebusch
Digen Zhang

Manis Valuations
and Prüfer Extensions I

A New Chapter in Commutative Algebra

 Springer

Authors

Manfred Knebusch
Digen ZHANG

Department of Mathematics
University of Regensburg
Universitätstr. 31
93040 Regensburg
Germany

e-mail:
Manfred.Knebusch@mathematik.uni-regensburg.de
Digen.Zhang@mathematik.uni-regensburg.de

Cover: "A good mathematician needs no counting rod",
       Lao Tse in Dao De Jing, chapter 27

Cataloging-in-Publication Data applied for.

Die Deutsche Bibliothek - CIP-Einheitsaufnahme

Knebusch, Manfred:
Manis valuations and Prüfer extensions / Manfred Knebusch ; Digen Zhang. -
Berlin ; Heidelberg ; New York ; Barcelona ; Hong Kong ; London ; Milan ;
Paris ; Tokyo : Springer
1. A new chapter in commutative algebra. - 2002
  (Lecture notes in mathematics ; 1791)
  ISBN 3-540-43951-X

Mathematics Subject Classification (2000):
PRIMARY 13A15, 13A18, 13F05, SECONDARY 13B02, 14A05, 14P10, 14P05

ISSN 0075-8434
ISBN 3-540-43951-x Springer-Verlag Berlin Heidelberg New York

Springer-Verlag Berlin Heidelberg New York a member of BertelsmannSpringer
Science + Business Media GmbH

http://www.springer.de

© Springer-Verlag Berlin Heidelberg 2002
Printed in Germany

Typesetting: Camera-ready TeX output by the author

SPIN: 10884676   41/3142/ du - 543210 - Printed on acid-free paper

Contents

Chapter III: PM-valuations and valuations of weaker type 177

Appendix 251

Introduction

We feel that new chapters of commutative algebra are needed to cope with the extensions and derivates of algebraic geometry over prominent fields like $\mathbb{R}$, $\mathbb{Q}_p$, ... and abstract versions of such geometries. In particular, the real algebra associated with abstract semialgebraic geometry, say the theory of real closed spaces (cf. [Sch], [Sch$_1$]), is still very much in its infancy. The present book is meant as a contribution to such a commutative algebra.

Typically, the commutative rings naturally occurring in geometries as above are not noetherian, but in compensation have other properties, which make them manageable. These properties may look strange from the viewpoint of classical commutative algebra with its center in polynomial rings over fields, but can be quite beautiful for an eye accustomed to them.

Valuations seem to play a much more dominant role here than in classical commutative algebra. For example, if F is any ordered field, then all the convex subrings of F are Krull valuation rings with quotient field F for a trivial reason (cf. e.g. [BCR, p.249], [KS, p.55]), and this fact is of primary importance in real algebra. Many of the mysteries of real algebra seem to be related to the difficulty of controlling the value group and the residue class field of such convex rings. They are almost never noetherian.

Let us stay a little with real algebra. If F is a formally real field, it is well-known that the intersection H of the real valuation rings of F is a *Prüfer domain*, and that H has quotient field F. {A valuation ring is called real if its residue class field is formally real.} H is the so called *real holomorphy ring* of F, cf. [B, §2], [S], [KS, Chap.III §12]. If F is the function field $k(V)$ of an algebraic variety V over a real closed field k (e.g. $k = \mathbb{R}$), suitable overrings of H in R can tell us a lot about the algebraic and the semialgebraic geometry of $V(k)$.

These rings, of course, are again Prüfer domains. A very interesting and – in our opinion – still mysterious role is played by some of these rings which are related to the *orderings of higher level* of F, cf. e.g. [B$_2$], [B$_3$]. Here we meet a remarkable phenomenon. For orderings of level 1 (i.e. orderings in the classical sense) the usual procedure is to observe first that the convex subrings of ordered fields are valuation rings, and then to go on to Prüfer domains as intersections of such valuation rings, cf. e.g. [B], [S], [KS]. But for higher levels, up to now, the best method is to directly construct a Prüfer domain A in F from a "torsion preordering" of F, and then to obtain the valuation rings necessary for analyzing the preordering as localizations $A_{\mathfrak{p}}$ of A, cf. [B$_2$, p.1956 f], [B$_3$]. Thus there is two way traffic between valuations and Prüfer domains.

Up to now, less is done for the function field $F = k(V)$ of an algebraic variety V over a p-adically closed field k (e.g. $k = \mathbb{Q}_p$). But work of Kochen and Roquette (cf. §6 and §7 in the book [PR] by Prestel and Roquette) gives ample evidence, that also here Prüfer domains play a prominent role. In particular, every formally p-adic field F contains a "*p-adic holomorphy ring*", called the *Kochen ring*, in complete analogy with the formally real case [PR, §6]. Actually, the Kochen ring has been discovered and studied much earlier than the real holomorphy ring ([Ko], [R$_1$]).

If R is a commutative ring (with 1) and k is a subring of R, then we can still define a real holomorphy ring $H(R/k)$ consisting of those elements a of R which can be bounded by elements of k on the real spectrum of R (cf. [BCR], [B$_1$], [KS]). {If R is a formally real field F and k the prime ring of F, this ring coincides with the real holomorphy ring H from above.} These rings $H(R/k)$ have proven to be very useful in real semialgebraic geometry. In particular, N. Schwartz and M. Prechtel have used them for completing a real closed space and, more generally, to turn a morphism between real closed spaces into a proper one in a universal way ([Sch, Chap V, §7], [Pt]).

The algebra of these holomorphy rings turns out to be particularly good-natured if we assume that $1 + \Sigma R^2 \subset R^*$, i.e. that all elements $1 + a_1^2 + \cdots + a_n^2$ $(n \in \mathbb{N}, a_i \in R)$ are units in R. This is a natural condition in real algebra. The rings used by Schwartz and Prechtel, consisting of abstract semialgebraic functions, fulfill the condition automatically. More generally, if A is any commutative ring (always with 1) then the localization $S^{-1}A$ with respect to the multiplicative set $S = 1 + \Sigma A^2$ is a ring R fulfilling the condition, and R has the same real spectrum as A. Thus for many problems in real geometry we may replace A by R.

Now, V. Powers has proved that if $1 + \Sigma R^2 \subset R^*$, then the real holomorphy ring $H(R/k)$, with respect to any subring k, is an R-Prüfer ring, as defined by Griffin in 1973 [G$_2$].[*] More generally she proved that if $1 + \Sigma R^{2d} \subset R^*$, for some even number $2d$, then every subring A of R, containing the elements $\frac{1}{1+q}$ with $q \in \Sigma R^{2d}$, is R-Prüfer ([P, Th.1.7], cf. also [BP]).

An R-Prüfer ring is related to *Manis valuations* on R in much the same way a Prüfer domain is related to valuations of its quotient field. Why shouldn't we try to repeat the success story of Prüfer domains and real valuations on the level of relative Prüfer rings and Manis valuations? Already Marshall in his important paper [Mar] has followed such a program. He worked with "Manis places" in a ring R with $1 + \Sigma R^2 \subset R^*$, and related them to the points of the real spectrum Sper R.

We mention that Marshall's notion of Manis places is slightly misleading. According to his definition, these places do not correspond to Manis valuations, but to a broader class of valuations which we call "special valuations", cf.I, §1 below. But then V. Powers (and independently one of us, D.Z.) observed that in the case $1 + \Sigma R^2 \subset R^*$, the places of Marshall in fact do correspond to the Manis valuations of R [P]. {In Chapter I, §1 below we prove that every special valuation on R is Manis under a much weaker condition on R, cf. Theorem 1.1.}

The program to study Manis valuations and relative Prüfer rings in rings of real functions has gained new impetus and urgency from the fact, that the theory of orderings of higher level has recently been pushed from fields to rings, leading to *real spectra of higher level*. These spectra in turn have already proven to be useful for ordinary real semialgebraic geometry. We mention an opus magnum by Ralph Berr [Be], where spectra of higher level are used in a fascinating way to classify the singularities of real semialgebraic functions.

It seems that p-adic semialgebraic geometry is accessible as well. L. Bröcker and H.-J. Schinke [BS] have brought the theory of p-adic spectra to a rather satisfactory level by studying the "*L-spectrum*"

---

[*] The definition by Griffin needs a slight modification, cf. Def.1 in I, §5 below.

L-spec A of a commutative ring A with respect to a given non-archimedian local field L (e.g. $L = \mathbb{Q}_p$). There seems to be no major obstacle in sight which prevents us from defining and studying rings of semialgebraic functions on a constructible (or even proconstructible) subset X of L-spec A. Here "semialgebraic" means definable in a model-theoretic sense plus satisfying a suitable continuity condition. Relative Prüfer subrings of such rings should be quite interesting.

The present book is devoted to the study of relative Prüfer rings and Manis valuations, with an eye towards applications in real and p-adic geometry.

Currently, there already exists a rich theory of *"Prüfer rings with zero divisors"*, also started by Griffin [G₁], cf. the books [LM], [Huc], and the literature cited there. But this theory does not seem to be taylored to geometric needs. A Prüfer ring with zero divisors A is the same as an R-Prüfer ring with $R = \text{Quot } A$, the total quotient ring of A. While this is a reasonable notion from the viewpoint of ring theory, it may be artificial coming from a geometric direction. A typical situation in real geometry is the following: R is the ring of (continuous) semialgebraic functions on a semialgebraic set M over a real closed field k or, more generally, the set of abstract semialgebraic functions on a proconstructible subset X of a real spectrum (cf. [Sch], [Sch₁]). Although the ring R has very many zero divisors, we have experience that in some sense R behaves nearly as well as a field, cf. e.g. our notion of "convenient ring extensions" in I, §6 below. Now, if A is a subring of R, then it is natural and interesting from a geometric viewpoint to study the R-Prüfer rings $B \supset A$, while the total quotient rings $\text{Quot } A$ and $\text{Quot } B$ seem to bear little geometric relevance.

Except for a paper by P.L. Rhodes from 1991 [Rh], little seems to be done on Prüfer extensions in general,[*] and in the original paper of Griffin the proofs of important facts [G₂, Prop.6, Th.7] are omitted. Moreover, the paper by Rhodes has a gap in the proof of his main theorem. {[Rh, Th.2.1], condition (5b) is apparently not a characterization of Prüfer extensions. Any algebraic field extension

---

[*] The important work of Gräter [Gr-Gr₃] and of Alajbegović and Močkoř [Al-M] will be discussed in Part II of the book.

is a counterexample.} Thus we have been careful about a foundation of this theory.

In Chapter I we develop the basics of a theory of Prüfer extensions and give some examples. Then in Chapter II we explicate the multiplicative ideal theory related to Prüfer extensions. At the end of that chapter (§11, §12) we give a picture of what has been attained so far in the case of noetherian rings. Here life is much less demanding but also less interesting than in the general theory. Nevertheless the reader may get the idea where we have to go and to work in the general setting.

In Chapter III, the last one of the present volume, we take a closer look at Manis valuations. We single out the all-important subclass of PM-valuations (= Prüfer-Manis valuations) and study relations between PM-valuations and other Manis valuations.

Anything else, in particular a more thorough working on examples, has been left to Part II of the book. We cannot hope there to reach the level of sophistication nowadays present in the theory of Prüfer *domains* and documented for example by Fontana, Huckaba and Papick in the recent book [FHP], but what can be done will be enough to intrigue the persistent reader.

We have been forced to change some of the terminology used by ring theorists, say in the books of Larsen-McCarthy [LM] and of Huckaba [Huc]. While these authors mean by a valuation on a ring a Manis valuation, we use the word "valuation" in the much broader sense of Bourbaki [Bo, Chap.VI, §3]. It is common belief that Manis valuations are the right valuations for computations. But the central notion is the Bourbaki valuation, since only with these valuations one can build an honest spectral space, the valuation spectrum [HK]. Valuation spectra have already proven to be immensely useful both in algebraic geometry (cf. [HK]) and rigid analytic geometry (e.g. [Hu₁], [Hu₂]). The closely related real valuation spectra (cf. [Hu₃, §1]) seem to be the natural basic spaces for endeavours in real algebra concerning valuations and Prüfer extensions.

The common belief just mentioned has been modified in the present book. Manis valuations can show some pathologies, which are absent for the valuations occuring in Prüfer extensions. These are the PM-valuations mentioned above. We devote a good part of Chapter III to their study.

Fortunately, PM-valuations seem to suffice for an understanding of the major part of the commutative algebra we have in mind (see above). Thus our message is that in many situations the PM-valuations, not the more general Manis valuations, are the really useful ones. Nevertheless we also define "tight valuations" in Chapter III. They form a broader class than the PM-valuations but are still Manis. We introduce tight valuations in order to get a better understanding of the good nature of PM-valuations. We do not exclude the possibility that they deserve interest on their own.

In Chapter III we also construct and study various *valuation hulls*. Given a ring A and a prime ideal $\mathfrak{p}$ of A we find a unique maximal ring extension $A \subset C$, such that there exists a (unique) PM-valuation v on C with $A = \{x \in C \mid v(x) \geq 0\}$ and $\mathfrak{p} = \{x \in C \mid v(x) > 0\}$. We call C the *PM-hull* of the pair $(A, \mathfrak{p})$. Similarly we construct a "TV-hull" of $(A, \mathfrak{p})$, which does the same for tight valuations instead of PM-valuations. Finally, in a more restricted setting, we construct a "Manis valuation hull" of $(A, \mathfrak{p})$.

These valuation hulls may serve as a good illustration, beyond Prüfer extensions, that it can be interesting to study valuations on rings instead of fields. Every valuation $v \colon R \to \Gamma \cup \infty$ on a ring R can be interpreted as a valuation $\hat{v} \colon k(\mathfrak{q}) \to \Gamma \cup \infty$ on the residue class field $k(\mathfrak{q})$ of a certain prime ideal $\mathfrak{q}$ of R, namely the support of v (cf. I, §1 below). Thus valuations on rings are in some sense nothing new, compared to valuations on fields. But the point of view is different. If we take the ring R into account instead of the field $k(\mathfrak{q})$, we have the possibility to ask new questions and to look for new objects, such as for example valuation hulls.

Some notations. In this book a "ring" always means a commutative ring with 1. For a ring A, we denote the group of units of A by A^*. We denote the total quotient ring of A by $\operatorname{Quot} A$. For $\mathfrak{p}$ a prime ideal of A we denote the field $\operatorname{Quot}(A/\mathfrak{p})$ by $k(\mathfrak{p})$.

$\mathbb{N} = \{1, 2, 3, \dots\}$, $\mathbb{N}_0 = \mathbb{N} \cup \{0\}$. If A and B are sets, then $A \subset B$ means that A is a subset of B, and $A \subsetneq B$ means that A is a proper subset of B. If two subsets M and N of some set X are given, then $M \setminus N$ denotes the complement of $M \cap N$ in M. The abbreviation "iff" means "if and only if".

Summary

We call a commutative ring extension $A \subset R$ Prüfer, if A is an R-Prüfer ring in the sense of Griffin (Can. J. Math. 26 (1974)). These ring extensions relate to Manis valuations in much the same way as Prüfer domains relate to Krull valuations. In the Introduction we tried to explain why Prüfer extensions and Manis valuations deserve attention from a geometric viewpoint. In Chapter I we develop a basic theory of Prüfer extensions and give some examples. Then in Chapter II we explicate the multiplicative ideal theory related to Prüfer extensions. Finally, in Chapter III we take a closer look at Manis valuations. We single out the all-important subclass of PM-valuations (= Prüfer-Manis valuations) and study relations between PM-valuations and other Manis valuations.

Prüfer extensions may be viewed as families of PM-valuations. This viewpoint will dominate Part II of the book.

An earlier version of the Introduction and Chapter I has been published in the electronic journal Documenta Mathematica 1 (1996), 149-197.

Acknowledgements. We thank Roland Huber and Niels Schwartz for many helpful comments, and Rosi Bonn for very efficiently typing countless versions of the manuscript.

Summary

Chapter I:

Basics on Manis valuations and Prüfer extensions

Summary:

In §1 and §2 we gather what we need about Manis valuations. Then in §3 and §4 we develop an auxiliary theory of "weakly surjective" ring homomorphisms. These form a class of epimorphisms in the category of commutative rings close to the flat epimorphisms studied by D. Lazard and others in the sixties, cf. [L], [Sa$_1$], [A]. In §5 the up to then independent theories of Manis valuations and weakly surjective homomorphisms are brought together to study Prüfer extensions. {We call a ring extension $A \subset R$ Prüfer, if A is R-Prüfer in the sense of Griffin.} It is remarkable that, although Prüfer extensions are defined in terms of Manis valuations (cf. §5, Def.1 below), they can be characterized entirely in terms of weak surjectivity. Namely, a ring extension $A \subset R$ is Prüfer iff every subextension $A \subset B$ is weakly surjective (cf. Th.5.2 below). A third way to characterize Prüfer extensions is by multiplicative ideal theory, as we will explicate in Chapter II.

Our first major result on Prüfer extensions is Theorem 5.2 giving various characterizations of these extensions which sometimes make it easy to recognize a given ring extension as Prüfer, cf. the examples in §6. We then establish various permanence properties of the class of Prüfer extensions. For example we prove for Prüfer extensions $A \subset B$ and $B \subset C$ that $A \subset C$ is again Prüfer (Th.5.6), a result already due to Rhodes [Rh].

At the end of §5 we prove that any commutative ring A has a universal Prüfer extension $A \subset P(A)$ which we call the *Prüfer hull* of A. Every other Prüfer extension $A \hookrightarrow R$ can be embedded into $A \hookrightarrow P(A)$ in a unique way. The Prüfer rings with zero divisors are just the rings A with $P(A)$ containing the total quotient ring Quot A. Prüfer hulls mean new territory leading to many new open questions. We will pursue some of them in part II of the book.

In §6 we prove theorems which give us various examples of Manis valuations and Prüfer extensions. We illustrate how naturally they come up in algebraic geometry over a field k which is not algebraically closed (§6, Example 7, Th.6.13, Th.6.17), and in real algebraic and semialgebraic geometry (§6, Examples 3 and 13). Perhaps our best result here is Theorem 6.16 giving a far-reaching generalization of an old lemma by A. Dress (cf. [D, Satz 2′]). This lemma states for F a field, in which -1 is not a square, that the subring of F generated by the elements $1/(1 + a^2)$, $a \in F$, is Prüfer in F. Dress's innocent looking lemma seems to have inspired generations of real algebraists (cf. e.g. [La, p.86], [KS, p.163]) and also ring theorists, cf. [Gi₁].

We finally prove in §7 for various Prüfer extensions $A \subset R$ that, if $\mathfrak{a}$ is a finitely generated A-submodule of R with $R\mathfrak{a} = R$, then some power $\mathfrak{a}^d$ (with d specified) is principal. Our main result here (Theorem 7.8) is a generalization of a theorem by P. Roquette [R, Th.1] which states this for R a field (cf. also [Gi₁]). Roquette used his theorem to prove by general principles that the Kochen ring of a formally p-adic field is Bezout [loc.cit]. Similar applications should be possible in p-adic semialgebraic geometry. Roquette's paper has been an inspiration for our whole work since it indicates well the ubiquity of Prüfer domains in algebraic geometry over a non algebraically closed field.

§1 Valuations on rings

Let R be a ring and Γ an (additive) totally ordered abelian group. We extend Γ to an ordered monoid $\Gamma \cup \infty := \Gamma \cup \{\infty\}$ by the rules $\infty + x = x + \infty = \infty$ for all $x \in \Gamma \cup \infty$ and $x < \infty$ for all $x \in \Gamma$. We use the notations $\Gamma_+ := \{x \in \Gamma \mid x \geq 0\}$ and $\Gamma_- := \{x \in \Gamma, x \leq 0\}$.

Definition 1 (Bourbaki [Bo, VI. 3.1]).
A *valuation on R* with values in Γ is a map $v: R \to \Gamma \cup \infty$ such that:

(1) $v(xy) = v(x) + v(y)$ for all $x, y \in R$.
(2) $v(x + y) \geq \min\{v(x), v(y)\}$ for all $x, y \in R$.
(3) $v(1) = 0$ and $v(0) = \infty$.

If $v(R) = \{0, \infty\}$ then v is said to be *trivial*, otherwise v is called *non-trivial*.

We recall some very basic facts[1] about valuations on rings and fix notations.

Let $v\colon R \to \Gamma \cup \infty$ be a valuation on R.

The subgroup of Γ generated by $v(R) \setminus \{\infty\}$ is called the *value group* of v and is denoted by Γ_v. The set $v^{-1}(\infty)$ is a prime ideal of R. It is called the *support* of v and is denoted by $\operatorname{supp} v$. v induces a valuation $\hat{v}\colon k(\operatorname{supp} v) \to \Gamma \cup \infty$ on the quotient field $k(\operatorname{supp} v)$ of $R/\operatorname{supp} v$. We denote the valuation ring of $k(\operatorname{supp} v)$ corresponding to $\hat{v}$ by $\mathfrak{o}_v$, its maximal ideal by $\mathfrak{m}_v$, and its residue class field by $\kappa(v)$, $\kappa(v) := \mathfrak{o}_v/\mathfrak{m}_v$. Notice that $\hat{v}(\mathfrak{o}_v) = (\Gamma_v)_+ \cup \{\infty\}$.

We further denote the set $\{x \in R \mid v(x) \geq 0\}$ by A_v and the set $\{x \in R \mid v(x) > 0\}$ by $\mathfrak{p}_v$. Clearly A_v is a subring of R and $\mathfrak{p}_v$ is a prime ideal of A_v. We call A_v the *valuation ring* of v and $\mathfrak{p}_v$ the *center* of v.

Definition 2. Two valuations v, w on R are said to be *equivalent*, in short, $v \sim w$, if the following equivalent conditions are satisfied:

(1) There is an isomorphism $f\colon \Gamma_v \cup \{\infty\} \to \Gamma_w \cup \{\infty\}$ of ordered monoids with $w(x) = f(v(x))$ for all $x \in R$.
(2) $v(a) \geq v(b) \Longleftrightarrow w(a) \geq w(b)$ for all $a, b \in R$.
(3) $\operatorname{supp} v = \operatorname{supp} w$ and $\mathfrak{o}_v = \mathfrak{o}_w$.

By abuse of language we will often regard equivalent valuations as "equal".

Definition 3. a) The *characteristic subgroup* $c_v(\Gamma)$ of Γ with respect to v is the smallest convex subgroup of Γ (convex with respect to the total ordering of Γ) which contains all elements $v(x)$ with $x \in R$, $v(x) \leq 0$. Clearly $c_v(\Gamma)$ is the set of all $\gamma \in \Gamma$ such that $v(x) \leq \gamma \leq -v(x)$ for some $x \in R$ with $v(x) \leq 0$.
b) v is called *special*,[2] if $c_v(\Gamma_v) = \Gamma_v$. (We replaced Γ by Γ_v.)

If H is any convex subgroup of Γ containing $c_v(\Gamma)$ then we obtain from v a new valuation $v|H\colon R \to \Gamma_\infty$ putting $(v|H)(x) = v(x)$ if $v(x) \in H$ and $v(x) = \infty$ else. Taking $H = c_v\Gamma$ we obtain from v a special valuation $w = v|c_v\Gamma$. Notice that $A_w = A_v$, $\mathfrak{p}_w = \mathfrak{p}_v$.

---

[1] For this we refer to [Bo, VI.3.1] and [HK, §1]

[2] The word "special" alludes to the fact that such a valuation has no proper primary specialization in the valuation spectrum of R, cf. [HK, §1].

Definition 4 (cf. [M]). v is called a *Manis valuation* on R, if $v(R) = \Gamma_v \cup \infty$.[3]

Manis valuation will be in the focus of the present paper. Notice that every Manis valuation is special, but that the converse is widely false.

Example. Let R be the polynomial ring $k[x]$ in one variable x over some field k. Consider the valuation $v\colon R \to \mathbb{Z} \cup \infty$ with $v(f) = -\deg f$ for any $f \in R \setminus \{0\}$. This valuation is special but definitely not Manis.   □

One of our primary observations is that nevertheless there are many interesting rings, on which every special valuation is Manis. For example this holds if for every $x \in R$ the element $1 + x^2$ is a unit in R. More generally we have the following theorem.

Theorem 1.1. Let k be a subring of R. Assume that for every $x \in R \setminus k$ there exists some monic polynomial $F(T) \in k[T]$ (one variable T) with $F(x) \in R^*$. Then every special valuation v on R with $A_v \supset k$ is Manis.

Proof. We may assume that v is non trivial. Let $x \in R$ be given with $v(x) \neq 0, \infty$. We have to find some $y \in R$ with $v(y) = -v(x)$. Since v is special there exists some $a \in R$ with $v(ax) < 0$. Let $F(T) = T^d + c_1 T^{d-1} + \cdots + c_d$ be a polynomial with $c_1, \ldots, c_d \in k$ and $F(ax) \in R^*$. Since $v(ax) < 0$, but $v(c_i) \geq 0$ for $i = 1, \ldots, d$, we have $v(F(ax)) = dv(ax)$. The element $y := \frac{a^d x^{d-1}}{F(ax)}$ does the job.[4]   □

We return to valuations in general. Up to the end of this section we will keep the following

Notations. $v\colon R \to \Gamma \cup \infty$ *is a valuation on some ring* R, $A := A_v$, $\mathfrak{p} := \mathfrak{p}_v$, $\mathfrak{q} := \operatorname{supp} v$, $\bar{R} := R/\mathfrak{q}$, $\bar{A} := A/\mathfrak{q}$, $\bar{\mathfrak{p}} := \mathfrak{p}/\mathfrak{q}$. $\pi\colon R \to \bar{R}$ *is the evident epimorphism from* R *to* $\bar{R}$. *We have a unique valuation* $\bar{v}\colon \bar{R} \to \Gamma \cup \infty$ *on* $\bar{R}$ *such that* $\bar{v} \circ \pi = v$.

---

[3] Since we often identify equivalent valuations we have slightly altered the definition in [M]. Manis demands that $v(R){=}\Gamma\cup\infty$.

[4] We are indebted to Roland Huber for this simple argument.

We have $A_{\bar{v}} = \bar{A}$, $\mathfrak{p}_{\bar{v}} = \bar{\mathfrak{p}}$, $\operatorname{supp}\bar{v} = \{0\}$, $\Gamma_{\bar{v}} = \Gamma_v$, $\mathfrak{o}_v = \mathfrak{o}_{\bar{v}}$. It is evident that v is special iff $\bar{v}$ is special, and that v is Manis iff $\bar{v}$ is Manis. Looking at the valuation $\hat{v}$ on the quotient field $k(\mathfrak{q})$ of $\bar{R}$ (which extends $\bar{v}$) one now obtains by an easy exercise

Proposition 1.2.
  a) v is Manis iff $k(\mathfrak{q}) = \bar{R} \cdot \mathfrak{o}_v^*$.
  b) v is special iff $k(\mathfrak{q}) = \bar{R} \cdot \mathfrak{o}_v$.
Here $\bar{R} \cdot \mathfrak{o}_v^*$ (resp. $\bar{R} \cdot \mathfrak{o}_v$) denotes the set of products xy with $x \in \bar{R}$, $y \in \mathfrak{o}_v^*$ (resp. $\mathfrak{o}_v$). The set $\bar{R} \cdot \mathfrak{o}_v$ is also the subring of $k(\mathfrak{q})$ generated by $\bar{R}$ and $\mathfrak{o}_v$. $\qquad\square$

Definition 5. v is called *local* if the pair $(A, \mathfrak{p})$ is local, i.e. $\mathfrak{p}$ is the unique maximal ideal of A.

Proposition 1.3 (cf. [G₂, Prop. 5]). The following are equivalent.
  i) v is Manis and local.
  ii) The pair $(R, \mathfrak{q})$ is local.
  iii) v is local and $\mathfrak{q}$ is a maximal ideal of R.

Proof. i) $\Rightarrow$ ii): Let $x \in R \setminus \mathfrak{q}$ be given. Since v is Manis there exists some $y \in R$ with $v(xy) = 0$. Since v is local this implies that xy is a unit of A, hence also a unit of R. Thus x is a unit of R.
ii) $\Rightarrow$ i): $\bar{v}$ is a valuation of the field $\bar{R}$. Thus $\bar{v}$ is Manis, which implies that v is Manis. Let $x \in A \setminus \mathfrak{p}$ be given. Then x is a unit in R. We have $v(x^{-1}) = -v(x) = 0$. Thus $x^{-1} \in A$, $x \in A^*$.
i), ii) $\Rightarrow$ iii): trivial.
iii) $\Rightarrow$ i): $\bar{v}$ is a valuation of the field $\bar{R}$. From this we conclude again that v is Manis. $\qquad\square$

If S is any multiplicative subset of R with $S \cap \mathfrak{q} = \emptyset$ then we denote by v_S the unique "extension" of v to a valuation on $S^{-1}R$, defined by

$$v_S\left(\frac{a}{s}\right) = v(a) - v(s) \qquad (a \in R, s \in S).$$

For $w = v_S$ we have $\Gamma_w = \Gamma_v$ and $c_w(\Gamma) \supset c_v(\Gamma)$. Thus if v is Manis then v_S is Manis and if v is special then v_S is special. v_S has the support $S^{-1}\mathfrak{q}$.

We now consider the special case $S = A \setminus \mathfrak{p}$. Then

$$v_S\left(\frac{a}{s}\right) = v(a) \qquad (a \in R, s \in S).$$

Thus for $w = v_S$ we now have $A_w = S^{-1}A = A_\mathfrak{p}$ and $\mathfrak{p}_w = S^{-1}\mathfrak{p} = \mathfrak{p}_\mathfrak{p}$, and we see that v_S is a local valuation. Moreover $A \setminus \mathfrak{p}$ is the smallest saturated multiplicative subset S of R such that v_S is local. We write $S^{-1}R = R_\mathfrak{p}$.

Definition 6. The valuation v_S with $S = A \setminus \mathfrak{p}$ is called *the localization of v*, and is denoted by $\tilde{v}$.

We have $\tilde{v}(R_\mathfrak{p}) = v(R)$, $\Gamma_{\tilde{v}} = \Gamma_v$, $c_v\Gamma = c_{\tilde{v}}\Gamma$. Thus v is Manis iff $\tilde{v}$ is Manis, and v is special iff $\tilde{v}$ is special. Applying Proposition 3 [5] to $\tilde{v}$ we obtain

Proposition 1.4. The following are equivalent.
  i) v is Manis.
 ii) $\mathfrak{q}$ is the unique ideal of R which is maximal among all ideals of R which do not meet $A \setminus \mathfrak{p}$.
iii) $\mathfrak{q}$ is maximal among all ideals of R which do not meet $A \setminus \mathfrak{p}$. $\square$

If S is a (non empty) multiplicative subset of R then we denote by $\mathrm{Sat}_R(S)$ the set of all elements of R which divide some element of S ("saturum of S in R"). Recall from basic commutative algebra that, if T is a second multiplicative subset of R, then $S^{-1}R = T^{-1}R$ iff $\mathrm{Sat}_R(S) = \mathrm{Sat}_R(T)$.

The following characterization of Manis valuations can be deduced from Proposition 4, but we will give an independent proof.

Proposition 1.5. The following are equivalent.
  i) v is Manis.
 ii) $\mathrm{Sat}_R(A \setminus \mathfrak{p}) = R \setminus \mathfrak{q}$.
iii) $R_\mathfrak{p} = R_\mathfrak{q}$.

Proof. The multiplicative set $R \setminus \mathfrak{q}$ is saturated. Thus the equivalence ii) $\Longleftrightarrow$ iii) is evident from what has been said above.

---

[5] Reference to Prop.1.3 in this section. In later sections we will refer to this proposition as "Prop.1.3", instead of "Prop.3".

i) $\Longleftrightarrow$ ii): v is Manis $\Longleftrightarrow$ For every $x \in R\backslash\mathfrak{q}$ there exists some $y \in R$ with $v(x) + v(y) = 0$, i.e. with $xy \in A \setminus \mathfrak{p} \Longleftrightarrow R \setminus \mathfrak{q} = \mathrm{Sat}_R(A \setminus \mathfrak{p})$. $\qquad\square$

Proposition 1.6. If v is Manis then $\mathfrak{o}_v = \bar{A}_{\bar{\mathfrak{p}}}$.

Proof. We may pass from v to $\bar{v}$. Thus we assume without loss of generality that $\mathfrak{q} = 0$. We have $\mathfrak{o}_v = \mathfrak{o}_{\tilde{v}}$ and v is Manis iff $\tilde{v}$ is Manis. Thus we may assume without loss of generality that v is also local. Now R is a field (cf. Prop. 3), and $\mathfrak{o}_v = A = A_{\mathfrak{p}}$. $\qquad\square$

Definition 7. We say that v *has maximal support* if $\mathfrak{q}$ is a maximal ideal of R.

Proposition 1.7. v has maximal support iff $\bar{v}$ is local and Manis. Then v is also a Manis valuation on R.

Proof. If v has maximal support, then $\bar{v}$ is a valuation on the field $\bar{R}$. Thus $\bar{v}$ is certainly Manis and local. Since $\bar{v}$ is Manis, also v is Manis.

If $\bar{v}$ is local and Manis then, applying Proposition 3 to $\bar{v}$, we learn that the pair $(\bar{R}, \{0\})$ is local. This means that $\mathfrak{q}$ is a maximal ideal of R. $\qquad\square$

Definition 8. An additive subgroup M of R is called v-*convex*, if for any elements $x \in M$, $y \in R$ with $v(x) \leq v(y)(\leq v(0) = \infty)$ it follows that $y \in M$.

If M is a v-convex additive subgroup of R, then certainly $ax \in M$ for any $a \in A$, $x \in M$, i.e. M is an A-submodule of R. We now have a closer look at the v-convex ideals of A.

Clearly $\mathfrak{q}$ is a v-convex ideal of A and is contained in any other v-convex ideal of A. Also $\mathfrak{p}$ is v-convex and $I \subset \mathfrak{p}$ for every v-convex ideal $I \neq A$.

Proposition 1.8. If v has maximal support then every A-submodule of R containing $\mathfrak{q}$ is v-convex.

Proof. Let I be an A-submodule of R containing $\mathfrak{q}$, and $\bar{I} := I/\mathfrak{q}$. It is easy to see that I is v-convex iff $\bar{I}$ is $\bar{v}$-convex. Since v has maximal support, $\bar{v}$ is a valuation on the field $\bar{R} := R/\mathfrak{q}$. From classical valuation theory we conclude that $\bar{I}$ is $\bar{v}$-convex. □

Corollary 1.9. If v is a local Manis valuation then every A-submodule of R containing $\mathfrak{q}$ is v-convex.

Proof. By Proposition 3 we know that v has maximal support. □

Proposition 1.10. [M, Prop. 3]. Assume that the valuation v is Manis. Then a prime ideal $\mathfrak{r}$ of A is v-convex iff $\mathfrak{q} \subset \mathfrak{r} \subset \mathfrak{p}$.

Proof. Replacing v by $\bar{v}$ we assume without loss of generality that $\mathfrak{q} = 0$. Since $v(A \setminus \mathfrak{p}) = \{0\}$ it is evident that the v-convex prime ideals $\mathfrak{r}$ of A correspond uniquely with the $\tilde{v}$-convex prime ideals $\mathfrak{r}'$ of $A_\mathfrak{p}$ via $\mathfrak{r}' = \mathfrak{r}_\mathfrak{p}$. Thus we may pass from v to $\tilde{v}$ and assume without loss of generality that v is local. All prime ideals (in fact, all ideals) of A are v-convex (Cor. 9). □

Proposition 1.11. Assume that v is a non-trivial Manis valuation. The following are equivalent.

i) Every ideal I of A with $\mathfrak{q} \subset I \subset \mathfrak{p}$ is v-convex.
ii) Any two ideals I, J of A with $\mathfrak{q} \subset I \subset \mathfrak{p}$ and $\mathfrak{q} \subset J \subset \mathfrak{p}$ are comparable by inclusion.
iii) $\bar{A}$ is a (Krull)valuation domain.
iv) $\mathfrak{p}$ is the unique maximal ideal of A containing $\mathfrak{q}$.
v) v has maximal support.
vi) Every ideal I of A containing $\mathfrak{q}$ is v-convex.
vii) $IR = R$ for every ideal $I \supsetneqq \mathfrak{q}$ of A.

Proof. For proving the equivalence of the conditions i) – vi) we find it convenient to assume that $\mathfrak{q} = \{0\}$, which we can do without loss of generality. Now R is an integral domain.

i) $\Rightarrow$ ii) is evident, since for any two v-convex ideals I and J of A we have $I \subset J$ or $J \subset I$. (This holds more generally for v-convex additive subgroups I, J of R.)

ii) $\Rightarrow$ iii): We verify: If $x \in A$, $y \in A$ then $Ax \subset Ay$ or $Ay \subset Ax$. This will imply that A is a valuation domain. We assume without loss of generality that $v(x) \leq v(y)$. If $x \in \mathfrak{p}$ then also $y \in \mathfrak{p}$. The ideals

Ax and Ay are comparable by our assumption ii). There remains the case that $x \notin \mathfrak{p}$. We choose an element $c \neq 0$ in $\mathfrak{p}$. Then $xc \in \mathfrak{p}$ and $v(xc) \leq v(yc)$. As we have proved this implies $Ayc \subset Axc$ or $Axc \subset Ayc$. Since R is a domain we conclude that $Ay \subset Ax$ or $Ax \subset Ay$.

iii) $\Longrightarrow$ iv): Trivial. iv) $\Longrightarrow$ v) is evident by Proposition 7, and v) $\Longrightarrow$ vi) is evident by Proposition 8. Clearly vi) $\Rightarrow$ i).

v) $\Rightarrow$ vii): IR is an ideal of R with $\mathfrak{q} \subsetneq IR$. Thus $IR = R$.
vii) $\Rightarrow$ v): Let $x \in R \setminus \mathfrak{q}$ be given. We verify that $\mathfrak{q} + Rx = R$. Then we will know that $\mathfrak{q}$ is a maximal ideal of R. Now $\mathfrak{q} + Rx$ is not contained in A, since $\mathfrak{q}$ is the conductor of A in R. We choose $y \in R$ with $\mathfrak{q} + yx \notin A$, hence $yx \notin A$. Then we choose $z \in \mathfrak{p}$ with $zyx \in A \setminus \mathfrak{p}$. This is possible since v is Manis. $\mathfrak{q} + Azyx$ is not contained in $\mathfrak{p}$. Thus certainly $\mathfrak{q} + Azyx \subsetneq \mathfrak{q}$. By our assumption vii) we conclude that $\mathfrak{q} + Rzyx = R$. A fortiori $\mathfrak{q} + Rx = R$.   □

Definition 9. A valuation $w: R \to \Gamma' \cup \infty$ is called *coarser* than v (or a *coarsening* of v) if there exists an order preserving homomorphism[6] $f: \Gamma_v \to \Gamma_w$ such that, for all $x \in R$, $w(x) = f(v(x))$ (put $f(\infty) = \infty$).

If H is a convex subgroup of Γ then the quotient Γ/H is a totally ordered abelian group in such a way that the natural projection from Γ to Γ/H is an order preserving homomorphism. We have $(\Gamma/H)_+ = (\Gamma_+ + H)/H$. From v we obtain a coarsening $w: R \to (\Gamma/H) \cup \infty$ putting $w(x) := x + H$ for all $x \in R$. (Read $\infty + H = \infty$.) This valuation w is denoted by v/H.

Remarks 1.12. a) v/H has the center $\mathfrak{p}_H := \{x \in R \mid v(x) > H\}$, and this is a v-convex prime ideal of A. $\{v(x) > H$ means $v(x) > \gamma$ for every $\gamma \in H\}$. If $\Gamma_+ \subset v(R)$ (e.g. v is Manis and $\Gamma = \Gamma_v$) then the v-convex prime ideals $\mathfrak{r}$ of A correspond uniquely with the convex subgroups H of Γ via $\mathfrak{r} = \mathfrak{p}_H$. {If necessary, we denote the ideal $\mathfrak{p}_H$ more precisely by $\mathfrak{p}_{v,H}$.}

b) Assume (without loss of generality) that $\Gamma = \Gamma_v$. The coarsenings w of v correspond, up to equivalence, uniquely with the convex subgroups H of Γ via $w = v/H$. We have $A \subset A_w$, $\mathfrak{p} \supset \mathfrak{p}_w$, $\operatorname{supp} w = \mathfrak{q}$,

---

[6] This means f is a homomorphism of abelian groups with $f(\alpha) \geq f(\beta)$ if $\alpha \geq \beta$. The homomorphism f is necessarily surjective.

$\hat{w} = \hat{v}/H$, $\bar{w} = \bar{v}/H$, $\tilde{w} = (\tilde{v}/H)^{\sim}$. If S is a multiplicative subset of R with $S \cap \mathfrak{q} = \emptyset$ then $v_S/H = (v/H)_S$. If v is special then v/H is special. If v is Manis then v/H is Manis.

All this is either trivial or can be verified in a straightforward way.

How do we obtain the ring A_w from $A_v = A$ if $w = v/H$? In order to give a satisfactory answer, at least in special cases, we need a definition which will be widely used also later on.

Definition 10. Let B be a subring of R, let S be a multiplicative subset of B and let $j_S: R \to S^{-1}R$ denote the localization map $x \mapsto \frac{x}{1}$ of R with respect to S. For any B-submodule M of R we define

$$M_{[S]} := j_S^{-1}(S^{-1}M).$$

Clearly $M_{[S]}$ is the set of all $x \in R$ such that $sx \in M$ for some $s \in S$. We call $M_{[S]}$ the *saturation of M (in R) by S*.[7] In the case $S = B \setminus \mathfrak{r}$ with $\mathfrak{r}$ a prime ideal of B we usually write $j_\mathfrak{r}$ and $M_{[\mathfrak{r}]}$ instead of j_S, $M_{[S]}$.

Notice that $B_{[S]}$ is a subring of R and $M_{[S]}$ is a $B_{[S]}$-submodule of R. If M is an ideal of B then $M_{[S]}$ is an ideal of $B_{[S]}$. If M is a prime ideal of B with $M \cap S = \emptyset$ then $M_{[S]}$ is a prime ideal of $B_{[S]}$.

Proposition 1.13. Let S be a multiplicative subset of $A \setminus \mathfrak{q}$, and let H denote the convex subgroup of Γ generated by $v(S)$, i.e. the smallest convex subgroup of Γ containing $v(S)$. Let $w := v/H$ and $\mathfrak{r} := \mathfrak{p}_H$. Then

$$A_w = A_{[S]} = A_{[\mathfrak{r}]},$$
$$\mathfrak{p}_w = \mathfrak{r} = \{x \in R \mid v(x) > v(S)\}.$$

Proof. We already stated above that $\mathfrak{p}_w = \mathfrak{p}_H = \mathfrak{r}$. This ideal coincides with the set of all $x \in R$ with $v(x) > v(S)$. It is evident that $A_{[S]} \subset A_w$. Let now $x \in A_w$ be given. There exists some element $\gamma \in H_+$ with $v(x) \geq -\gamma$, and some element $s \in S$ with

---

[7] $M_{[S]}$ is called the "S-component of M" in [LM]. To be more precise, we sometimes write $M_{[S]}^R$ instead of $M_{[S]}$.

$\gamma \leq v(s)$. We obtain $v(xs) \geq 0$, i.e. $xs \in A$. This proves that $A_w = A_{[S]}$. We have $S \subset A \setminus \mathfrak{r}$, thus $A_{[S]} \subset A_{[\mathfrak{r}]}$. Let $x \in A_{[\mathfrak{r}]}$ be given. We choose $y \in A \setminus \mathfrak{r}$ with $xy \in A$. There exists some $\gamma \in H_+$ with $v(y) \leq \dot{\gamma}$ and some $s \in S$ with $\gamma \leq v(s)$. We have

$$0 \leq v(x) + v(y) \leq v(x) + v(s) = v(sx).$$

Thus $sx \in A$, $x \in A_{[S]}$. This proves $A_{[S]} = A_{[\mathfrak{r}]}$. $\square$

Remark. If v is Manis then a converse to Proposition 13 holds, cf. Theorem 2.6.ii below.

Corollary 1.14. Assume that $\Gamma_+ \subset v(R)$ (e.g. v Manis and $\Gamma_v = \Gamma$). Let H be a convex subgroup of Γ, $w := v/H$ and $\mathfrak{r} := \mathfrak{p}_H$. We have $A_w = A_{[\mathfrak{r}]}$ and $\mathfrak{p}_w = \mathfrak{r}$.

Proof. Apply Prop. 13 to the set $S := \{x \in R \mid v(x) \in H_+\}$. $\square$

Proposition 1.15. Let I be an A-submodule of R with $\mathfrak{q} \subset I$. Assume that v is Manis. Then I is v-convex iff $I = I_{[\mathfrak{p}]}$.

Proof. Assume first that I is v-convex. We have $I \subset I_{[\mathfrak{p}]}$. Let $x \in I_{[\mathfrak{p}]}$ be given. We choose $d \in A \setminus \mathfrak{p}$ with $dx \in I$. We have $v(x) = v(dx)$. Since I is v-convex this implies $x \in I$. Thus $I = I_{[\mathfrak{p}]}$.

Assume now that $I = I_{[\mathfrak{p}]}$. This means $I = j_{\mathfrak{p}}^{-1}(I_{\mathfrak{p}})$ with $j_{\mathfrak{p}}$ the localization map from R to $R_{\mathfrak{p}}$. As always, let $\tilde{v}: R_{\mathfrak{p}} \to \Gamma \cup \infty$ denote the localization of v. We have $A_{\tilde{v}} = A_{\mathfrak{p}}$, $\operatorname{supp} \tilde{v} = \mathfrak{q}_{\mathfrak{p}}$. Since $\tilde{v}$ is local and Manis, every $A_{\mathfrak{p}}$-submodule of $R_{\mathfrak{p}}$ containing $\mathfrak{q}_{\mathfrak{p}}$ is $\tilde{v}$-convex (Cor.9). In particular $I_{\mathfrak{p}}$ is $\tilde{v}$-convex. Since $I = j_{\mathfrak{p}}^{-1}(I_{\mathfrak{p}})$ and $v = \tilde{v} \circ j_{\mathfrak{p}}$, we conclude that I is v-convex. $\square$

We briefly discuss a process of restriction, which gives us special valuations on subrings of R.

Let B be a subring of R. The restriction $u = v|B: B \to \Gamma \cup \infty$ of the map $v: R \to \Gamma \cup \infty$ is a valuation on B. Let $\Delta := c_u(\Gamma)$ and $w := u|\Delta$. Then $w: B \to \Delta \cup \infty$ is a special valuation on B.

Definition 11. We call w the *special restriction* of v to B, and denote this valuation by $v|_B$.

For $w = v|_B$ we have $A_w = A \cap B$, $\mathfrak{p}_w = \mathfrak{p} \cap B$, $\operatorname{supp} w \supset \mathfrak{q} \cap B$. Notice also that $v|_B = (v|c_v\Gamma)|_B$. Thus in essence our restriction process deals with special valuations.

In the case that v is Manis the question arises, under which conditions on B the special restriction $v|_B$ is again Manis. We need an easy lemma.

Lemma 1.16. If $v \colon R \to \Gamma \cup \infty$ is special and $(\Gamma_v)_+ \subset v(R)$, then v is Manis.

Proof. This is a consequence of Proposition 2. By that proposition $k(\mathfrak{q}) = \bar{R}\mathfrak{o}_v$. From $(\Gamma_v)_+ \subset v(R) = \bar{v}(\bar{R})$ we conclude that $\mathfrak{o}_v \subset \bar{R}\mathfrak{o}_v^*$, hence $k(\mathfrak{q}) = \bar{R}\mathfrak{o}_v^*$, and this means that v is Manis. □

Proposition 1.17. Assume that v is Manis and that B is a subring of R containing $\mathfrak{p} = \mathfrak{p}_v$. Then the special restriction $v|_B \colon B \to \Delta \cup \infty$ of v is again Manis. If v is surjective (i.e. $\Gamma = \Gamma_v$) then $v|_B$ is surjective.

Proof. We assume without loss of generality that v is surjective. Let $u := v|B$ and $w := v|_B$. Let $\gamma \in \Delta$ be given with $\gamma > 0$. There exists some $a \in \mathfrak{p}_v$ with $v(a) = \gamma$. Since $\mathfrak{p}_v \subset B$ we have $a \in B$, hence $v(a) = u(a) = w(a)$. {Recall that for any $x \in B$ with $u(x) \in \Delta$ we have $w(x) = u(x)$.} This proves that $\Delta_+ \subset w(B)$. Now Lemma 16 tells us that w is Manis. □

Scholium 1.18. Let $v \colon R \to \Gamma \cup \infty$ be a Manis valuation and H a convex subgroup of Γ. Let $w := v/H$ and $B := A_w$. We have[8]

$$A_w = \{x \in R \mid v(x) \geq h \quad \text{for some} \quad h \in H\} =: A_H$$
$$\mathfrak{p}_w = \{x \in R \mid v(x) > h \quad \text{for all} \quad h \in H\} =: \mathfrak{p}_H.$$

Let $v_H \colon B \to \Delta \cup \infty$ denote the special restriction $v|_B$ of v. Here $\Delta = c_{v|B}(\Gamma) \subset H$. v_H has support $\mathfrak{p}_H$, hence gives us a Manis

---

[8] If necessary, we will write more precisely $A_{v,H}$ and $\mathfrak{p}_{v,H}$ instead of A_H and $\mathfrak{p}_H$.

valuation $\overline{v_H}: A_H/\mathfrak{p}_H \to \bar{\Delta} \cup \infty$ of support zero. If v is surjective then $\Delta = H$.

The proof of all this is a straightforward exercise. Later we will prove a converse to these statements (Prop. 2.8).

Using Lemma 16 from above we can prove a converse to Proposition 6.

Proposition 1.19. Assume that the valuation v on R is special and that $\mathfrak{o}_v = \bar{A}_{\bar{\mathfrak{p}}}$ (cf. notations above). Then v is Manis.

Proof. Replacing A by $\bar{A} = A/\mathfrak{q}$ and v by $\bar{v}$ we assume without loss of generality that $\mathfrak{q} = 0$. Now R is an integral domain, and $A \subset R \subset K$ with K the quotient field of R. We also assume without loss of generality that $\Gamma = \Gamma_v$. The valuation $v: R \to \Gamma \cup \infty$ extends to the valuation $\hat{v}: K \twoheadrightarrow \Gamma \cup \infty$, and $\hat{v}$ has the valuation ring $\mathfrak{o}_v$. We have $v(A \setminus \mathfrak{p}) = \{0\}$, hence $v(A) = \hat{v}(A_\mathfrak{p}) = \hat{v}(\mathfrak{o}_v) = \Gamma_+$. By Lemma 16 we conclude that v is Manis. $\qquad\square$

We turn to a construction of "direct limits" of Manis valuations. This construction will not be used seriously before Chapter III, §5, thus may safely be skipped by the reader until then.

Let $(B_i \mid i \in I)$ be an inductive system of subrings of a given ring R, i.e. I is a partially ordered index set, such that for any two indices $i, j \in I$ there exists some $k \in I$ with $i \leq k$, $j \leq k$, and $B_i \subset B_j$ if $i \leq j$. Assume that for every $i \in I$ there is given a surjective Manis valuation $v_i: B_i \twoheadrightarrow \Gamma_i \cup \infty$, such that $v_j|_{B_i}$ is equivalent to v_i if $i < j$. Then we construct a valuation v on the ring $C := \bigcup_{i \in I} B_i$ as follows.

Construction 1.20. If $i < j$ we have a unique monomorphism $\sigma_{ji}: \Gamma_i \to \Gamma_j$ of ordered abelian groups with convex image $\sigma_{ji}(\Gamma_i) \subset \Gamma_j$, such that $v_j(x) = \sigma_{ji} \circ v_i(x)$ for every $x \in B_i$ with $v_i(x) \neq \infty$. The σ_{ji} fit together to an inductive system $(\Gamma_i \mid i \in I)$ of ordered abelian groups with transition maps σ_{ji} for $i < j$. Let $\Gamma := \varinjlim \Gamma_i$.

This is again an ordered abelian group and comes with order preserving monomorphisms $\sigma_i: \Gamma_i \to \Gamma$, such that $\sigma_i = \sigma_j \circ \sigma_{ji}$ if $i < j$. The subgroups $\sigma_i(\Gamma_i)$ are convex in Γ.

We define a map $v: C \to \Gamma \cup \infty$ as follows. Let $x \in C$. If $x \notin A$, we choose some $i \in I$ with $x \in B_i$ and put $v(x) := \sigma_i(v_i(x))$. If $x \in A$ and $v_i(x) = \infty$ for every $i \in I$, we put $v(x) = \infty$. Otherwise we choose some $i \in I$ with $v_i(x) \neq \infty$ and put again $v(x) := \sigma_i(v_i(x))$. One easily verifies that v is a well defined valuation with support $\operatorname{supp} v = \bigcap_{i \in I} \operatorname{supp} v_i$ and $v(C \setminus \operatorname{supp} v) = \Gamma$. Thus v is Manis. One further checks that $v|_{B_i} = \sigma_i \circ v_i$ for every $i \in I$. Notice that v and all the v_i have the same valuation ring $A_v = A_{v_i}$ and the same center $\mathfrak{p}_v = \mathfrak{p}_{v_i}$. $\qquad\qquad\square$

§2 Valuation subrings and Manis pairs

As before let R be a ring (commutative, with 1).

Definition 1. a) A *valuation subring of R* is a subring A of R such that there exists some valuation $v: R \to \Gamma \cup \infty$ with $A = A_v$. A *valuation pair in R* (also called "R-valuation pair") is a pair $(A, \mathfrak{p})$ consisting of a subring A of R and a prime ideal $\mathfrak{p}$ of A such that $A = A_v$, $\mathfrak{p} = \mathfrak{p}_v$ for some valuation v of R.
b) We speak of a *Manis subring A of R* and a *Manis pair $(A, \mathfrak{p})$ in R* respectively if here v can be chosen as a Manis valuation of R.

Two bunches of questions come to mind immediately. 1) How can a valuation subring or a Manis subring of R be characterized ring theoretically? Ditto for pairs.
2) How far is a valuation v determined by the associated ring A_v or pair $(A_v, \mathfrak{p}_v)$?

As stated in §1 the pair $(A_v, \mathfrak{p}_v)$ does not change if we pass from v to the associated special valuation $v|c_v\Gamma$. Thus we will concentrate on special valuations.

If $A = R$ then a special valuation v with $A_v = A$ must be trivial, and any prime ideal $\mathfrak{p}$ of R occurs as the center (= support) of such a valuation v. The valuation v is completely determined by $(R, \mathfrak{p})$ and is Manis. These pairs $(R, \mathfrak{p})$ are called the *trivial Manis pairs* in R.

If $A \neq R$ and A is a valuation subring of R then clearly $R \setminus A$ is a multiplicatively closed subset of R. P. Samuel started an investigation of such subrings of R. We quote one of his very remarkable results.

Definition 2. Let A be a subring of R with $A \neq R$ and $S := R \setminus A$ multiplicatively closed. We define the following subsets $\mathfrak{p}_A$ and $\mathfrak{q}_A$ of A. $\mathfrak{p}_A$ is the set of all $x \in A$ such that there exists some $s \in S$ with $sx \in A$, and $\mathfrak{q}_A$ is the set of all $x \in A$ with $sx \in A$ for all $s \in R \setminus A$.

Clearly $\mathfrak{q}_A \subset \mathfrak{p}_A$. Also $\mathfrak{q}_A = \{x \in R \mid rx \in A \text{ for all } r \in R\}$. Thus $\mathfrak{q}_A$ is the biggest ideal of R contained in A, called the *conductor* of A in R.[1]

Theorem 2.1. [Sa, Th.1 and Th.2]. Let A be a proper subring of R with $R \setminus A$ multiplicatively closed.
 i) $\mathfrak{p}_A$ is a prime ideal of A and $\mathfrak{q}_A$ is a prime ideal both of A and R.
 ii) A is integrally closed in R.
 iii) If R is a field then A is a valuation domain, and R is the quotient field of A. $\qquad\square$

If v is a special nontrivial valuation then the support of v is determined by the ring A_v alone. More precisely we have the following proposition, whose proof is an easy exercise.

Proposition 2.2. Let v be a non-trivial valuation on R and $A := A_v$. Then $\mathfrak{q}_A \supset \operatorname{supp} v$. The valuation v is special iff $\mathfrak{q}_A = \operatorname{supp} v$. $\qquad\square$

We cannot expect that a special valuation v is determined up to equivalence by the pair $(A, \mathfrak{p}) := (A_v, \mathfrak{p}_v)$, as is already plausible from the example in §1. (We will give a counterexample later, III.7.2.) But this holds if v is Manis. Indeed, if v is also non-trivial, then we see from Prop. 2 and Prop.1.6 that $\mathfrak{o}_v = \bar{A}_{\bar{\mathfrak{p}}}$ with $\bar{A} = A/\mathfrak{q}_A$, $\bar{\mathfrak{p}} = \mathfrak{p}/\mathfrak{q}_A$. Even more is true. The following proposition implies that v is determined up to equivalence by A alone. The proof is again an easy exercise.

Proposition 2.3. Let v be a non-trivial valuation on R and $A := A_v$. Then $\mathfrak{p}_A \subset \mathfrak{p}_v$. If v is Manis then $\mathfrak{p}_A = \mathfrak{p}_v$. $\qquad\square$

---

[1] If necessary, we more precisely write $\mathfrak{p}_A^R, \mathfrak{q}_A^R$ instead of $\mathfrak{p}_A, \mathfrak{q}_A$.

We have the following important characterization of Manis pairs.

Theorem 2.4 ([M, Prop. 1], or [Huc, Th. 5.1]). Let A be a subring of R and $\mathfrak{p}$ a prime ideal of A. The following are equivalent.

  i) $(A, \mathfrak{p})$ is a Manis pair in R.
  ii) If B is a subring of R and $\mathfrak{q}$ a prime ideal of B with $A \subset B$ and $\mathfrak{q} \cap A = \mathfrak{p}$ then $A = B$. [2]
  iii) For every $x \in R \setminus A$ there exists some $y \in \mathfrak{p}$ with $xy \in A \setminus \mathfrak{p}$. $\square$

For the proof we refer to the literature [loc.cit.]. There also exists a satisfying characterization of the valuation subrings of R in ring theoretic terms, due to Samuel and Griffin [e.g.Huc, Th.5.5], but we do not need this here.

We give a characterization of local Manis pairs (i.e. Manis pairs corresponding to local Manis valuations) in a classical style.

Theorem 2.5. Let $A \subset R$ be a ring extension, $A \neq R$.
i) The following are equivalent
(1) Every $x \in R \setminus A$ is a unit in R and $x^{-1} \in A$.
(2) A has a unique maximal ideal $\mathfrak{p}$ (hence is local) and $(A, \mathfrak{p})$ is Manis in R.
ii) If (1), (2) hold, then R is a local ring with maximal ideal $\mathfrak{q} := \mathfrak{q}_A$, and $A_\mathfrak{q} = R_\mathfrak{p} = R$. Moreover, $\mathfrak{p} = \mathfrak{q} \cup \{x^{-1} | x \in R \setminus A\}$.

Proof. Assume that (1) holds. Then $R \setminus A$ is closed under mulptiplication. Indeed, let $x, y \in R \setminus A$ be given. Then $(xy)y^{-1} \in R \setminus A$, but $y^{-1} \in A$, hence $xy \in R \setminus A$. We introduce the prime ideals $\mathfrak{p} := \mathfrak{p}_A$ and $\mathfrak{q} := \mathfrak{q}_A$ (cf. Def. 2). If $\mathfrak{M}$ is any maximal ideal of R then $\mathfrak{M} \cap (R \setminus A) = \emptyset$, since $R \setminus A \subset R^*$. Thus $\mathfrak{M} \subset A$. It follows that $\mathfrak{M}$ is contained in the conductor $\mathfrak{q}$ of A in R, and we conclude that $\mathfrak{M} = \mathfrak{q}$. Thus $\mathfrak{q}$ is the only maximal ideal of R. Let K denote the field $R/\mathfrak{q}$ and $\overline{A}$ the subring $A/\mathfrak{q}$ of K. For every $z \in K \setminus \overline{A}$ the inverse z^{-1} is contained in $\overline{A}$. Thus $\overline{A}$ is a valuation domain with quotient field K. We conclude that A is Manis in R, and then, that $(A, \mathfrak{p})$ is a Manis pair in R (cf. Prop. 3). Since $(R, \mathfrak{q})$ is local we learn from Proposition 1.3 that $(A, \mathfrak{p})$ is local.

---

[2] In [M] and [Huc] it is not assumed that $\mathfrak{q}$ is a prime ideal. It can be proved easily that their condition can be changed to our condition (ii).

Now assume that (2) holds. We know from Proposition 1.3 that R is local with maximal ideal $\mathfrak{q} := \mathfrak{q}_A$. Thus $R \setminus A \subset R \setminus \mathfrak{q} = R^*$. Since $(A, \mathfrak{p})$ is Manis in R we have $x^{-1} \in \mathfrak{p} \subset A$ for every $x \in R \setminus A$, and it is also clear that $\mathfrak{p} = \mathfrak{q} \cup \{x^{-1} | x \in R \setminus A\}$.

Since $A \setminus \mathfrak{q} \subset R^*$, we have $A_{\mathfrak{q}} \subset R$. If $x \in R \setminus A$ then $x = \frac{1}{y}$ with $y \in A \setminus \mathfrak{q}$. Thus $x \in A_{\mathfrak{q}}$. This proves that $A_{\mathfrak{q}} = R$. Since $A \setminus \mathfrak{p} = A^*$, it is trivial that $R_{\mathfrak{p}} = R$. $\qquad\qquad\square$

Let $v : R \longrightarrow \Gamma \cup \infty$ and w be valuations on R. We have called w coarser than v if w is equivalent to v/H for some convex subgroup H of v (§1, Def. 9 and Remark 1.12). How can the coarsening relation be expressed in terms of the pairs $(A_v, \mathfrak{p}_v)$, $(A_w, \mathfrak{p}_w)$ if both v and w are Manis?

Theorem 2.6 (cf. [M, Prop.4] for a weaker statement). Assume that $v : R \longrightarrow \Gamma \cup \infty$ and w are two non-trivial Manis valuations of R, and (without loss of generality) that $\Gamma = \Gamma_v$.

i) The following are equivalent:
(1) w is coarser than v.
(2) $\mathrm{supp}\,(v) = \mathrm{supp}\,(w)$ and $\mathfrak{o}_v \subset \mathfrak{o}_w$.
(3) $A_v \subset A_w$ and $\mathfrak{p}_w \subset \mathfrak{p}_v$.
(4) $\mathfrak{p}_w$ is an ideal of A_v contained in $\mathfrak{p}_v$.

ii) Let $A := A_v$, $\mathfrak{p} := \mathfrak{p}_v$, and let $\mathfrak{r}$ be a prime ideal of A with $\mathrm{supp}\,v \subset \mathfrak{r} \subset \mathfrak{p}$. There exists a convex subgroup H of Γ with $\mathfrak{r} = \mathfrak{p}_H$ and $v(A \setminus \mathfrak{r}) = H_+$. For the valuation $w := v/H$ we have $\mathfrak{p}_w = \mathfrak{r}$ and $A_w = A_{[\mathfrak{r}]} = A_H$.[3] Also $A_w = \{x \in R \mid x\mathfrak{r} \subset \mathfrak{r}\}$.

Proof: i): (1) $\Longleftrightarrow$ (2): We may assume in advance that $\mathrm{supp}\,v = \mathrm{supp}\,w$. It is now evident that w is coarser than v iff $\hat{w}$ is coarser than $\hat{v}$. By classical valuation theory this holds iff the valuation ring $\mathfrak{o}_v$ of $\hat{v}$ is contained in $\mathfrak{o}_w$.

(2) $\Longrightarrow$ (3): Replacing R by $R/\mathrm{supp}\,v$ we assume without loss of generality that $\mathrm{supp}\,v = \mathrm{supp}\,w = \{0\}$. In the quotient field K of R we have $\mathfrak{o}_v \cap R = A_v$, $\mathfrak{o}_w \cap R = A_w$, $\mathfrak{m}_v \cap R = \mathfrak{p}_v$ and $\mathfrak{m}_w \cap R = \mathfrak{p}_w$. By assumption $\mathfrak{o}_v \subset \mathfrak{o}_w$. This implies $\mathfrak{m}_v \supset \mathfrak{m}_w$. We conclude that $A_v \subset A_w$ and $\mathfrak{p}_v \supset \mathfrak{p}_w$.

---

[3] Recall the notations from 1.12 and 1.18.

(3) $\Longrightarrow$ (2): We verify first that $\operatorname{supp}(v) = \operatorname{supp}(w)$. We know that $\operatorname{supp}(v) = \{x \in R \mid xR \subset A_v\}$ and $\operatorname{supp}(w) = \{x \in R \mid xR \subset A_w\}$ (cf. Proposition 2). Using the assumption $A_v \subset A_w$ we conclude $\operatorname{supp} v \subset \operatorname{supp} w$. Since v, w are Manis valuations, it is also evident that $\operatorname{supp}(v) = \{x \in R \mid xR \subset \mathfrak{p}_v\}$ and $\operatorname{supp}(w) := \{x \in R \mid xR \subset \mathfrak{p}_w\}$. Using the assumption $\mathfrak{p}_v \supset \mathfrak{p}_w$ we conclude that $\operatorname{supp} v \supset \operatorname{supp} w$. Thus indeed $\operatorname{supp}(v) = \operatorname{supp}(w)$.

In order to prove that $\mathfrak{o}_v \subset \mathfrak{o}_w$ we may replace R by $R/\operatorname{supp} v$. Thus we may assume that $\operatorname{supp} v = \operatorname{supp} w = \{0\}$. Now we know from Proposition 1.6 that $\mathfrak{o}_v = (A_v)_{\mathfrak{p}_v}$ and $\mathfrak{o}_w = (A_w)_{\mathfrak{p}_w}$. The inclusions $A_v \subset A_w$ and $\mathfrak{p}_v \supset \mathfrak{p}_w$ imply that $\mathfrak{o}_v \subset \mathfrak{o}_w$.

(3) $\Longrightarrow$ (4): trivial.

(4) $\Longrightarrow$ (3): Since w is Manis we have $A_w = \{x \in R \mid x\mathfrak{p}_w \subset \mathfrak{p}_w\}$. Now $\mathfrak{p}_w$ is an ideal of A_v. Thus $A_v \subset A_w$.

ii): We know from Prop.1.10 that the ideal $\mathfrak{r}$ is v-convex. Thus there exists a unique convex subgroup H of Γ such that $\mathfrak{r} = \{x \in R \mid v(x) > H\} = \mathfrak{p}_H$. {Here we use that $\Gamma = \Gamma_v$.} Since $v(A) = \Gamma_+ \cup \infty$, it follows that $v(A \setminus \mathfrak{r}) = H_+$. For $w := v/H$ we have $\mathfrak{p}_w = \mathfrak{p}_H$ and $A_w = A_H$ (cf.1.18). Since $\mathfrak{r} = \mathfrak{p}_w$ and w is Manis it is obvious that A_w is the set of all $x \in R$ with $x\mathfrak{r} \subset \mathfrak{r}$.

It remains to prove that $B := A_w$ coincides with $A_{[\mathfrak{r}]}$. Let $x \in A_{[\mathfrak{r}]}$ be given. We choose some $d \in A \setminus \mathfrak{r}$ with $dx \in A$. Since $A \subset A_w$ and $\mathfrak{r} = \mathfrak{p}_w$, we have $w(dx) \geq 0$, $w(d) = 0$, hence $w(x) \geq 0$, i.e. $x \in B$. This proves that $A_{[\mathfrak{r}]} \subset B$. Let now $x \in B$ be given. Suppose that $x \notin A_{[\mathfrak{r}]}$. Since $x \notin A$ there exists some $x' \in \mathfrak{p}$ with $xx' \in A \setminus \mathfrak{p} \subset A \setminus \mathfrak{r} \subset A$. Since $x \notin A_{[\mathfrak{r}]}$ we have $x' \in \mathfrak{r}$. Thus $x\mathfrak{r} \not\subset \mathfrak{r}$. This is a contradiction, since $\mathfrak{r}$ is an ideal of B and $x \in B$. Thus $x \in A_{[\mathfrak{r}]}$. We have proved $B = A_{[\mathfrak{r}]}$. $\qquad\square$

Corollary 2.7. Let $v : R \to \Gamma \cup \infty$ be a Manis valuation and $A := A_v$, $\mathfrak{p} = \mathfrak{p}_v$. The coarsenings w of v correspond uniquely, up to equivalence, with the prime ideals $\mathfrak{r}$ of A between $\operatorname{supp} v$ and $\mathfrak{p}$ via $\mathfrak{r} = \mathfrak{p}_w$. Also $A_{[\mathfrak{r}]} = A_w$.

Proof. If v is trivial then $\operatorname{supp} v = \mathfrak{p}$, and all assertions are evident. Assume now that v is not trivial. For the trivial coarsening t of v we have $\mathfrak{p}_t = \operatorname{supp} t = \operatorname{supp} v$ and $A_{[\mathfrak{p}_t]} = R$. If w is a non-trivial coarsening of v then $\mathfrak{p}_w$ is an ideal of A with $\operatorname{supp} v \subsetneqq \mathfrak{p}_w \subset \mathfrak{p}$ (cf. Th.6.i). This ideal is prime in A since it is prime in the ring $A_w \supset A$. Conversely, if $\mathfrak{r}$ is a prime ideal of A with $\operatorname{supp} v \subsetneqq \mathfrak{r} \subset \mathfrak{p}$ then, by Theorem 6.ii, there exists a coarsening w of v with $\mathfrak{p}_w = \mathfrak{r}$, $A_w = A_{[\mathfrak{r}]}$, and w is not trivial. Finally, if w and w' are two nontrivial coarsenings of v with $\mathfrak{p}_w = \mathfrak{p}_{w'} = \mathfrak{r}$, then $A_w = \{x \in R \mid x\mathfrak{r} \subset \mathfrak{r}\} = A_{w'}$, and we learn from (3) in Theorem 6.i (or by a direct argument) that $w \sim w'$. $\qquad\square$

We establish a converse to the construction 1.18.

Proposition 2.8. Let w be a non-trivial Manis valuation on R and u a Manis valuation on $A_w/\mathfrak{p}_w$. Let A and $\mathfrak{p}$ denote the preimages of A_u and $\mathfrak{p}_u$ in A_w under the natural homomorphism $\varphi: A_w \to A_w/\mathfrak{p}_w$.

i) $(A, \mathfrak{p})$ is a Manis pair in R iff $\operatorname{supp} u = \{0\}$.
ii) If this holds, let $v: R \longrightarrow\!\!\!\!\!\rightarrow \Gamma \cup \infty$ be a surjective valuation with $A_v = A$, $\mathfrak{p}_v = \mathfrak{p}$. Then Γ has a convex subgroup H, uniquely determined by w and u, such that w is equivalent to v/H. The valuation u is equivalent to $\overline{v_H}$ (cf. 1.18).

Proof. We have $\mathfrak{p}_w \subset \mathfrak{p} \subset A \subset A_w \subset R$.

a) We assume that $\operatorname{supp} u = \{0\}$ and prove that the pair $(A, \mathfrak{p})$ is Manis in R. Let $x \in R \setminus A$ be given. By Theorem 4 we are done if we find some $y \in \mathfrak{p}$ with $xy \in A \setminus \mathfrak{p}$.

Case 1: $x \in A_w$. Since $\varphi(x) \notin A_u$ there exists some $y \in \mathfrak{p}$ with $\varphi(x)\varphi(y) \in A_u \setminus \mathfrak{p}_u$, hence $xy \in A \setminus \mathfrak{p}$.

Case 2: $x \in R \setminus A_w$. Since w is Manis there exists some $y \in \mathfrak{p}_w$ with $xy \in A_w \setminus \mathfrak{p}_w$. We have $\varphi(xy) \neq 0$. Since u has support zero there exists some $z \in A_w$ with $\varphi(xy)\varphi(z) \in A_u \setminus \mathfrak{p}_u$, hence $xyz \in A \setminus \mathfrak{p}$. Clearly $yz \in \mathfrak{p}_w \subset \mathfrak{p}$.

b) Assume now that $(A, \mathfrak{p})$ is Manis in R, and that $v: R \longrightarrow\!\!\!\!\!\rightarrow \Gamma \cup \infty$ is a surjective valuation with $A_v = A$, $\mathfrak{p}_v = \mathfrak{p}$. We verify that u has support zero and prove the second part of the proposition. Since w is not trivial, we know from Theorem 6 that w is a coarsening of v. There is a unique convex subgroup H of Γ with $w \sim v/H$, and $A_w = A_H$, $\mathfrak{p}_w = \mathfrak{p}_H$ (notations from 1.18). We obtain from v and H a Manis valuation $v_H: A_w \longrightarrow\!\!\!\!\!\rightarrow H \cup \infty$ with support $\mathfrak{p}_w$, as explained in 1.18. The pair associated to v_H is $(A, \mathfrak{p})$. Thus $v_H \sim u \circ \varphi$ and $\overline{v_H} \sim u$. In particular $\operatorname{supp} u = \operatorname{supp} \overline{v_H} = \{0\}$. $\qquad\square$

We now consider the following situation: *A is a subring of R and* $\mathfrak{p}$ *is a prime ideal of A.* We are looking for criteria that the pair $(A_{[\mathfrak{p}]}, \mathfrak{p}_{[\mathfrak{p}]})$ (cf. §1, Def. 10) is Manis.

We need an easy lemma.

Lemma 2.9. a) $R_{\mathfrak{p}} = R_{(\mathfrak{p}_{[\mathfrak{p}]})}$.
b) If M is an A-submodule of R then $M_{\mathfrak{p}} = (M_{[\mathfrak{p}]})_{\mathfrak{p}_{[\mathfrak{p}]}}$.
c) If M is an A-submodule of R and $\mathfrak{r}$ is a prime ideal of A contained in $\mathfrak{p}$, then

$$M_{[\mathfrak{r}]} = (M_{[\mathfrak{p}]})_{[\mathfrak{r}_{[\mathfrak{p}]}]}.$$

Proof. We have $R_{\mathfrak{p}} = S^{-1}R$ and $R_{(\mathfrak{p}_{[\mathfrak{p}]})} = T^{-1}R$ with $S = A \setminus \mathfrak{p}$, $T = A_{[\mathfrak{p}]} \setminus \mathfrak{p}_{[\mathfrak{p}]}$. Notice that $S \subset T$. Let $x \in T$ be given. Choose some $d \in S$ with $dx \in A$. Then $dx \in A \setminus \mathfrak{p} = S$. This proves that $\operatorname{Sat}_R(S) = \operatorname{Sat}_R(T)$, and we conclude that $S^{-1}R = T^{-1}R$.

If M is an A-submodule of R, then $M_{[\mathfrak{p}]}$ is an $A_{[\mathfrak{p}]}$-submodule of R, and $M_{\mathfrak{p}} = S^{-1}M$, $(M_{[\mathfrak{p}]})_{\mathfrak{p}_{[\mathfrak{p}]}} = T^{-1}M_{[\mathfrak{p}]}$. Clearly $S^{-1}M \subset T^{-1}M_{[\mathfrak{p}]}$. (N.B. Both are subsets of $S^{-1}R = T^{-1}R$.) Also $T^{-1}M_{[\mathfrak{p}]} = S^{-1}M_{[\mathfrak{p}]}$. Let $z \in S^{-1}M_{[\mathfrak{p}]}$ be given. Write $z = \frac{x}{s}$ with $x \in M_{[\mathfrak{p}]}$, $s \in S$. We choose some $d \in S$ with $dx = m \in M$. We have $z = \frac{m}{sd} \in M_{\mathfrak{p}}$. This proves part b) of the lemma. The last statement c) follows from the obvious equality $M_{\mathfrak{r}} = (M_{\mathfrak{p}})_{\mathfrak{r}_{\mathfrak{p}}}$ by taking preimages under the various localization maps. $\qquad\square$

Proposition 2.10. $(A_{[\mathfrak{p}]}, \mathfrak{p}_{[\mathfrak{p}]})$ is a Manis pair in R iff $(A_{\mathfrak{p}}, \mathfrak{p}_{\mathfrak{p}})$ is a Manis pair in $R_{\mathfrak{p}}$. In this case, if $(A_{[\mathfrak{p}]}, \mathfrak{p}_{[\mathfrak{p}]})$ comes from the Manis valuation v on R, then $(A_{\mathfrak{p}}, \mathfrak{p}_{\mathfrak{p}})$ comes from the localization $\tilde{v}$ of v defined in §1 (Def. 6). {Recall from the lemma that $A_{\mathfrak{p}} = (A_{[\mathfrak{p}]})_{\mathfrak{p}_{[\mathfrak{p}]}}, \mathfrak{p}_{\mathfrak{p}} = (\mathfrak{p}_{[\mathfrak{p}]})_{\mathfrak{p}_{[\mathfrak{p}]}}$.} With $\mathfrak{q} := A \cap \operatorname{supp} v$ we have $\operatorname{supp} v = \mathfrak{q}_{[\mathfrak{p}]}, \operatorname{supp} \tilde{v} = \mathfrak{q}_{\mathfrak{p}}$.

Proof. a) Assume first that there exists a Manis valuation $v: R \to \Gamma \cup \infty$ with $A_v = A_{[\mathfrak{p}]}, \mathfrak{p}_v = \mathfrak{p}_{[\mathfrak{p}]}$. Let $\tilde{v}: R_{\mathfrak{p}_v} \to \Gamma \cup \infty$ denote the localization of v. Then $\tilde{v}$ is again Manis and $A_{\tilde{v}} = (A_v)_{\mathfrak{p}_v}, \mathfrak{p}_{\tilde{v}} = (\mathfrak{p}_v)_{\mathfrak{p}_v}, \operatorname{supp} \tilde{v} = (\operatorname{supp} v)_{\mathfrak{p}_v}$ (cf. §1). By part a) of the lemma above we have $R_{\mathfrak{p}_v} = R_{\mathfrak{p}}$, $A_{\tilde{v}} = A_{\mathfrak{p}}, \mathfrak{p}_{\tilde{v}} = \mathfrak{p}_{\mathfrak{p}}$. Let $\mathfrak{q} := A \cap \operatorname{supp} v$. Certainly $\mathfrak{q}_{[\mathfrak{p}]} \subset \operatorname{supp} v$. Let $x \in \operatorname{supp} v$ be given. We have $x \in A_v = A_{[\mathfrak{p}]}$. We choose some $d \in A \setminus \mathfrak{p}$ with $dx \in A$. Then $v(dx) = \infty$, thus $dx \in A \cap \operatorname{supp} v = \mathfrak{q}$, $x \in \mathfrak{q}_{[\mathfrak{p}]}$. This proves $\operatorname{supp} v = \mathfrak{q}_{[\mathfrak{p}]}$. Using part b) of the lemma we obtain $\operatorname{supp} \tilde{v} = \mathfrak{q}_{\mathfrak{p}}$.

b) Assume finally that $w: R_{\mathfrak{p}} \to \Gamma \cup \infty$ is a Manis valuation with $A_w = A_{\mathfrak{p}}$, $\mathfrak{p}_w = \mathfrak{p}_{\mathfrak{p}}$. Let $j_T: R \to R_{\mathfrak{p}}$ denote the localization map of R with respect to $T := A_{[\mathfrak{p}]} \setminus \mathfrak{p}_{[\mathfrak{p}]}$. Let v denote the valuation $w \circ j_T$ on R. We have $v(T) = \{0\}$. Thus $v(R) = w(R_{\mathfrak{p}}) = \Gamma_w$, and we conclude that v is Manis. Also $A_v = j_T^{-1}(A_w) = A_{[\mathfrak{p}]}$, $\mathfrak{p}_v = j_T^{-1}(\mathfrak{p}_w) = \mathfrak{p}_{[\mathfrak{p}]}$, and w coincides with the localization $\tilde{v}$ of v. $\qquad\square$

Proposition 2.11. Let $\mathfrak{r}$ be a prime ideal of A contained in $\mathfrak{p}$. Assume that $v: R \to \Gamma \cup \infty$ is a valuation with $A_v = A_{[\mathfrak{p}]}$, $\mathfrak{p}_v = \mathfrak{p}_{[\mathfrak{p}]}$, $A \cap \operatorname{supp} v \subset \mathfrak{r}$. Let H denote the convex subgroup of Γ generated by $v(A \setminus \mathfrak{r})$ and let $w := v/H$. Then $A_w = A_{[\mathfrak{r}]}$, $\mathfrak{p}_w = \mathfrak{r}_{[\mathfrak{p}]} = \mathfrak{r}_{[\mathfrak{r}]}$. Thus, if $(A_{[\mathfrak{p}]}, \mathfrak{p}_{[\mathfrak{p}]})$ is a Manis pair in R the same holds for $(A_{[\mathfrak{r}]}, \mathfrak{r}_{[\mathfrak{r}]})$.

Proof. By the last statement in Prop. 10 we have $\operatorname{supp} v \subset \mathfrak{r}_{[\mathfrak{p}]}$. It follows from Proposition 1.13 and part c) of lemma 9 above that $A_w = A_{[\mathfrak{r}]}$, $\mathfrak{p}_w = \mathfrak{r}_{[\mathfrak{p}]}$. It is evident that $\mathfrak{r}_{[\mathfrak{p}]} \subset \mathfrak{r}_{[\mathfrak{r}]} \subset \mathfrak{p}_w$. Thus $\mathfrak{r}_{[\mathfrak{p}]} = \mathfrak{r}_{[\mathfrak{r}]}$. $\qquad\square$

We state a criterion which will play a key role for the theory of Prüfer extensions in §5.

Theorem 2.12. Assume that A is integrally closed in R. The following are equivalent.

i) $(A_{[\mathfrak{p}]}, \mathfrak{p}_{[\mathfrak{p}]})$ is a Manis pair in R.
ii) For each $x \in R$ there exists some polynomial $F[T] \in A[T] \setminus \mathfrak{p}[T]$ with $F(x) = 0$.

Proof. i) $\Rightarrow$ ii): We first consider the case that $x \in A_{[\mathfrak{p}]}$. We choose some $s \in A \setminus \mathfrak{p}$ with $sx = a \in A$. The polynomial $F(T) := sT - a$ fulfills the requirements. Let now $x \in R \setminus A_{[\mathfrak{p}]}$. Since $(A_{[\mathfrak{p}]}, \mathfrak{p}_{[\mathfrak{p}]})$ is a Manis pair there exists some $y \in \mathfrak{p}_{[\mathfrak{p}]}$ with $xy \in A_{[\mathfrak{p}]} \setminus \mathfrak{p}_{[\mathfrak{p}]}$. We choose elements s and t in $A \setminus \mathfrak{p}$ with $ty \in \mathfrak{p}$, $sxy \in A$. We have $sxy \in A \setminus \mathfrak{p}$. Put $a_0 := sty \in \mathfrak{p}$, $a_1 := -stxy \in A \setminus \mathfrak{p}$. The polynomial $F(T) := a_0 T + a_1$ fulfills the requirements.

ii) $\Rightarrow$ i): We verify the property (iii) in Theorem 4. Let $x \in R \setminus A_{[\mathfrak{p}]}$ be given. We look for an element $y \in \mathfrak{p}_{[\mathfrak{p}]}$ with $xy \in A_{[\mathfrak{p}]} \setminus \mathfrak{p}_{[\mathfrak{p}]}$. Let

$$F(T) := a_0 T^n + a_1 T^{n-1} + \cdots + a_n$$

be a polynomial of minimal degree $n \geq 1$ in $A[T] \setminus \mathfrak{p}[T]$ with $F(x) = 0$. From $F(x) = 0$ we deduce that $b := a_0 x$ is integral over A. Thus $b \in A$. Since $x \notin A_{[\mathfrak{p}]}$ we conclude that $a_0 \in \mathfrak{p}$. Suppose that $n > 1$. We put $G(T) := a_0 T - b$ in the case $b \notin \mathfrak{p}$, and

$$G(T) := (b + a_1) T^{n-1} + a_2 T^{n-2} + \cdots + a_n$$

in the case $b \in \mathfrak{p}$. In both cases $G(T) \in A[T] \setminus \mathfrak{p}[T]$ and $G(x) = 0$. This contradicts the minimality of n. Thus $n = 1$, $F(T) = a_0 T + a_1$. Since $a_0 \in \mathfrak{p}$, certainly $a_1 \in A \setminus \mathfrak{p}$. For $y := a_0$ we have $y \in \mathfrak{p}_{[\mathfrak{p}]}$, $xy \in A_{[\mathfrak{p}]} \setminus \mathfrak{p}_{[\mathfrak{p}]}$. $\qquad\square$

Essentially as a consequence of Theorems 4 and 12 we derive still another criterion for a pair $(A_{[\mathfrak{p}]}, \mathfrak{p}_{[\mathfrak{p}]})$ to be Manis in R. In the case of Krull valuation rings (i.e. R a field) such a criterion had been observed by Gilmer [Gi, Th. 19.15]. We need (a special case of) an easy lemma.

Lemma 2.13. Let $(B, \mathfrak{r})$ be a Manis pair in R. Let I be a B-submodule of R with $I \cap B \subset \mathfrak{r}$. Then $I \subset \mathfrak{r}$.

Proof. Suppose there exists an $x \in I$ with $x \notin \mathfrak{r}$, hence $x \notin B$. Since $(B, \mathfrak{r})$ is Manis there exists some $y \in B$ with $xy \in B \setminus \mathfrak{r}$. Then $xy \notin I$. On the other hand $x \in I$ and $y \in B$, a contradiction. $\qquad\square$

Theorem 2.14 (cf. [Gi, Th. 19.15] for R a field). Assume that A is integrally closed in R, and let $\mathfrak{p}$ be a prime ideal of A. The following are equivalent.

i) $(A_{[\mathfrak{p}]}, \mathfrak{p}_{[\mathfrak{p}]})$ is a Manis pair in R.

ii) If B is a subring of R containing $A_{[\mathfrak{p}]}$ and $\mathfrak{q}, \mathfrak{q}'$ are prime ideals of B with $\mathfrak{q} \subset \mathfrak{q}'$ and $\mathfrak{q} \cap A_{[\mathfrak{p}]} = \mathfrak{q}' \cap A_{[\mathfrak{p}]} \subset \mathfrak{p}_{[\mathfrak{p}]}$, then $\mathfrak{q} = \mathfrak{q}'$.

ii') If B is a subring of R containing $A_{[\mathfrak{p}]}$ and $\mathfrak{q} \subset \mathfrak{q}'$ are prime ideals of B lying over $\mathfrak{p}_{[\mathfrak{p}]}$, then $\mathfrak{q} = \mathfrak{q}'$.

iii) If B is a subring of R containing A and $\mathfrak{q}, \mathfrak{q}'$ are prime ideals of B with $\mathfrak{q} \subset \mathfrak{q}'$ and $\mathfrak{q} \cap A = \mathfrak{q}' \cap A \subset \mathfrak{p}$ then $\mathfrak{q} = \mathfrak{q}'$.

iii') If B is a subring of R containing A and $\mathfrak{q} \subset \mathfrak{q}'$ are prime ideals of B lying over $\mathfrak{p}$ then $\mathfrak{q} = \mathfrak{q}'$.

iv) There exists only one Manis pair $(B, \mathfrak{q})$ in R over $(A, \mathfrak{p})$, i.e. with $A \subset B$ and $\mathfrak{q} \cap A = \mathfrak{p}$.

v) For every subring B of R containing A there exists at most one prime ideal $\mathfrak{q}$ of B over $\mathfrak{p}$.

vi) For every Manis pair $(B, \mathfrak{q})$ in R over $(A, \mathfrak{p})$ the field extension $k(\mathfrak{p}) \subset k(\mathfrak{q})$ is algebraic.

Proof. The implication i) $\Rightarrow$ ii) is evident by the preceding lemma. The implications ii) $\Rightarrow$ ii') and iii) $\Rightarrow$ iii') are trivial. ii') $\Rightarrow$ iii'): If $\mathfrak{q}$ and $\mathfrak{q}'$ are prime ideals of B over $\mathfrak{p}$ with $\mathfrak{q} \subset \mathfrak{q}'$, then $\mathfrak{q}_{[\mathfrak{p}]}$ and $\mathfrak{q}'_{[\mathfrak{p}]}$ are prime ideals of $B_{[\mathfrak{p}]}$ over $\mathfrak{p}_{[\mathfrak{p}]}$ with $\mathfrak{q}_{[\mathfrak{p}]} \subset \mathfrak{q}'_{[\mathfrak{p}]}$. Thus $\mathfrak{q}_{[\mathfrak{p}]} = \mathfrak{q}'_{[\mathfrak{p}]}$. Intersecting with B we obtain $\mathfrak{q} = \mathfrak{q}'$. ii) $\Rightarrow$ iii): The proof is similar.

iii') $\Rightarrow$ i): Suppose that $(A_{[\mathfrak{p}]}, \mathfrak{p}_{[\mathfrak{p}]})$ is not Manis in R. By Theorem 12 there exists some $x \in R$ such that $F(x) \neq 0$ for every polynomial $F(T) \in A[T] \backslash \mathfrak{p}[T]$. We introduce the subring $B := A[x]$ of R and the surjective ring homomorphism $\varphi \colon A[T] \longrightarrow B$ over A with $\varphi(T) = x$. The kernel of φ is contained in $\mathfrak{p}[T]$. This implies that the ideals $\mathfrak{q}$ and $\mathfrak{q}'$ of B defined by

$$\mathfrak{q} := \varphi(\mathfrak{p}[T]) = \mathfrak{p}[x] = \mathfrak{p}B, \quad \mathfrak{q}' := \varphi(\mathfrak{p} + TA[T]) = \mathfrak{p} + xB = \mathfrak{q} + xB,$$

both are prime and lie over $\mathfrak{p}$. Since $\mathfrak{q} \neq \mathfrak{q}'$ this contradicts the assumption iii'). Thus $(A_{[\mathfrak{p}]}, \mathfrak{p}_{[\mathfrak{p}]})$ is Manis in R.

i) $\Rightarrow$ iv): Let $(B, \mathfrak{q})$ be a Manis pair in R over $(A, \mathfrak{p})$. It is easily verified that $(B, \mathfrak{q})$ is a pair over $(A_{[\mathfrak{p}]}, \mathfrak{p}_{[\mathfrak{p}]})$. Since $(A_{[\mathfrak{p}]}, \mathfrak{p}_{[\mathfrak{p}]})$ is Manis in R we conclude by Theorem 4 that $(B, \mathfrak{q}) = (A_{[\mathfrak{p}]}, \mathfrak{p}_{[\mathfrak{p}]})$.

iv) $\Rightarrow$ v): Assume that B is a subring of R containing A and $\mathfrak{q}_1, \mathfrak{q}_2$ are prime ideals of B over $\mathfrak{p}$. We extend the pairs $(B, \mathfrak{q}_1)$ and $(B, \mathfrak{q}_2)$ to maximal pairs $(C, \mathfrak{q}'_1)$ and $(D, \mathfrak{q}'_2)$ in R. These pairs are Manis in R by Theorem 4. They both lie over $(A, \mathfrak{p})$, hence $(C, \mathfrak{q}'_1) = (D, \mathfrak{q}'_2)$. Intersecting with B we obtain $\mathfrak{q}_1 = \mathfrak{q}_2$.

v) $\Rightarrow$ iii'): trivial.

i) $\Rightarrow$ vi): Since (i) and (iv) hold we know that $(B, \mathfrak{q}) := (A_{[\mathfrak{p}]}, \mathfrak{p}_{[\mathfrak{p}]})$ is the only Manis pair in R over $(A, \mathfrak{p})$. We have $k(\mathfrak{p}) = k(\mathfrak{q})$.

vi) $\Rightarrow$ iii')): Suppose that $(B, \mathfrak{q}_1)$ and $(B, \mathfrak{q}_2)$ are pairs in R over $(A, \mathfrak{p})$ with $\mathfrak{q}_1 \subsetneqq \mathfrak{q}_2$. We choose a maximal pair $(C, \mathfrak{r})$ in R over $(B, \mathfrak{q}_1)$. Then $(C, \mathfrak{r})$ is Manis, hence $k(\mathfrak{r})$ is algebraic over $k(\mathfrak{p})$. It follows that $k(\mathfrak{q}_1)$ is algebraic over $k(\mathfrak{p})$. We choose an element $x \in \mathfrak{q}_2 \setminus \mathfrak{q}_1$. Since $k(\mathfrak{q}_1)$ is algebraic over $k(\mathfrak{p})$ we have a relation

$$(*) \qquad \sum_{i=0}^{n} a_i x^i = b$$

with $a_0, a_1, \ldots, a_n \in A$, $a_n \notin \mathfrak{p}$, $b \in \mathfrak{q}_1$. Let B' denote the subring $A[b, a_n x]$ of B, and $\mathfrak{q}'_1 := \mathfrak{q}_1 \cap B'$, $\mathfrak{q}'_2 := \mathfrak{q}_2 \cap B'$. We have $\mathfrak{q}'_1 \subsetneqq \mathfrak{q}'_2$, since $a_n x \in \mathfrak{q}'_2 \setminus \mathfrak{q}'_1$. But $\mathfrak{q}'_1 \cap A = \mathfrak{q}'_2 \cap A = \mathfrak{p}$. We learn from the relation $(*)$ that $B'/\mathfrak{q}'_1$ is integral over $A/\mathfrak{p}$. But the ring $B'/\mathfrak{q}'_1$ contains the prime ideal $\mathfrak{q}'_2/\mathfrak{q}'_1 \neq \{0\}$ with $(\mathfrak{q}'_2/\mathfrak{q}'_1) \cap A/\mathfrak{p} = \{0\}$. Such a situation is impossible in an integral ring extension (cf. [Bo, V §2, n° 1]). Thus (iii') is valid. $\qquad \square$

§3 Weakly surjective homomorphisms

In section §5 we will start our theory of "Prüfer extensions". In the terminology developed there the Prüfer rings (with zero divisors) of the classical literature (e.g. [LM], [Huc]) are those commutative rings A which are Prüfer in their total quotient rings Quot A. In the present section and the following one we develop an auxiliary theory of "weakly surjective" ring extensions. It will turn out later (cf.Th.5.2 below) that the Prüfer rings (with zero divisors) are precisely those commutative rings A, such that for every subring B of Quot A containing A the inclusion mapping $A \hookrightarrow B$ is weakly surjective.

Definition 1. i) Let $\varphi: A \to B$ be a ring homomorphism. We call φ *locally surjective* if for every prime ideal $\mathfrak{q}$ of B the induced homomorphism $\varphi_{\mathfrak{q}}: A_{\varphi^{-1}(\mathfrak{q})} \to B_{\mathfrak{q}}$ is surjective. We call φ *weakly surjective* (abbreviated: ws) if for every prime ideal $\mathfrak{p}$ of A with $\mathfrak{p}B \neq B$ the induced homomorphism $\varphi_{\mathfrak{p}}: A_{\mathfrak{p}} \to B_{\mathfrak{p}}$ is surjective.
ii) If A is a subring of a ring B, then we say that A is *locally surjective in B* (resp. *weakly surjective in B*) if the inclusion mapping $A \hookrightarrow B$ is locally surjective (resp. ws).

At first glance "locally surjective" seems to be a more natural notion than "weakly surjective", but it is the latter notion which will be needed below.

Of course, a surjective homomorphism is both weakly surjective and locally surjective. We now prove that weak surjectivity is a stronger property than local surjectivity.

Proposition 3.1. If $\varphi\colon A \to B$ is weakly surjective then φ is locally surjective.

This follows from

Lemma 3.2. Let $\varphi\colon A \to B$ be a ring homomorphism. Let $\mathfrak{q}$ be a prime ideal of B and $\mathfrak{p}:= \varphi^{-1}(\mathfrak{q})$. Assume that $\varphi_{\mathfrak{p}}\colon A_{\mathfrak{p}} \to B_{\mathfrak{p}}$ is surjective. Then the natural map $B_{\mathfrak{p}} \to B_{\mathfrak{q}}$ is an isomorphism, in short, $B_{\mathfrak{p}} = B_{\mathfrak{q}}$. Furthermore $\mathfrak{p}B_{\mathfrak{p}} = \mathfrak{p}B_{\mathfrak{q}} = \mathfrak{q}B_{\mathfrak{q}}$.

Proof of the lemma. One easily retreats to the case that A is a subring of B and φ is the inclusion $A \hookrightarrow B$. Now $\mathfrak{p} = \mathfrak{q} \cap A$ and $A_{\mathfrak{p}} = B_{\mathfrak{p}}$. We have $\mathfrak{p}A_{\mathfrak{p}} = \mathfrak{p}B_{\mathfrak{p}} \subset \mathfrak{q}B_{\mathfrak{p}}$. Since $\mathfrak{p}A_{\mathfrak{p}}$ is the maximal ideal of $A_{\mathfrak{p}}$ and $(\mathfrak{q}B_{\mathfrak{p}})\cap B = \mathfrak{q}$, hence $\mathfrak{q}B_{\mathfrak{p}} \neq B_{\mathfrak{p}}$, we have $\mathfrak{p}B_{\mathfrak{p}} = \mathfrak{q}B_{\mathfrak{p}}$. The natural homomorphism $B \to B_{\mathfrak{p}}$ maps $B \setminus \mathfrak{q}$ into the group of units of $B_{\mathfrak{p}}$, hence factors through a homomorphism from $B_{\mathfrak{q}}$ to $B_{\mathfrak{p}}$. This homomorphism is inverse to the natural map from $B_{\mathfrak{p}}$ to $B_{\mathfrak{q}}$. $\square$

Example 3.3. If S is a multiplicative subset of a ring A then the localization map $A \to S^{-1}A$ is weakly surjective. $\square$

Example 3.4. Let K be a field. The diagonal homomorphism $K \to K \times K$, $x \mapsto (x, x)$, is locally surjective but not weakly surjective, as is easily verified.

Proposition 3.5. If $\varphi\colon A \to B$ is locally surjective and B is an integral domain then φ is weakly surjective.

Proof. Let $\mathfrak{p}$ be a prime ideal of A with $\mathfrak{p}B \neq B$. We choose a prime ideal $\mathfrak{q}$ of B containing $\mathfrak{p}B$. Let $\mathfrak{r}:= \varphi^{-1}(\mathfrak{q})$. We have a natural commuting triangle

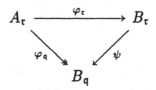

$\varphi_{\mathfrak{q}}$ is surjective, since φ is locally surjective. On the other hand ψ is injective since B is a domain. Thus ψ is bijective and $\varphi_{\mathfrak{r}}$ is surjective. (We have $B_{\mathfrak{r}} = B_{\mathfrak{q}}$, $\varphi_{\mathfrak{r}} = \varphi_{\mathfrak{q}}$.) Since $\mathfrak{p} \subset \mathfrak{r}$ also $\varphi_{\mathfrak{p}}$ is surjective. $\qquad\square$

Proposition 3.6. Every weakly surjective homomorphism is an epimorphism in the category of rings (commutative, with 1).[*)]

Proof. Assume that $\varphi: A \to B$ is ws, and that $\psi_1: B \to C$, $\psi_2: B \to C$ are two ring homomorphisms with $\psi_1 \circ \varphi = \psi_2 \circ \varphi$. We have to prove that $\psi_1 = \psi_2$. By general principles (cf. [Bo, Chap.II, §3]) it suffices to verify that $j \circ \psi_1 = j \circ \psi_2$ with $j: C \to C_{\mathfrak{r}}$ the localisation map for an arbitrary prime ideal $\mathfrak{r}$ of C. Thus we may assume in advance that the ring C is local. Let $\mathfrak{m}$ denote the maximal ideal of C, and let $\mathfrak{q}_1 := \psi_1^{-1}(\mathfrak{m})$, $\mathfrak{q}_2 := \psi_2^{-1}(\mathfrak{m})$, $\mathfrak{p} := \varphi^{-1}(\mathfrak{q}_1) = \varphi^{-1}(\mathfrak{q}_2)$. By Lemma 2 we know that the natural maps $B_{\mathfrak{p}} \to B_{\mathfrak{q}_1}$ and $B_{\mathfrak{p}} \to B_{\mathfrak{q}_2}$ are isomorphisms. This implies that the prime ideals $\mathfrak{q}_1$ and $\mathfrak{q}_2$ coincide, $\mathfrak{q}_1 = \mathfrak{q}_2 =: \mathfrak{q}$. Again by Lemma 2 we know that the localisation $\varphi_{\mathfrak{p}}: A_{\mathfrak{p}} \to B_{\mathfrak{p}} = B_{\mathfrak{q}}$ of φ with respect to $\mathfrak{p}$ is surjective. We have natural commuting diagrams $(j = 1, 2)$

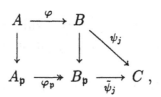

and $\tilde{\psi}_1 \circ \varphi_{\mathfrak{p}} = \tilde{\psi}_2 \circ \varphi_{\mathfrak{p}}$, since $\psi_1 \circ \varphi = \psi_2 \circ \varphi$. From this we conclude that $\tilde{\psi}_1 = \tilde{\psi}_2$ and then that $\psi_1 = \psi_2$. $\qquad\square$

We now verify that this class of epimorphisms has pleasant formal properties.

[*)] In [KZ, p.170] we erroneously stated that locally surjective homomorphisms are epimorphisms. Example 4 above is a counterexample.

Proposition 3.7. Let $\varphi\colon A \to B$ and $\psi\colon B \to C$ be ring homomorphisms.
a) If both φ and ψ are ws then $\psi \circ \varphi$ is ws.
b) If $\psi \circ \varphi$ is ws then ψ is ws.

Proof. a): Let $\mathfrak{p}$ be a prime ideal of A with $\mathfrak{p}C \neq C$. We choose a prime ideal $\mathfrak{r}$ of C containing $\mathfrak{p}C$. Let $\mathfrak{q} := \psi^{-1}(\mathfrak{r})$ and $\tilde{\mathfrak{p}} := \varphi^{-1}(\mathfrak{q})$. The map $\varphi_{\tilde{\mathfrak{p}}}\colon A_{\tilde{\mathfrak{p}}} \to B_{\tilde{\mathfrak{p}}}$ is surjective. By Lemma 2 we know that $B_{\mathfrak{q}} = B_{\tilde{\mathfrak{p}}}$. Thus also $C_{\tilde{\mathfrak{p}}} = C \otimes_A A_{\tilde{\mathfrak{p}}} = C \otimes_B (B \otimes_A A_{\tilde{\mathfrak{p}}}) = C \otimes_B B_{\tilde{\mathfrak{p}}} = C \otimes_B B_{\mathfrak{q}} = C_{\mathfrak{q}}$, and $\psi_{\tilde{\mathfrak{p}}} = \psi_{\mathfrak{q}}$, which is surjective. We conclude that $(\psi \circ \varphi)_{\tilde{\mathfrak{p}}} = \psi_{\tilde{\mathfrak{p}}} \circ \varphi_{\tilde{\mathfrak{p}}}$ is surjective. Since $\mathfrak{p} \subset \tilde{\mathfrak{p}}$, also $(\psi \circ \varphi)_{\mathfrak{p}}$ is surjective.
b): Let $\mathfrak{q}$ be a prime ideal of B with $\mathfrak{q}C \neq C$. Let $\mathfrak{p} := \varphi^{-1}(\mathfrak{q})$. The map $\psi_{\mathfrak{p}} \circ \varphi_{\mathfrak{p}} = (\psi \circ \varphi)_{\mathfrak{p}}$ is surjective. Thus $\psi_{\mathfrak{p}}$ is surjective. Since $\varphi(A \setminus \mathfrak{p}) \subset B \setminus \mathfrak{q}$, also $\psi_{\mathfrak{q}}$ is surjective. $\qquad\square$

Proposition 3.8. If $\varphi\colon A \to B$ and $\psi\colon B \to C$ are ring homomorphisms and φ is ws then $\psi\varphi(A)$ is ws in $\psi(B)$.

Proof. We have a commuting square

$$
\begin{array}{ccc}
A & \xrightarrow{\varphi} & B \\
{\scriptstyle p}\downarrow & & \downarrow{\scriptstyle q} \\
\psi\varphi(A) & \xhookrightarrow{\ i\ } & \psi(B)
\end{array}
$$

with i an inclusion mapping and surjections p and q. Since φ and q are ws, the composite $q \circ \varphi = i \circ p$ is ws. Thus also i is ws. $\qquad\square$

Corollary 3.9. Let $\varphi\colon A \to B$ a ring homomorphism. φ is ws iff $\varphi(A)$ is ws in B.

Proof. Applying Proposition 8 with $\psi = id_B$ we see that weak surjectivity of φ implies weak surjectivity of the inclusion mapping $i\colon \varphi(A) \hookrightarrow B$. Conversely, if i is ws, then φ is ws, since $\varphi = i \circ p$ with p a surjection. $\qquad\square$

It is also easy to verify the corollary directly by using Definition 1.

Proposition 3.10. Let

$$
\begin{array}{ccc}
A & \xrightarrow{\varphi} & B \\
\alpha \downarrow & & \downarrow \beta \\
C & \xrightarrow[\psi]{} & D
\end{array}
$$

be a commuting square of ring homomorphisms. Assume that φ is ws and $D = \beta(B) \cdot \psi(C)$. Then ψ is ws.

Proof. Let $q \in \operatorname{Spec} C$ be given with $\psi(q)D \neq D$, and let $p := \alpha^{-1}(q)$. The commuting square above "extends" to a commuting square

$$
\begin{array}{ccc}
A_p & \xrightarrow{\tilde\varphi} & B_p \\
\tilde\alpha \downarrow & & \downarrow \tilde\beta \\
C_q & \xrightarrow[\tilde\psi]{} & D_q
\end{array}
$$

with $\tilde\varphi = \varphi_p$, $\tilde\psi = \psi_q$. We have $pB \neq B$. The map $\tilde\varphi$ is surjective. We are done, if we verify that $\tilde\psi$ is surjective.

Let $\xi \in D_q$ be given. Write $\xi = \frac{x}{s}$ with $x \in D$, $s \in C \setminus q$. Since $D = \beta(B)\psi(C)$ we have an equation

$$
x = \sum_{i \in I} \beta(b_i)\psi(c_i)
$$

with finite index set I, $b_i \in B$, $c_i \in C$. This equation gives us

$$
\xi = \sum_{i \in I} \tilde\beta\left(\frac{b_i}{1}\right)\tilde\psi\left(\frac{c_i}{s}\right).
$$

Since $\tilde\varphi$ is surjective we have elements $a_i \in A$ $(i \in I)$ and an element $t \in A \setminus p$ with $\frac{b_i}{1} = \tilde\varphi\left(\frac{a_i}{t}\right)$ for every $i \in I$. Then

$$
\xi = \tilde\psi\left(\frac{y}{s\alpha(t)}\right)
$$

with $y := \sum_{a \in I} \alpha(a_i)c_i$. This proves that $\tilde\psi$ is surjective. □

In order to understand weakly surjective homomorphisms it suffices by Cor. 9 to analyse weakly surjective ring extensions.

In the following R is a ring and A is a subring of R.

Definition 2. An *R-overring of* A is a subring B of R with $A \subset B$.

Proposition 3.11.
  a) Let B_1 and B_2 be R-overrings of A. If A is ws both in B_1 and B_2 then A is ws in $B_1 B_2$.
  b) There exists a unique R-overring $M(A, R)$ of A such that A is ws in $M(A, R)$ and $M(A, R)$ contains every R-overring of A in which A is ws.

Proof. a) Since $A \hookrightarrow B_1$ is ws, the inclusion $B_2 \hookrightarrow B_1 B_2$ is ws, as follows from Proposition 10. Since also $A \hookrightarrow B_2$ is ws, the composite $A \hookrightarrow B_2 \hookrightarrow B_1 B_2$ is ws (Prop. 7).
b) Let $\mathfrak{A}$ denote the set of all R-overrings of A in which A is ws. Then $\mathfrak{A}$ is an upward directed system of subrings of R. Let $M(A, R)$ denote the union of all these subrings, which is again a subring of R. A is ws in $M(A, R)$ by the following general remark, which is immediate from Definition 1.

Remark 3.12. Let $(B_i | i \in I)$ be an upward directed system of R-overrings of A. If A is ws in each B_i then A is ws in $\bigcup_{i \in I} B_i$.

Definition 3. We call $M(A, R)$ the *weakly surjective hull* of A in R.

We now derive criteria for a homomorphism to be weakly surjective. Without essential loss of generality we concentrate on ring extensions. Let R be a ring and A a subring of R. Recall from §2 that for $\mathfrak{p}$ a prime ideal of A we denote by $A_{[\mathfrak{p}]}$ the preimage of $A_\mathfrak{p}$ under the localization map $R \to R_\mathfrak{p}$.

Notation. If $x \in R$ then $(A \colon x)$ denotes the ideal of A consisting of all $a \in A$ with $ax \in A$.

Theorem 3.13 (cf. [G_1, Prop. 10] in the case $R = \operatorname{Quot} A$). Let B be an R-overring of A. The following are equivalent.
  (1) A is weakly surjective in B.
  (2) $B_{[\mathfrak{q}]} = A_{[\mathfrak{q} \cap A]}$ for every prime ideal $\mathfrak{q}$ of B.
  (2') $B_{[\mathfrak{q}]} = A_{[\mathfrak{q} \cap A]}$ for every maximal ideal $\mathfrak{q}$ of B.

(3) $B \subset A_{[\mathfrak{p}]}$ for every prime ideal $\mathfrak{p}$ of A with $\mathfrak{p}B \neq B$.
(4) $(A\colon x)B = B$ for every $x \in B$.
(5) $(A\colon x)B = (B\colon x)$ for every $x \in R$.

Proof. (1) $\Longleftrightarrow$ (3): We verify the following: For any $\mathfrak{p} \in \mathrm{Spec}A$

$$B \subset A_{[\mathfrak{p}]} \Longleftrightarrow B_{\mathfrak{p}} = A_{\mathfrak{p}}.$$

Then we will be done according to Def. 1.

$\Rightarrow$: If $B \subset A_{[\mathfrak{p}]}$, then $B_{\mathfrak{p}} \subset (A_{[\mathfrak{p}]})_{\mathfrak{p}} = A_{\mathfrak{p}}$, hence $B_{\mathfrak{p}} = A_{\mathfrak{p}}$.
$\Leftarrow$: If $B_{\mathfrak{p}} = A_{\mathfrak{p}}$ then the preimage $A_{[\mathfrak{p}]}$ of $A_{\mathfrak{p}}$ under the localization map $R \to R_{\mathfrak{p}}$ contains B.

(3) $\Rightarrow$ (2): Let $\mathfrak{q} \in \mathrm{Spec}B$ and $\mathfrak{p} := \mathfrak{q} \cap A$. Of course, $A_{[\mathfrak{p}]} \subset B_{[\mathfrak{q}]}$. In order to prove the converse inclusion we first remark that $\mathfrak{p}B \subset \mathfrak{q}$, hence $\mathfrak{p}B \neq B$. By hypothesis $B \subset A_{[\mathfrak{p}]}$. Let $x \in B_{[\mathfrak{q}]}$ be given. Choose $b \in B \setminus \mathfrak{q}$ with $bx =: b_1 \in B$. We then have elements a, a_1 in $A \setminus \mathfrak{p}$ with $ab \in A$, $a_1 b_1 \in A$. Since $a \in B \setminus \mathfrak{q}$, also $ab \in B \setminus \mathfrak{q}$, hence $ab \in A \cap (B \setminus \mathfrak{q}) = A \setminus \mathfrak{p}$. Also $a_1 ab \in A \setminus \mathfrak{p}$. From $(a_1 ab)x = a(a_1 bx) = a(a_1 b_1) \in A$ we see that $x \in A_{[\mathfrak{p}]}$.

(2) $\Rightarrow$ (2'): trivial.

(2') $\Rightarrow$ (4): Let $x \in B$ be given. Suppose that $(A\colon x)B \neq B$. We choose a maximal ideal $\mathfrak{q}$ of B containing $(A\colon x)B$. Let $\mathfrak{p} := \mathfrak{q} \cap A$. Then $(A\colon x) \subset \mathfrak{p}$. But it follows from (2') that $x \in A_{[\mathfrak{p}]}$, i.e. $(A\colon x) \not\subset \mathfrak{p}$. This contradiction proves that $(A\colon x)B = B$.

(4) $\Rightarrow$ (3): Let $\mathfrak{p}$ be a prime ideal of A with $\mathfrak{p}B \neq B$. Suppose there exists some $x \in B$ with $x \notin A_{[\mathfrak{p}]}$. Then $(A\colon x) \subset \mathfrak{p}$. Thus $(A\colon x)B \subset \mathfrak{p}B \subsetneqq B$. This contradicts the assumption (4). We conclude that $B \subset A_{[\mathfrak{p}]}$.

(5) $\Rightarrow$ (4): trivial.

(4) $\Rightarrow$ (5): Let $x \in R$ be given. Of course, $(A\colon x)B \subset (B\colon x)$. We pick some $b \in (B\colon x)$ and verify that $b \in (A\colon x)B$. Let $u := bx \in B$. By assumption we have $(A\colon b)B = B$ and $(A\colon u)B = B$, hence $(A\colon b)(A\colon u)B = B$, which implies $[(A\colon b) \cap (A\colon u)]B = B$. We thus have a relation

$$1 = \sum_{i=1}^{N} a_i c_i$$

with $a_i \in A$, $c_i \in B$, $a_i b \in A$, $a_i u \in A$ for every $i \in \{1, \ldots, N\}$. This gives us $b = \sum_{i=1}^{N} (ba_i)c_i$. Now $ba_i \in A$ and $ba_i x = a_i u \in A$, hence $ba_i \in (A\!:\!x)$. We conclude that indeed $b \in (A\!:\!x)B$. $\qquad\square$

Remarks. In the case of domains Richman [Ri, §2] has studied the properties (3), (4) under the name "good extensions". If $A \subset B$ and B is a domain then good means the same as weakly surjective and as locally surjective. Theorem 13 has a close relation to work of Lazard [L, Chap. IV] and Akiba [A], cf. Theorem 4.4 in the next section.

Definition 4. [Lb, §2.3]. a) An ideal $\mathfrak{a}$ of a ring C is called *dense in C* if its annulator ideal $\text{Ann}_C(\mathfrak{a})$ is zero.

b) A *ring of quotients of A* is a ring $B \supset A$ such that $(A\!:\!x)B$ is dense in B for every $x \in B$.

We recall the following important fact from Lambek's book [Lb, §2.3]. For any ring A there exists a ring of quotients $Q(A)$ of A, explicitly constructed in [Lb], such that for any other ring of quotients B of A there exists a unique homomorphism from B to $Q(A)$ over A. Every such homomorphism is injective. $Q(A)$ is called the *complete ring of quotients* of A. Of course $Q(A)$ contains the total quotient ring $\text{Quot}\,(A)$ {also called the "classical" quotient ring of A}. For A noetherian it is known that $\text{Quot}\,A = Q(A)$, cf. [A, Prop. 1], but in general these two extensions of A may be different.

From condition (4) in Theorem 13 it is clear that, if $A \subset B$ is a weakly surjective ring extension, then B is a ring of quotients of A. Thus every weakly surjective ring extension of A embeds into $Q(A)$ in a unique way.

Definition 5. The *weakly surjective hull* $M(A)$ of A is defined as the ws hull $M(A, Q(A))$ of A in $Q(A)$.

From our discussion of the hulls $M(A, R)$ above the following is evident.

Proposition 3.14. For every weakly surjective ring extension $A \subset B$ there exists a unique homomorphism $B \to M(A)$ over A, and this is a monomorphism. $\qquad\square$

Thus, without serious abuse, we may regard any ws extension $A \subset B$ as a subextension of $A \subset M(A)$. In particular, $A \subset \text{Quot } A \subset M(A)$.

Remark 3.15. If C is any subring of $M(A)$ containing A then $M(C) = M(A)$. In particular, $MM(A) = M(A)$.

Proof. Since C is ws in $M(A)$ we have embeddings $C \subset M(A) \subset M(C)$. Now A is ws in $M(A)$ and $M(A)$ is ws in $M(C)$, hence A is ws in $M(C)$. Due to the maximality of $M(A)$ we have $M(C) = M(A)$. $\square$

Caution. In general, if C is a subring of $M(A)$ containing A, then A is not necessarily ws in C (cf. §5).

Corollary 3.16. Let $A \subset B_1$ and $A \subset B_2$ be weakly surjective extensions. Then there exists at most one homomorphism $\lambda: B_1 \to B_2$ over A, and λ is injective.

Proof. We have unique homomorphisms $\mu_i: B_i \to M(A)$ over A ($i = 1, 2$), and they both are injective. If $\lambda: B_1 \to B_2$ is a homomorphism over A, this implies that $\mu_2 \circ \lambda = \mu_1$. Thus λ is injective and is uniquely determined by μ_1 and μ_2. $\square$

Of course, the uniqueness of λ is a priori clear, since $A \hookrightarrow B_1$ is an epimorphism (Prop. 6).

We briefly discuss relations between weakly surjective extensions and integral extensions.

Proposition 3.17 (cf. [G$_1$, Prop. 11]). If a ring homomorphism $\varphi: A \to B$ is both weakly surjective and integral then φ is surjective.

Proof. Replacing A by $\varphi(A)$ we assume without loss of generality that $A \subset B$ and φ is the inclusion mapping. We have to prove that $A = B$.

Suppose there exists an element $x \in B \setminus A$. Then $(A : x)$ is a proper ideal of A. Since B is integral over A, this implies that $(A : x)B \neq B$. This contradicts property (4) in Theorem 13. Thus $A = B$. $\square$

Proposition 3.18. ([Ri, §4] for R a field, [G_1, Prop. 11] for $R = $ Quot A). Assume that $A \subset B \subset R$ are ring extensions, and that A is weakly surjective in B. For the integral closures $\tilde{A}$ and $\tilde{B}$ of A and B in R the following holds.

i) $\tilde{B} = \tilde{A} \cdot B$.

ii) $\tilde{A}$ is weakly surjective in $\tilde{B}$.

Proof. The argument in [Ri] (p.797, proof of Prop.1) extends to our more general situation. We repeat this argument for the convenience of the reader. Of course $\tilde{B} \supset \tilde{A}B$. Let $x \in \tilde{B}$ be given. We have an equation

$$(*) \qquad\qquad x^n + b_1 x^{n-1} + \cdots + b_n = 0$$

with $b_1, \ldots, b_n \in B$. Let $\mathfrak{a}_i := (A : b_i) \quad (1 \leq i \leq n)$ and $\mathfrak{a} := \mathfrak{a}_1 \cap \cdots \cap \mathfrak{a}_n$.

By Theorem 13 we have $\mathfrak{a}_i B = B$ for every i. This implies

$$(\mathfrak{a}_1 \ldots \mathfrak{a}_n)B = B$$

and then $\mathfrak{a}B = B$. Given an element $a \in \mathfrak{a}$ we multiply the relation $(*)$ by a^n and learn that ax is integral over A. Thus $\mathfrak{a}x \subset \tilde{A}$, and $xB = x\mathfrak{a}B \subset \tilde{A}B$, i.e. $x \in \tilde{A}B$. This proves $\tilde{A}B = \tilde{B}$. Using Prop.10 we conclude that $\tilde{A}$ is ws in $\tilde{B}$. $\qquad\square$

§4 More on weakly surjective extensions

Having set the stage we discuss some properties of weakly surjective ring extensions. We are mainly interested in functorial properties and the behavior of ideals.

Proposition 4.1. Every weakly surjective ring extension $A \subset B$ is flat (i.e., B is a flat A-module).

Proof. Let $\alpha : M' \to M$ be an injective homomorphism of A-modules.

We verify that $\alpha \otimes_A B \colon M' \otimes_A B \to M \otimes_A B$ is again injective. Let $\mathfrak{q}$ be a prime ideal of B and $\mathfrak{p} := \mathfrak{q} \cap A$. Then $A_{\mathfrak{p}} = B_{\mathfrak{q}}$, thus

$$(\alpha \otimes_A B)_{\mathfrak{q}} = (\alpha \otimes_A B) \otimes_B B_{\mathfrak{q}} = \alpha \otimes_A B_{\mathfrak{q}} = \alpha \otimes_A A_{\mathfrak{p}}.$$

Since $A \to A_{\mathfrak{p}}$ is flat the homomorphism $(\alpha \otimes_A B)_{\mathfrak{q}}$ is injective. Since this holds for every $\mathfrak{q} \in \operatorname{Spec} B$ we conclude that $\alpha \otimes_A B$ is injective. $\square$

Proposition 4.2. Let $A \subset B_1$ and $A \subset B_2$ be weakly surjective ring extensions.

  a) Then the natural map $A \to B_1 \otimes_A B_2$ is injective and weakly surjective, hence may be regarded as a ws extension.
  b) If both $A \subset B_1$ and $A \subset B_2$ are subextensions of a ring extension $A \subset R$, then the natural map $B_1 \otimes_A B_2 \to B_1 B_2$ is an isomorphism, in short, $B_1 \otimes_A B_2 = B_1 B_2$.

Proof. a) Since B_1 is flat over A the natural map $B_1 \to B_1 \otimes_A B_2$ is injective. Also $B_2 \to B_1 \otimes_A B_2$ and $A \to B_1 \otimes B_2$ are injective. We regard A, B_1, B_2 as subrings of $B_1 \otimes_A B_2$ and conclude from Propositions 3.7.a and 3.10. that A is ws in $B_1 \otimes B_2$.
b) In the situation $B_1 \subset R$, $B_2 \subset R$ the ring A is also ws in $B_1 B_2$. The natural map $\lambda \colon B_1 \otimes_A B_2 \to B_1 B_2$ is a surjective homomorphism over A. By Cor.3.16 λ is also injective, hence is an isomorphism. $\square$

Example 4.3. If $\varphi \colon A \to B$ is a weakly surjective homomorphism then the natural map $B \otimes_A B \longrightarrow B, x \otimes y \longmapsto xy$, is an isomorphism.

This follows from the proposition since $B \otimes_A B = B \otimes_{\varphi(A)} B$. The statement is just a reformulation of the fact, already known to us (Prop. 3.6), that φ is an epimorphism, cf. e.g. [St, p. 380] or Appendix A below.

We now invoke the important work of Lazard in his thesis [L] and of Akiba [A]. We have seen that every injective weakly surjective homomorphism is a flat epimorphism (in the category of rings). By [L, IV. Prop. 2.4] or [A, Th.1] the converse also holds.

Theorem 4.4 (Lazard, Akiba). An injective homomorphism φ is weakly surjective iff φ is a flat epimorphism. $\square$

For the convenience of the reader we will reproduce Lazard's proof in Appendix A. Up to very minor points also the results to follow, up to Proposition 10, are contained in Lazard's thesis [L], and many more. We give short proofs in the present frame work.

We assume, up to Proposition 10, that $A \subset B$ is a ws ring extension.

Proposition 4.5. (cf. [L, IV, Cor.3.2]) Let C be a subring of B containing A. Then $A \subset C$ is weakly surjective iff C is flat over A.

Proof. We know already that weak surjectivity of $A \hookrightarrow C$ implies flatness. In order to prove the converse we look at the following commuting square of natural maps:

$$
\begin{array}{ccc}
C \otimes_A C & \xrightarrow{m_1} & C \\
\big\downarrow{\scriptstyle j} & & \big\uparrow{\scriptstyle i} \\
B \otimes_A B & \xrightarrow{m_2} & B .
\end{array}
$$

Here i denotes the inclusion map from C to B and j the induced map from $C \otimes_A C$ to $B \otimes_A B$. Finally m_1 and m_2 denote the "multiplication maps" $x \otimes y \mapsto xy$. We know from 4.3 that m_2 is an isomorphism. Assume that C is flat over A. Then j is injective. This implies that m_1 is injective, hence an isomorphism. It follows that the two maps $C \rightrightarrows C \otimes_A C$, $x \mapsto x \otimes 1$, $x \mapsto 1 \otimes x$, are equal, which means that $A \hookrightarrow C$ is an epimorphism. Now Theorem 4 tells us that $A \hookrightarrow C$ is ws. $\square$

As before we are given a ws extension $A \subset B$.

Proposition 4.6. Let $\mathfrak{b}$ be an ideal of B and $\mathfrak{a} := \mathfrak{b} \cap A$. Then $\mathfrak{b} = \mathfrak{a}B$.

Proof. Let $\mathfrak{c} := \mathfrak{a}B$. Then $\mathfrak{c} \subset \mathfrak{b}$ and $\mathfrak{c} \cap A = \mathfrak{a}$. We have a commuting triangle of natural homomorphisms

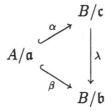

with α and β injective (and λ surjective). Both α and β are ws. Thus λ is injective (hence an isomorphism) by Cor. 3.16. This means that $\mathfrak{c} = \mathfrak{b}$. $\qquad\square$

The nil radical of a ring C will be denoted by $\operatorname{Nil} C$.

Example 4.7. $\operatorname{Nil} B = (\operatorname{Nil} A)B$.

Indeed, we have $(\operatorname{Nil} B) \cap A = \operatorname{Nil} A$. $\qquad\square$

Theorem 4.8. Let $\mathfrak{p}$ be a prime ideal of A with $\mathfrak{p}B \neq B$. Then $\mathfrak{q} := \mathfrak{p}B$ is a prime ideal of B. This is the unique prime ideal of B lying over $\mathfrak{p}$. If B is given as a subextension of an extension $A \subset R$, then $B \subset A_{[\mathfrak{p}]}$ and $\mathfrak{q} = \mathfrak{p}_{[\mathfrak{p}]} \cap B$.

Proof. We have $A_\mathfrak{p} = B_\mathfrak{p}$. Thus $\mathfrak{p}B_\mathfrak{p}$ is the unique maximal ideal of $B_\mathfrak{p}$. Let $\mathfrak{q}$ denote the preimage of $\mathfrak{p}B_\mathfrak{p}$ under the localization map $B \to B_\mathfrak{p}$. From the natural commuting triangle

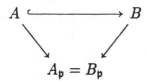

$$A \lhook\joinrel\longrightarrow B$$
$$A_\mathfrak{p} = B_\mathfrak{p}$$

we read off that $\mathfrak{q} \cap A = \mathfrak{p}$. By Prop. 6 we have $\mathfrak{p}B = \mathfrak{q}$. Thus $\mathfrak{p}B$ is a prime ideal. Now assume that $A \subset B \subset R$. Then $B \subset A_{[\mathfrak{p}]}$ by Theorem 3.13. $\mathfrak{q}' := \mathfrak{p}_{[\mathfrak{p}]} \cap B$ is a prime ideal of B with $\mathfrak{q}' \cap A = \mathfrak{p}_{[\mathfrak{p}]} \cap A = \mathfrak{p}$. Thus $\mathfrak{q}' = \mathfrak{q}$. $\qquad\square$

Remark 4.9. If $\mathfrak{p}B = B$ then certainly $\mathfrak{p}B \neq \mathfrak{p}_{[\mathfrak{p}]} \cap B$.

Let $X(B/A)$ denote the image of the restriction map $\mathfrak{q} \mapsto \mathfrak{q} \cap A$ from $\operatorname{Spec} B$ to $\operatorname{Spec} A$. We endow $X(B/A)$ with the subspace topology in $\operatorname{Spec} A$. It follows from Theorem 8 that $X(B/A)$ is the set of all $\mathfrak{p} \in \operatorname{Spec} A$ with $\mathfrak{p}B \neq B$.

Proposition 4.10. The restriction map $\operatorname{Spec} B \to \operatorname{Spec} A$ is a homeomorphism from $\operatorname{Spec} B$ to $X(B/A)$. The set $X(B/A)$ is proconstructible and dense in $\operatorname{Spec} A$. It is closed under generalizations in $\operatorname{Spec} A$.

Proof. We use the framework of spectral spaces, cf. [Ho] or e.g. [KS, Chap. III]. The restriction map $\operatorname{Spec}B \to \operatorname{Spec}A$ is spectral. Thus $X(B/A)$ is proconstructible in $\operatorname{Spec}A$, hence is itself a spectral space. Again by Theorem 8 the restriction map $r: \operatorname{Spec}B \to X(B/A)$ is bijective. If $x, y \in \operatorname{Spec}B$ and $r(y)$ is a specialization of $r(x)$ then y is a specialization of x. Since r is spectral this implies that r is a homeomorphism.

Since A is a subring of B, the image $X(B/A)$ of the restriction map contains all minimal prime ideals of A and is dense in $\operatorname{Spec}A$. If $\mathfrak{p} \in \operatorname{Spec}A$ and $\mathfrak{p}B \neq B$, then $\mathfrak{r}B \neq B$ for the prime ideals $\mathfrak{r}$ of A contained in $\mathfrak{p}$. Thus $X(B/A)$ is closed under generalizations. {This already follows from the fact that $A \hookrightarrow B$ is flat, hence the "going down theorem" holds for prime ideals.} $\qquad\square$

We finally look again at the relation between the notions "weakly surjective" and "locally surjective" (cf. §3, Def.1).

Proposition 4.11 (N. Schwartz [Sch₄]). Given a ring homomorphism $\varphi: A \to B$, the following are equivalent:
(1) φ is weakly surjective.
(2) φ is locally surjective and the map $\operatorname{Spec}(\varphi): \operatorname{Spec}B \to \operatorname{Spec}A$ is a homeomorphism onto its image.

Proof. φ has a factorisation $A \xrightarrow{\pi} A/\mathfrak{a} \xhookrightarrow{\bar\varphi} B$ with $\mathfrak{a}$ the kernel of φ and π the natural map from A to $A/\mathfrak{a}$. For π both the properties (1) and (2) are true. Thus it suffices to prove everything for $\bar\varphi$ instead of φ. In the following we assume that A is a subring of B and $\varphi = i$ is the inclusion map from A to B. We know already from Proposition 10 and Proposition 3.1 that (1) implies (2).

Assume now that A is locally surjective in B and $\operatorname{Sper}(i)$ is a homeomorphism of $\operatorname{Spec}B$ onto its image in $\operatorname{Spec}A$. Let $\mathfrak{p}$ be a prime ideal of A with $\mathfrak{p}B \neq B$. We have to prove that the localised homomorphism $i_{\mathfrak{p}}: A_{\mathfrak{p}} \to B_{\mathfrak{p}}$ is an isomorphism. We choose a prime ideal $\mathfrak{q} \supset \mathfrak{p}B$ of B. Let $\tilde{\mathfrak{p}} := \mathfrak{q} \cap A$. Then $\tilde{\mathfrak{p}} \supset \mathfrak{p}$. It suffices to verify that $i_{\tilde{\mathfrak{p}}}$ is an isomorphism, since $i_{\mathfrak{p}}$ is a localisation of $\varphi_{\tilde{\mathfrak{p}}}$. Thus we may assume without loss of generality that $\mathfrak{q} \cap A = \mathfrak{p}$. We have a commuting triangle of natural maps

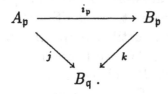

Here j is an isomorphism, since i is locally surjective. Of course, $i_{\mathfrak{p}}$ is injective. Thus we will be done, if we verify that k is injective.

Let $f = \operatorname{Sper}(\varphi)\colon \operatorname{Spec}B \to \operatorname{Spec}A$ and $X := \operatorname{Im} f$. As usual we regard $\operatorname{Spec}B_{\mathfrak{p}}$ and $\operatorname{Spec}A_{\mathfrak{p}}$ as topological subspaces of $\operatorname{Spec}B$ and $\operatorname{Spec}A$ respectively. Then $\operatorname{Spec}A_{\mathfrak{p}}$ is the set of generalisations of the point $\mathfrak{p} \in \operatorname{Spec}A$. We have $f(\operatorname{Spec}B_{\mathfrak{p}}) = X \cap \operatorname{Spec}A_{\mathfrak{p}}$, and $f(\mathfrak{q}) = \mathfrak{p}$. Since f maps $\operatorname{Spec}B_{\mathfrak{p}}$ homeomorphically onto $X \cap \operatorname{Spec}A_{\mathfrak{p}}$, we conclude that $\operatorname{Spec}B_{\mathfrak{p}}$ consists of the generalisations of the point $\mathfrak{q}B_{\mathfrak{p}} \in \operatorname{Spec}(B_{\mathfrak{p}})$. This means that $B_{\mathfrak{p}}$ is a local ring with the maximal ideal $\mathfrak{q}B_{\mathfrak{p}}$. It is now evident that the map k in the triangle above even is an isomorphism. $\qquad\square$

§5 Basic theory of Prüfer extensions

Let R be a ring and A a subring of R.

Definition 1 [G_2, §4] [*] A is called an *R-Prüfer ring*, or a *Prüfer subring of* R, if $(A_{[\mathfrak{p}]}, \mathfrak{p}_{[\mathfrak{p}]})$ is a Manis pair in R for every maximal ideal $\mathfrak{p}$ of A. We then also say that A is *Prüfer in* R, or that R is *Prüfer over* A, or that R is a *Prüfer extension* of A.

N.B. According to Prop. 2.10 this holds iff $(A_{\mathfrak{p}}, \mathfrak{p}_{\mathfrak{p}})$ is a Manis pair in $R_{\mathfrak{p}}$ for every maximal ideal $\mathfrak{p}$ of A. In particular, if R is a field, we arrive at the classical notion of a Prüfer domain.

Proposition 5.1. Assume that A is Prüfer in R.

---

[*] It turned out that Griffin's definition is not quite "correct". He only demands that the $A_{[\mathfrak{p}]}$ are Manis subrings of R. For a reasonable theory it is necessary to include a condition on the $\mathfrak{p}_{[\mathfrak{p}]}$, cf. also [Gr, p.285].

i) For every prime ideal $\mathfrak{p}$ of A the pair $(A_{[\mathfrak{p}]}, \mathfrak{p}_{[\mathfrak{p}]})$ is Manis in R, and the pair $(A_\mathfrak{p}, \mathfrak{p}_\mathfrak{p})$ is Manis in $R_\mathfrak{p}$.

ii) If S is a multiplicative subset of A then $S^{-1}A$ is Prüfer in $S^{-1}R$.

iii) The following are equivalent.

(1) A is a Manis subring of R.
(2) A is a valuation subring of R.
(3) $R \setminus A$ is multiplicatively closed, i.e. $(R \setminus A)(R \setminus A) \subset R \setminus A$.

Moreover, if $A \neq R$ and (1) – (3) hold, then $(A, \mathfrak{p}_A)$ is a Manis pair of R. $\{\mathfrak{p}_A$ had been defined in §2, Def.2.$\}$

Proof. Let $\mathfrak{p}$ be a prime ideal of A. We choose a maximal ideal $\mathfrak{m} \supset \mathfrak{p}$. There exists a Manis valuation v on R with $A_v = A_{[\mathfrak{m}]}$, $\mathfrak{p}_v = \mathfrak{m}_{[\mathfrak{m}]}$. If $A \cap \operatorname{supp} v \not\subset \mathfrak{p}$, then we choose some $s \in (\operatorname{supp} v) \cap (A \setminus \mathfrak{p})$. We have $sR \subset A_{[\mathfrak{m}]} \subset A_{[\mathfrak{p}]}$, and we conclude that $A_{[\mathfrak{p}]} = R$. Thus $(A_{[\mathfrak{p}]}, \mathfrak{p}_{[\mathfrak{p}]})$ is certainly Manis in R in this case. Assume now that $A \cap \operatorname{supp} v \subset \mathfrak{p}$. Then it follows from Prop.2.11 that $(A_{[\mathfrak{p}]}, \mathfrak{p}_{[\mathfrak{p}]})$ is Manis in R, and Prop.2.10 tells us that $(A_\mathfrak{p}, \mathfrak{p}_\mathfrak{p})$ is Manis in $R_\mathfrak{p}$. This proves assertion (i). Assertion (ii) is an immediate consequence of (i).

In assertion (iii) the implications (1) $\Rightarrow$ (2) $\Rightarrow$ (3) are trivial. We prove (3) $\Rightarrow$ (1). We may assume $A \neq R$. Let $\mathfrak{p} := \mathfrak{p}_A$. Let $x \in A_{[\mathfrak{p}]}$ be given. There exists some $d \in A \setminus \mathfrak{p}$ with $dx \in A$. If $x \notin A$ this would imply $d \in \mathfrak{p}$ by definition of $\mathfrak{p} = \mathfrak{p}_A$. Thus $x \in A$. This proves $A_{[\mathfrak{p}]} \subset A$, i.e. $A_{[\mathfrak{p}]} = A$. Then $\mathfrak{p}_{[\mathfrak{p}]} \subset A$, hence $\mathfrak{p} = \mathfrak{p}_{[\mathfrak{p}]} \cap A = \mathfrak{p}_{[\mathfrak{p}]}$. Since A is Prüfer in R we conclude that the pair $(A, \mathfrak{p})$ is Manis in R. $\qquad\square$

The following theorem gives a bunch of criteria for a given ring extension $A \subset R$ to be Prüfer. It is here that the theory of Manis valuations and the theory of weakly surjective ring extensions, displayed in §1, §2 and in §3, §4 respectively, come together.

Theorem 5.2. The following are equivalent.

(1) A is an R-Prüfer ring.
(2) A is weakly surjective in every R-overring.
(2$'$) A is weakly surjective in $A[x]$ for every $x \in R$.
(3) If B is any R-overring of A then $(A : x)B = B$ for every $x \in B$.

(3') If B is any R-overring of A then $(A:x)B = (B:x)$ for every $x \in R$.

(4) Every R-overring of A is integrally closed in R.

(5) A is integrally closed in R, and $A[x] = A[x^n]$ for every $x \in R$ and $n \in \mathbb{N}$.

(5') A is integrally closed in R, and $A[x] = A[x^2]$ for every $x \in R$.

(6) A is integrally closed in R. For every $x \in R$ there exists a polynomial $F[T] = \sum_{i=0}^{d} a_i T^i$ with all $a_i \in A$, $a_j = 1$ for at least one index j, such that $F(x) = 0$.

(7) A is integrally closed in R. For every $x \in R$ and every maximal ideal $\mathfrak{p}$ of A there exists a polynomial $F_{x,\mathfrak{p}}(T) \in A[T] \setminus \mathfrak{p}[T]$ such that $F_{x,\mathfrak{p}}(x) = 0$.

(8) $(A:x) + x(A:x) = A$ for every $x \in R$.

(9) A is integrally closed in R. For every overring B of R the restriction map $\operatorname{Spec} B \to \operatorname{Spec} A$ is injective.

(9') A is integrally closed in R. If B is an R-overring of A and $\mathfrak{q} \subset \mathfrak{q}'$ are prime ideals of B with $\mathfrak{q} \cap A = \mathfrak{q}' \cap A$ then $\mathfrak{q} = \mathfrak{q}'$.

(10) A is integrally closed in R. For every prime ideal $\mathfrak{p}$ of A there exists a unique Manis pair $(B, \mathfrak{q})$ in R over $(A, \mathfrak{p})$, i.e. with $A \subset B$, $\mathfrak{q} \cap A = \mathfrak{p}$.

(11) For every R-overring B of A the inclusion map $A \hookrightarrow B$ is an epimorphism (in the category of rings).

(11') For every $x \in R$ the inclusion map $A \hookrightarrow A[x]$ is an epimorphism.   □

Remarks. The equivalence of (1), (2), (3), (4) had already been stated by Griffin [G$_2$, Prop.6, Th.7], but he made additional assumptions and did not present the proofs. On the other hand, Griffin weakened our hypothesis that the rings have unit elements. The equivalence of (1), (4), (8) has been proved by Eggert for $R = Q(A)$, the complete ring of quotients of A [Eg, Th.2]. The equivalence of (1) and any of the conditions (4) - (7) is a generalization of classical results for R a field (cf. e.g. [E, Th.11.10]). The equivalence of (1) and (11) for R a field has been proved by Storrer [St$_1$]. The equivalence of (1), (2), (4), (8) has been stated in full generality by Rhodes [Rh, Th.2.1]. Unfortunately his proof contains a gap (cf. our Introduction). E.D. Davis studied extensions $A \subset R$ with property (4) under the name "normal pairs". In the case of domains some of our results in this section can be read off from his paper [Da].

Proof. (1) $\Rightarrow$ (2): Let B be an R-overring of A and $\mathfrak{q}$ a prime ideal of B. Let $\mathfrak{p} := \mathfrak{q} \cap A$. We verify that $A_{[\mathfrak{p}]} = B_{[\mathfrak{q}]}$ and then will be done by Theorem 3.13. Of course, $A_{[\mathfrak{p}]} \subset B_{[\mathfrak{q}]}$. Let $x \in R \setminus A_{[\mathfrak{p}]}$ be given. We prove that $x \notin B_{[\mathfrak{q}]}$, and then will be done.

Since $(A_{[\mathfrak{p}]}, \mathfrak{p}_{[\mathfrak{p}]})$ is a Manis pair in R there exists an element y of $\mathfrak{p}_{[\mathfrak{p}]}$ with $xy \in A_{[\mathfrak{p}]} \setminus \mathfrak{p}_{[\mathfrak{p}]}$. We choose elements a and c in $A \setminus \mathfrak{p}$ with $a(xy) \in A$ and $cy \in \mathfrak{p}$. We have $a(xy) \in A \setminus \mathfrak{p}$. Suppose that $x \in B_{[\mathfrak{q}]}$. Then there exists some $b \in B \setminus \mathfrak{q}$ with $bx \in B$. We have $a(bx)(cy) \in \mathfrak{q}$. On the other hand, $a(bx)(cy) = bc(axy) \in B \setminus \mathfrak{q}$. This contradiction proves that $x \notin B_{[\mathfrak{q}]}$.

(2) $\Rightarrow$ (2'): trivial.

(2) $\Leftrightarrow$ (3) $\Leftrightarrow$ (3'): Clear from Th. 3.13.

(2') $\Rightarrow$ (3): Let $x \in B$. Then $(A : x)A[x] = A[x]$. A fortiori $(A : x)B = B$.

(2) $\Rightarrow$ (4): Let B be an R-overring of A, and let $C = \tilde{B}$ denote the integral closure of B in R. By (2) A is ws in C. Thus B is ws in C (Prop. 3.7.b). Prop. 3.17 tells us that $C = B$, i.e. B is integrally closed in R.

(4) $\Rightarrow$ (5): x is integral over $A[x^n]$. By assumption (4) the subring $A[x^n]$ is integrally closed in R. Thus $x \in A[x^n]$.

(5) $\Rightarrow$ (5'): trivial.

(5') $\Rightarrow$ (6): For every $x \in R$ we have a relation $x = \sum_{i=0}^{m} a_i x^{2i}$ with $m \in \mathbb{N}_0$, $a_i \in A$.

(6) $\Rightarrow$ (7): trivial.

(7) $\Rightarrow$ (1): Theorem 2.12 tells us that $(A_{[\mathfrak{p}]}, \mathfrak{p}_{[\mathfrak{p}]})$ is a Manis pair in R for every $\mathfrak{p} \in \operatorname{Spec} A$.

(1) $\Rightarrow$ (8): Suppose there exists some $x \in R$ with $I := (A : x) + x(A : x) \neq A$. We choose a maximal ideal $\mathfrak{m}$ of A containing I. Then $x \in R \setminus A_{[\mathfrak{m}]}$ since $(A : x) \subset \mathfrak{m}$. By (1) and Theorem 2.4 (iii) there exists some $x' \in \mathfrak{m}_{[\mathfrak{m}]}$ with $xx' \in A_{[\mathfrak{m}]} \setminus \mathfrak{m}_{[\mathfrak{m}]}$. We then choose some $d \in A \setminus \mathfrak{m}$ with $dx' \in \mathfrak{m}$ and $dxx' \in A \setminus \mathfrak{m}$. It follows that $dx' \in (A : x)$ and $dxx' \in x(A : x) \subset \mathfrak{m}$, a contradiction. Thus (8) holds.

(8) $\Rightarrow$ (1): We prove for a given prime ideal $\mathfrak{p}$ of A that the pair $(A_{[\mathfrak{p}]}, \mathfrak{p}_{[\mathfrak{p}]})$ is Manis in R by verifying condition (iii) in Theorem 2.4.

Let $x \in R \setminus A_{[\mathfrak{p}]}$. Then $(A : x) \subset \mathfrak{p}$. By (8) we know that $x(A : x) \not\subset \mathfrak{p}$. Thus there exists some $x' \in (A : x) \subset \mathfrak{p}$ with $xx' \in A \setminus \mathfrak{p} \subset A_{[\mathfrak{p}]} \setminus \mathfrak{p}_{[\mathfrak{p}]}$.

The equivalence of (1), (9), (9'), (10) is evident from Theorem 2.14. The implication $(2') \Rightarrow (11')$ follows from the fact that every weakly surjective map is an epimorphism (cf. Prop.3.6).

$(11') \Rightarrow (11)$: Suppose there exists an R-overring B of A such that the inclusion map $A \hookrightarrow B$ is not an epimorphism. Then there exist two ring homomorphisms φ_1, φ_2 from B to some ring C with $\varphi_1 | A = \varphi_2 | A$ but $\varphi_1 \neq \varphi_2$. We choose some $x \in B$ with $\varphi_1(x) \neq \varphi_2(x)$. The restrictions $\varphi_1 | A[x]$ and $\varphi_2 | A[x]$ are different, but $\varphi_1 | A = \varphi_2 | A$. This contradicts the assumption $(11')$.

$(11) \Rightarrow (4)$: Let B be an R-overring of A, and let $x \in R$ be integral over B. We want to prove that $x \in B$. The inclusion $A \hookrightarrow B[x]$ is an epimorphism. Thus (for purely categorial reasons) also the inclusion $B \hookrightarrow B[x]$ is an epimorphism. By an easy proposition of Lazard [L, Chap. IV, Prop.1.7], $B[x] = B$. We will state and prove this proposition in Appendix A. $\qquad\square$

From condition (4) in this theorem one obtains immediately

Corollary 5.3. Let B be an R-overring of A. If A is Prüfer in R then B is Prüfer in R and A is Prüfer in B. $\qquad\square$

From condition (8) in the theorem we obtain

Corollary 5.4. If A is Prüfer in R then for any $x \in R$ the ideal $(A : x)$ is generated by two elements.

Indeed, we have elements a and b in $(A : x)$ with $1 = a + xb$. If $u \in (A : x)$ then $u = ua + (ux)b$. Thus $(A : x) = Aa + Ab$. $\qquad\square$

Theorem 2 contains the fact that every R-Prüfer ring is integrally closed in R. The reader might ask for a more direct proof of this statement. Indeed this follows from the definition of R-Prüfer rings and an elementary fact which holds *without any assumption* about our subring A of R.

Remark 5.5. If I is an A-submodule of R, then

$$I = \bigcap_{\mathfrak{p} \in M} (A_{[\mathfrak{p}]} I) = \bigcap_{\mathfrak{p} \in M} I_{[\mathfrak{p}]},$$

with M denoting the set of maximal ideals of A. In particular $A = \bigcap_{\mathfrak{p} \in M} A_{[\mathfrak{p}]}$.

Proof. Of course, $I \subset A_{[\mathfrak{p}]} I \subset I_{[\mathfrak{p}]}$ for every $\mathfrak{p} \in M$. Let $x \in \bigcap_{\mathfrak{p} \in M} I_{[\mathfrak{p}]}$ be given. Consider the ideal $\mathfrak{a} := \{a \in A \mid ax \in I\}$. For every $\mathfrak{p} \in M$ the intersection $\mathfrak{a} \cap (A \setminus \mathfrak{p})$ is not empty, i.e. $\mathfrak{a} \not\subset \mathfrak{p}$. Thus $\mathfrak{a} = A$, i.e. $x \in I$. $\qquad\square$

We now look for permanence properties of Prüfer extensions.

Theorem 5.6 [Rh, Prop.3.1.3]. Assume that A is a Prüfer subring of B and B is a Prüfer subring of C. Then A is Prüfer in C.

Proof (cf. [Rh, loc.cit]). We verify for a given prime ideal $\mathfrak{p}$ of A that the pair $(A_\mathfrak{p}, \mathfrak{p}_\mathfrak{p})$ is Manis in $C_\mathfrak{p}$. Replacing A, B, C by $A_\mathfrak{p}$, $B_\mathfrak{p}$, $C_\mathfrak{p}$ we assume without loss of generality (cf. Prop.1.ii) that A is local and $\mathfrak{p}$ is the maximal ideal of A. We will apply Theorem 2.5. By this theorem (or Prop.1.3) B is local, and the maximal ideal $\mathfrak{q}$ of B is contained in $\mathfrak{p}$. Let $x \in C \setminus A$ be given. If $x \in B$ then, by Theorem 2.5, $x \in B^*$ and $x^{-1} \in \mathfrak{p}$. If $x \notin B$ then, by the same theorem, $x \in C^*$ and $x^{-1} \in \mathfrak{q} \subset \mathfrak{p}$. Thus in both cases x is a unit in C and $x^{-1} \in A$. We conclude, again by Theorem 2.5, that $(A, \mathfrak{p})$ is Manis in C. $\qquad\square$

Proposition 5.7. Assume that A is a Prüfer subring of a ring R. Then, for any ring homomorphism $\varphi: R \to D$ the ring $\varphi(A)$ is Prüfer in $\varphi(R)$.

Proof. Let C' be a subring of $\varphi(R)$ containing $\varphi(A)$. We verify that $\varphi(A)$ is weakly surjective in C', and then will be done by condition (2) in Theorem 2. Indeed, $C := \varphi^{-1}(C')$ is a subring of R containing A. Thus A is weakly surjective in C. By Proposition 3.8 $\varphi(A)$ is weakly surjective in $\varphi(C) = C'$. $\qquad\square$

This proof relied on our study of ws extensions in §3. We give a second proof, which may be more direct. It relies on the criterion (8) in Theorem 5.2.

2. Proof. Let $\bar{x} \in \varphi(R)$ be given. We have to verify that $(\varphi(A):\bar{x}) + (\varphi(A):\bar{x})\bar{x} = \varphi(A)$. We choose some $x \in R$ with $\varphi(x) = \bar{x}$. We have $(A:x) + (A:x)x = A$, since A is Prüfer in R. Applying φ, we obtain $\varphi((A:x)) + \varphi((A:x))\varphi(x) = \varphi(A)$. If $a \in (A:x)$, i.e. $a \in A$ and $ax \in A$, then $\varphi(a)\varphi(x) \in \varphi(A)$, hence $\varphi(a) \in (\varphi(A):\bar{x})$. Thus $\varphi((A:x)) \subset (\varphi(A):\bar{x})$. It follows that $(\varphi(A):\bar{x}) + (\varphi(A):\bar{x})\bar{x} = \varphi(A)$, as desired. $\square$

Also for other claims to follow different proofs are possible, since Theorem 2 provides us with many criteria for a ring extension to be Prüfer.

Proposition 5.8 [Rh, Prop.3.1.1]. Let $A \subset R$ be a ring extension and I an ideal of R contained in A. Then A is Prüfer in R iff A/I is Prüfer in R/I.

Proof. If A is Prüfer in R then the preceding proposition tells us that A/I is Prüfer in R/I. Assume now that the latter holds. We verify condition (4) in Theorem 2 and then will be done.

Let B be an R-overring of A. Then B/I is an R/I-overring of A/I. Thus B/I is integrally closed in R/I. Let $x \in R$ be integral over B. Then $x + I \in B/I$. Since $I \subset B$ we conclude that $x \in B$. Thus B is integrally closed in R. $\square$

Theorem 5.9. Let $\varphi: R \to R'$ be an integral ring homomorphism. Let A be a Prüfer subring of R, and let A' denote the integral closure of $\varphi(A)$ in R'. Then A' is a Prüfer subring of R', and $R' = A' \cdot \varphi(R)$.

Proof. We verify condition (7) in Theorem 2. Let an element x of R' and a prime ideal $\mathfrak{q}$ of R' be given. Let $\mathfrak{p} := \varphi^{-1}(\mathfrak{q})$. We look for a polynomial $G(T) \in A[T] \setminus \mathfrak{p}[T]$ with $G^\varphi(x) = 0$, where $G^\varphi(T)$ denotes the polynomial obtained from $G(T)$ by applying φ to the coefficients.

We start with a polynomial

$$F(T) = T^n + u_1 T^{n-1} + \cdots + u_n \in R[T]$$

such that $F^\varphi(x) = 0$. Such a polynomial exists since φ is integral. Let $v: R \longrightarrow \Gamma \cup \infty$ denote the Manis valuation on R with $A_v = A_{[\mathfrak{p}]}$,

$\mathfrak{p}_v = \mathfrak{p}_{[\mathfrak{p}]}$. We choose an index $r \in \{1, \ldots, n\}$ with

$$v(u_r) = \text{Min}\{v(u_i) \mid 1 \le i \le n\}.$$

We distinguish two cases.

Case 1: $v(u_r) = \infty$. Now certainly $u_i \in A_{[\mathfrak{p}]}$ for $i = 1, 2, \ldots, n$. We choose some $d \in A \setminus \mathfrak{p}$ with $du_i \in A$ for all i. The polynomial $G(T) := dF(T)$ does the job.

Case 2: $v(u_r) < \infty$. We choose some $b \in R$ with $v(bu_r) = 0$. This is possible since v is Manis. We have $bu_i \in A_{[\mathfrak{p}]}$ for every $i \in \{1, \ldots, n\}$ and $bu_r \notin \mathfrak{p}_{[\mathfrak{p}]}$. We choose some $c \in A \setminus \mathfrak{p}$ with $cbu_i \in A$ for $i = 1, \ldots, n$. The polynomial $G(T) := cbF(T)$ does the job. Since $\varphi(A)$ is weakly surjective in $\varphi(R)$ (Prop.3.8), we conclude from Prop.3.18 that $R' = A' \cdot \varphi(R)$.   □

Theorem 5.10. Let A be a subring of R and B, C be two R-overrings of A. Assume that A is Prüfer in B and weakly surjective in C. Then C is Prüfer in BC.

Proof. We pick a prime ideal $\mathfrak{q}$ of C and verify that $(C_\mathfrak{q}, \mathfrak{q}_\mathfrak{q})$ is a Manis pair in $(BC)_\mathfrak{q}$.

Let $\mathfrak{p} := \mathfrak{q} \cap A$. Then $A_\mathfrak{p} = C_\mathfrak{p} = C_\mathfrak{q}$ and $\mathfrak{q}_\mathfrak{q} = \mathfrak{q}_\mathfrak{p} = \mathfrak{p}_\mathfrak{p}$ (cf. Lemma 3.2). Thus $C \setminus \mathfrak{q}$ is the saturum of the multiplicative set $A \setminus \mathfrak{p}$ in C. Notice also that $BC = B \otimes_A C$ (Prop. 4.2). Thus $(BC)_\mathfrak{p} = B_\mathfrak{p} \otimes_{A_\mathfrak{p}} C_\mathfrak{p} = B_\mathfrak{p}$. More precisely, the subrings $(BC)_\mathfrak{p}$ and $B_\mathfrak{p}$ of $R_\mathfrak{p}$ are equal. We conclude that $(C_\mathfrak{q}, \mathfrak{q}_\mathfrak{q}) = (A_\mathfrak{p}, \mathfrak{p}_\mathfrak{p})$ and $(BC)_\mathfrak{q} = (BC)_\mathfrak{p} = B_\mathfrak{p}$. Since A is Prüfer in B, the pair $(C_\mathfrak{q}, \mathfrak{q}_\mathfrak{q})$ is Manis in $(BC)_\mathfrak{q}$.   □

Corollary 5.11. Let A be a subring of R and B, C be two R-overrings of A. If A is Prüfer in B and in C, then A is Prüfer in BC.

Proof. This follows from theorems 10 and 6.   □

Counterexample 5.12. If $A \subset B$ is a Prüfer extension and $A \subset C$ is a flat ring extension then C is not necessarily Prüfer in $B \otimes_A C$.

Here is a simple example: Let A be a non trivial valuation ring of a field K. Then A is Prüfer in K, but the polynomial ring $A[T]$ in one variable T is not Prüfer in $K[T]$.

Indeed, let $\mathfrak{m}$ be the maximal ideal of A and let $M := \mathfrak{m} + TA[T]$, which is a maximal ideal of $C := A[T]$. In the extension $K[T]$ of C we have $C_{[M]} = C$, $M_{[M]} = M$, as is easily verified. The pair (C, M) is not Manis in $K[T]$.   □

Proposition 5.13. Let $(R_\alpha \mid \alpha \in I)$ be a direct system of rings with transition homomorphisms $\varphi_{\alpha\beta} \colon R_\alpha \to R_\beta$ $(\alpha, \beta \in I, \ \alpha < \beta)$. Assume that for every $\alpha \in I$ there is given a subring A_α of R_α which is Prüfer in R_α. Assume further that $\varphi_{\alpha\beta}(A_\alpha) \subset A_\beta$ if $\alpha < \beta$. Then $A := \varinjlim_\alpha A_\alpha$ is a Prüfer subring of $R := \varinjlim_\alpha R_\alpha$.

Proof. A is clearly a subring of R. We verify that the extension $A \subset R$ has the property $(5')$ in Theorem 2. For every $\alpha \in I$ we have a canonical homomorphism $\varphi_\alpha \colon R_\alpha \to R$. Let x be an element of R which is integral over A. We choose some $\alpha \in I$ and $x_\alpha \in R_\alpha$ with $\varphi_\alpha(x_\alpha) = x$. There exists an index $\beta > \alpha$ such that $\varphi_{\beta\alpha}(x_\alpha)$ is integral over A_β, as is easily verified. Since A_β is integrally closed in R_β we have $\varphi_{\beta\alpha}(x_\alpha) \in A_\beta$, and we conclude that $x = \varphi_\alpha(x_\alpha) = \varphi_\beta(\varphi_{\beta\alpha}(x_\alpha)) \in A$. Thus A is integrally closed in R.

Let now x be any element of R. Again we choose some $\alpha \in I$ and $x_\alpha \in R_\alpha$ with $\varphi_\alpha(x_\alpha) = x$. By Theorem 2 we have $x_\alpha \in A_\alpha[x_\alpha^2]$. Applying φ_α to this relation we see that $x \in A[x^2]$.   □

Remark 5.14. As a very special case of Proposition 13 we mention that, if $A \subset R$ is a ring extension and $(B_i \mid i \in I)$ is an upward directed family of R-overrings of A with A Prüfer in each B_i, then A is Prüfer in $\bigcup_{i \in I} B_i$.   □

We now have the means to establish a theory of "Prüfer hulls" analogous to the theory of weakly surjective hulls in §3.

Theorem 5.15. Let $A \subset R$ be a ring extension. Then there exists a unique R-overring $P(A, R)$ of A, such that A is Prüfer in $P(A, R)$ and $P(A, R)$ contains every R-overring of A in which A is Prüfer.

Proof. Let $\mathcal{P}$ denote the set of all R-overrings B of A with A Prüfer in B. We regard $\mathcal{P}$ as a partially ordered set, the ordering being given by the inclusion relation. By Remark 14 and Zorn's lemma it is clear that $\mathcal{P}$ has a maximal element C. Now, if B is any element of $\mathcal{P}$, then $BC \in \mathcal{P}$ by Corollary 11. Thus $BC = C$, i.e. $B \subset C$. $\square$

Definition 2. We call $P(A, R)$ the *Prüfer hull of A in R*.

Of course, $P(A, R)$ is contained in the weakly surjective hull $M(A, R)$ of A in R, and $P(A, R) = P(A, C)$ for every R-overring C with $C \supset P(A, R)$. More generally, if C is any R-overring then $P(A, C) = C \cap P(A, R)$. Also $P(A, R) = P(B, R)$, if B is any R-overring of A contained in $P(A, R)$.

Definition 3. For any ring A the *Prüfer hull $P(A)$ of A* is defined as the Prüfer hull of A in the complete quotient ring $Q(A)$ (cf. §3), $P(A) := P(A, Q(A))$.

Remarks 5.16. $P(A)$ is contained in the weakly surjective hull $M(A)$. The classical Prüfer rings (with zero divisors) are precisely the rings A with $\operatorname{Quot} A \subset P(A)$. If A' is a weakly surjective ring extension of A, contained in $M(A)$ without loss of generality, then $A' \cdot P(A) \subset P(A')$ by Theorem 10 above. If $A \subset B$ is any Prüfer extension, it is clear by Proposition 3.14 that there is a unique homomorphism $\varphi \colon B \to P(A)$ over A, and φ is injective.

Example 5.17. Assume that A is Prüfer domain (in the classical sense), i.e. A is an integral domain, and for every maximal ideal $\mathfrak{p}$ the ring $A_{\mathfrak{p}}$ is a Krull valuation ring of the quotient field $K = \operatorname{Quot} A$. Then $A \subset K$ is a Prüfer extension. Since we have $K = Q(A)$, we conclude that $P(A) = K$. In particular, if A is a Dedekind domain then $P(A) = Q(A) = \operatorname{Quot} A$. If A is any integral domain, then clearly A is a Prüfer domain *if and only if A is Prüfer in* $\operatorname{Quot} A$. $\square$

Example 5.18. Assume that A is a ring with Krull dimension $\dim A = 0$. Then $P(A) = A$.

This can be seen as follows. Let $R := P(A)$ and suppose that $R \neq A$. We choose a maximal ideal $\mathfrak{m}$ of A with $A_{\mathfrak{m}} \neq R_{\mathfrak{m}}$. Then we have a nontrivial Manis valuation v of $R_{\mathfrak{m}}$ with $A_v = A_{\mathfrak{m}}$. In $A_{\mathfrak{m}}$ we have the two different prime ideals $\operatorname{supp} v \underset{\neq}{\subset} \mathfrak{p}_v$. Thus $\dim A_\mathfrak{m} \geq 1$, contradicting $\dim A = 0$. It follows that $R = A$. $\square$

Example 5.19. Let V be an affine algebraic variety over some real closed field k. The ring R of (k-valued, continuous) semialgebraic functions on the set $V(k)$ of rational points of V is "Prüfer closed", i.e. $P(R) = R$. This has been proved recently by Niels Schwartz [Sch$_2$] within the framework of his theory of real closed rings (cf. [Sch$_2$, Example 5.13]. His proof would take us here too far afield.

Let d be a natural number. We will see that R is Prüfer over the subring $A = k\left[\frac{1}{1+x^{2d}} \,|\, x \in R\right]$ generated by k and the elements $\frac{1}{1+x^{2d}}$, $x \in R$, cf. below §6, Example 12. Thus $R = P(A)$. $\square$

We now briefly discuss the behavior of Prüfer extensions under formation of direct products. Let $(A_i \subset R_i \,|\, i \in I)$ be a family of ring extensions. This gives us a ring extension $A \subset R$ with $A := \prod_{i \in I} A_i$ and $R := \prod_{i \in I} R_i$.

Proposition 5.20. A is Prüfer in R iff A_i is Prüfer in R_i for every $i \in I$.

Proof. We use the criterion (8) in Theorem 2. Let $x = (x_i | i \in I) \in R$ be given. We clearly have $(A : x) = \prod_{i \in I} (A_i : x_i)$ and $x(A : x) = \prod_{i \in I} x_i(A_i : x_i)$. If A_i is Prüfer in R_i for every $i \in I$, then we have equations $1_i = u_i + x_i v_i$ with 1_i the unit element of A_i, and u_i, v_i elements of $(A_i : x_i)$. With $u := (u_i | i \in I)$ and $v := (v_i | i \in I)$ this gives us an equation $1 = u + xv$ with u and v elements of $(A : x)$. Thus A is Prüfer in R.

Conversely, if we have an equation $1 = u + xv$ with $u, v \in (A : x)$ then we obtain equations $1 = u_i + x_i v_i$ with $u_i, v_i \in (A_i : x_i)$ for every $i \in I$. Thus, if A is Prüfer in R, then A_i is Prüfer in R_i for every $i \in I$. (This is also clear from Proposition 7.) $\square$

Proposition 5.21. Let $(A_i | i \in I)$ be a family of rings. For every $i \in I$ we have the Prüfer hull $A_i \subset P(A_i)$. Then $A := \prod_{i \in I} A_i$ has the

Prüfer hull $P(A) = \prod\limits_{i \in I} P(A_i)$ (where, of course, we regard A as a subring of $\prod\limits_{i \in I} P(A_i)$ in the obvious way).

Proof. It is well known, that $Q(A) = \prod\limits_{i \in I} Q(A_i)$, cf. [Lb, p.41 and p.100]. By the preceding proposition the extension $A \subset \prod\limits_{i \in I} P(A_i)$ is Prüfer. Thus we have a chain of inclusions

$$A = \prod_{i \in I} A_i \subset \prod_{i \in I} P(A_i) \subset P(A) \subset \prod_{i \in I} Q(A_i).$$

Fixing some $j \in I$ let π_j denote the natural projection map from $\prod\limits_{i \in I} Q(A_i)$ to $Q(A_j)$. We have $\pi_j(A) = A_j$ and $\pi_j(\prod\limits_{i \in I} P(A_i)) = P(A_j)$, hence

$$A_j \subset P(A_j) \subset \pi_j(P(A)).$$

Now A_j is Prüfer in $\pi_j(P(A))$ by Proposition 7. Thus also $P(A_j)$ is Prüfer in $\pi_j(P(A))$ (cf. Cor.3). This forces $P(A_j) = \pi_j(P(A))$. Since this holds for every $j \in I$, we conclude that $\prod\limits_{i \in I} P(A_j) = P(A)$.

$\square$

We will say more about Prüfer hulls in II, §5 and in Part II of the book.

§6 Examples of Prüfer extensions and convenient ring extensions

In this section R is a ring and A is a subring of R. We are looking for handy criteria which guarantee that A is Manis or Prüfer in R, and we will discuss examples emanating from some of these criteria.

Theorem 6.1. Assume that A is integrally closed in R. Assume further that for every $x \in R \setminus A$ there exists a monic polynomial $F(T) \in A[T]$ and a unimodular polynomial $G(T) \in A[T]$ (i.e. the ideal of A generated by all coefficients of $G(T)$ is A), such that

$F(x) \in R^*$, $\deg G < \deg F$ and $G(x)/F(x) \in A$. Then A is Prüfer in R.

Proof. We verify that for a given element x of R and a given maximal ideal $\mathfrak{m}$ of A there exists a polynomial $H(T) \in A[T] \setminus \mathfrak{m}[T]$ with $H(x) = 0$, and then will be done by Theorem 5.2.

If $x \in A$ we take $H(T) = T - x$. Now let $x \in R \setminus A$. We choose polynomials $F(T)$, $G(T)$ as indicated in the theorem. We put $b := G(x)/F(x) \in A$ and take $H(T) := bF(T) - G(T)$. Then $H(x) = 0$. If $b \in \mathfrak{m}$ then $H(T) \notin \mathfrak{m}[T]$, since $G(T)$ is unimodular. If $b \notin \mathfrak{m}$ then again $H(T) \notin \mathfrak{m}[T]$, since $\deg G < \deg F$ and F is monic. $\qquad\square$

Definition 1. We call a valuation v on R a *Prüfer-Manis valuation* (or *PM-valuation*, for short), if v is Manis and A_v is Prüfer in R. We call a subring B of R a *Prüfer-Manis subring* of R if $B = A_v$ for some Prüfer-Manis valuation v on R. We then also say that the ring B is *Prüfer-Manis* (or *PM*, for short) *in* R, or that the extension $A \subset B$ is PM.

If A is Prüfer in R and B is an R-overring of A which is Manis in R, then, of course, B is PM in R. Thus the valuations which really matter in the theory of relative Prüfer rings are the PM-valuations and not just the Manis valuations. We defer a systematic theory of PM-subrings of R to Chapter III, but now look for examples of such rings.

Theorem 6.2. Assume that $A \neq R$ and the set $S := R \setminus A$ is multiplicatively closed. Assume further that for every $x \in R \setminus A$ there exists a monic polynomial $F(T) \in A[T]$ of degree ≥ 1 with $F(x) \in R^*$. Then A is PM in R.

Proof. We verify that A is Prüfer in R and then will be done by Prop. 5.1.iii. We know from Theorem 2.1 that A is integrally closed in R. Let $x \in R \setminus A$ be given. We choose a polynomial $F(T) \in A[T]$ as indicated in the theorem. Certainly $F(x) \in R \setminus A$, since A is integrally closed in R. We conclude from the equation $1 = F(x) \cdot F(x)^{-1}$ that $1/F(x) \in A$, since otherwise we would get the contradiction $1 \in R \setminus A$. Now Theorem 1 tells us that A is Prüfer. $\qquad\square$

Definition 2. a) Let k be a subring of R. We say that R is *convenient over k*, if every R-overring A of k which has a multiplicatively closed complement $R \setminus A$ is PM in R.
b) We call the ring R *convenient*, if R is convenient over its prime ring $\mathbb{Z} \cdot 1$.

Example 1. Every field is a convenient ring. □

The idea behind Definition 2 is that, as far as valuations are concerned, a convenient ring is nearly as "convenient" as a field. If R is only convenient over some subring k then at least this should be true for the (special) valuations v with $A_v \supset k$. In particular we expect that for a convenient ring extension $k \subset R$ we have a theory of R-Prüfer rings $A \supset k$ nearly as good as in the field case.

We now look first for examples of convenient rings, then for the – still interesting – case of convenient ring extensions.

Example 2 (Generalization of Example 1)**.** If R has Krull dimension zero then R is convenient.

Proof. Let A be a subring of R with $A \neq R$ and $R \setminus A$ multiplicatively closed. We prove that A is Prüfer in R. Then it will follow from Prop. 5.1.iii that A is also Manis in R.

The ring A is integrally closed in R by Theorem 2.1.ii. Given an element $x \in R$ we prove that there exists a unimodular polynomial $F(T) \in A[T]$ with $F(x) = 0$. Then we will be done by Theorem 5.2.

If $x \in A$ take $F(T) = T - x$. Now let $x \in R \setminus A$. There exists some $n \in \mathbb{N}$ and $y \in R$ with $x^{n+1}y = x^n$, cf. [Huc, Th.3.1]. Then $(xy)^{n+1} = (xy)^n$. Since A is integrally closed in R, this implies $xy \in A$. Since $R \setminus A$ is closed under multiplication we conclude that $y \in A$. The polynomial $F(T) = yT^{n+1} - T^n$ fits our needs. □

Example 3. Every ring R with $1 + \Sigma R^2 \subset R^*$ is convenient.[*]
Indeed, it suffices to know that $1 + R^2 \subset R^*$ in order to conclude that R is convenient. □

---

[*] $1 + \Sigma R^2$ denotes the set of elements $1 + x_1^2 + \cdots + x_n^2$ with $n \in \mathbb{N}, x_i \in A$.

Comment. This is the most important class of rings we have in mind for use in real algebra. Recall that for every ring A the localization $\Sigma^{-1}A$ with respect to the multiplicative set $\Sigma := 1 + \Sigma A^2$ is such a ring, and that A and $\Sigma^{-1}A$ have the same real spectrum. For many problems in real algebra we may replace A by $\Sigma^{-1}A$ and thus arrive at a convenient ring. {If A is not real, i.e. $-1 \in \Sigma A^2$, then $\Sigma^{-1}A$ is the null ring, but this does not bother us.} □

Subexample 3 bis. If A is any ring and X is a proconstructible subset of the real spectrum Sper A then the ring $CS(X, A)$ of abstract semialgebraic functions on X (i.e. the real closure of A on X, cf. [Sch] or [Sch$_1$]) is convenient, since in this ring R we have $1 + \Sigma R^2 \subset R^*$. In general $CS(X, A)$ has very many zero divisors. □

Example 4. If more generally R is a ring such that, for every $x \in R$, there exists a natural number d with $1 + x^d \in R^*$, then R is convenient. □

Such rings (with d even) seem to be important in the theory of orderings of higher level and higher real spectra (cf. e.g. [B$_2$], [B$_3$], [P], [BP], [Be]).

Using some more elementary commutative ring theory we can generalize Example 2 greatly. We start with a well known lemma.

Lemma 6.3. Assume that $\dim R = 0$ and Nil $R = 0$, i.e. R is von Neumann regular. For every $a \in R$ there exists some $c \in R$ such that $ac = 0$ and $a + dc \in R^*$ for every $d \in R^*$.

Proof. There exists some $b \in R$ such that $a^2 b = a$. Taking $c := 1 - ab$ we see that $ac = a - a^2 b = 0$. For every prime ideal $\mathfrak{m}$ of R either $a \in \mathfrak{m}$ or $c \in \mathfrak{m}$. If $a \in \mathfrak{m}$ then $ab \in \mathfrak{m}$, which yields $c \notin \mathfrak{m}$, hence $a + dc \notin \mathfrak{m}$ for every $d \in R^*$. If $a \notin \mathfrak{m}$, then $c \in \mathfrak{m}$, hence $a + dc \notin \mathfrak{m}$ for every $d \in R^*$. This proves $a + dc \in R^*$ for every $d \in R^*$. □

Let J denote the Jacobson radical of R.

Definition 3 (cf. [G$_2$]). We say that R *has large Jacobson radical*, if R/J has Krull dimension zero. □

Proposition 6.4 (G$_2$, Prop.19]. The following are equivalent (for R any ring).

(1) R has large Jacobson radical.
(2) For every $x \in R$ there exists some $y \in R$ such that $xy \in J$ and $x + dy \in R^*$ for every $d \in R^*$.
(3) For every $x \in R$ there exists some $y \in R$ with $xy \in J$ and $x + y \in R^*$.

Proof. (1) $\Rightarrow$ (2) follows from the preceding lemma, and (2) $\Rightarrow$ (3) is trivial.
(3) $\Rightarrow$ (1): Suppose there exist prime ideals $\mathfrak{p}, \mathfrak{m}$ of R with $J \subset \mathfrak{p} \subsetneq \mathfrak{m}$.
We choose some $x \in \mathfrak{m} \setminus \mathfrak{p}$ and have by assumption some $y \in R$ with $xy \in J$ and $x + y \in R^*$. Since $xy \in \mathfrak{p}$ but $x \notin \mathfrak{p}$, we have $y \in \mathfrak{p}$. This implies $x + y \in \mathfrak{m}$, a contradiction. Thus $\dim R/J = 0$. $\square$

Theorem 6.5. If R has large Jacobson radical then R is convenient.

Proof. Let A be a subring of R such that $R \setminus A$ is multiplicatively closed. Let $x \in R \setminus A$ be given. By Theorem 2 we are done if we find a monic polynomial $F[T] \in A[T]$ of degree ≥ 1 with $F(x) \in R^*$. By the preceding proposition 4 there exists some $y \in R$ such that $x + dy \in R^*$ for every $d \in R^*$.
Case 1: $y \in A$. Now the polynomial $F(T) = T + y$ fits our needs.
Case 2: $y \notin A$. Now $xy \notin A$, hence $1 + xy \notin A$. But $1 + xy \in R^*$ since $xy \in J$. Take $d := \frac{1}{1+xy} \in R^*$. We have $d \in A$ since $(1+xy)d = 1 \in A$, but $1 + xy \notin A$. From $xyd = 1 - d \in A$ it follows in the same way that $yd \in A$. The polynomial $F(T) = T + yd$ fits our needs.[*] $\square$

Examples 5. If R is semilocal, i.e. has only finitely many maximal ideals, then it is obvious that R has large Jacobson radical. Thus every semilocal ring is convenient. Slightly more generally, if R' is an integral extension of a semilocal ring R, then R' has large Jacobson radical, hence is convenient. In particular the infinite Galois extensions of R (cf. e.g.[K], there called infinite "coverings" of R) are convenient rings, as well as the finite ones. $\square$

Proposition 6.6. Assume there exists a family $(\varphi_\alpha : R_\alpha \to R \mid \alpha \in I)$ of homomorphisms from convenient rings R_α to R such that R is the union of the subrings $\varphi_\alpha(R_\alpha)$. Then R is convenient.

---

[*] In fact we proved that the ring extension $A \subset R$ is "additively regular", cf. Def.3 in Chapter III, §3 below.

Proof. Let A be a subring of R such that the set $R \setminus A$ is multiplicatively closed and not empty. We verify that A is Prüfer in R. Then we will know that A is also Manis in R by Prop.5.1.iii.

We will use criterion (5') in Theorem 5.2. We know from Theorem 2.1 that A is integrally closed in R. Let $x \in R$ be given. We choose some $\alpha \in I$ and $x_\alpha \in R_\alpha$ such that $\varphi_\alpha(x_\alpha) = x$. Let $A_\alpha := \varphi_\alpha^{-1}(A)$. Then $R_\alpha \setminus A_\alpha$ is closed under multiplication. Since R_α is convenient, it follows that A_α is Prüfer (even PM) in R_α. Thus $x_\alpha \in A_\alpha[x_\alpha^2]$ by Theorem 5.2. Applying φ_α to this relation we obtain $x \in A[x^2]$. This proves that A is Prüfer in R. ∎

Remark 6.7. In particular, every homomorphic image of a ring with large Jacobson radical is convenient.

Example 6. Let $(X_\alpha \mid \alpha \in I)$ be an inverse system of quasiprojective schemes over some field k with transition maps $f_{\alpha\beta} : X_\beta \to X_\alpha$ $(\alpha, \beta \in I, \ \alpha < \beta)$. Assume that for every $\alpha \in I$ we are given a finite set of points $S_\alpha \subset X_\alpha$ such that $f_{\alpha\beta}(S_\beta) \subset S_\alpha$ if $\alpha < \beta$. We introduce the semilocal rings $R_\alpha := \mathcal{O}_{X_\alpha, S_\alpha}$ $(\alpha \in I)$. If $\alpha, \beta \in I$ and $\alpha < \beta$ then $f_{\alpha\beta} : X_\beta \to X_\alpha$ gives us a ring homomorphism $\varphi_{\alpha\beta} : X_\alpha \to X_\beta$. Thus we obtain a direct system $(R_\alpha, \varphi_{\alpha\beta})$ of semilocal rings. Let $R := \varinjlim_\alpha R_\alpha$ be the associated direct limit. R is convenient by Theorem 5 and Proposition 6. Such rings may be useful in the business of resolution of singularities and related matters. ∎

We now give examples of convenient ring *extensions*. From Theorem 2 we extract

Scholium 6.8. Let k be a subring of R with the following property. (∗) For every $x \in R \setminus k$ there exists some monic polynomial $F_x(T) \in k[T]$, $F_x \neq 1$, with $F_x(x) \in R^*$. Then R is convenient over k. ∎

Example 7. Let A be an affine algebra over a field k which is not algebraically closed. Let $V(k)$ denote the set of rational points of the associated k-variety V. {We may identify $V(k) = \mathrm{Hom}_k(A, k)$.} Let U be a k-Zariski-open subset of $V(k)$. {In other words, U is open in the subspace topology of $V(k)$ in $\mathrm{Spec}\, A$.} Let finally S be the multiplicative set consisting of all $a \in A$ with $a(p) \neq 0$ for every $p \in U$. Then $S^{-1}A$ is convenient over k.

Proof. We choose a monic polynomial $F(T) \in k[T]$, $F \neq 1$, in one variable T which has no zeros in k. Let $x \in S^{-1}A$ be given. Write $x = \frac{a}{s}$ with $a \in A$, $s \in S$, and write $F(T) = T^d + c_1 T^{d-1} + \cdots + c_d$. We have $F(x) = \frac{b}{s^d}$ with $b = a^d + c_1 a^{d-1} s + \cdots + c_d s^d \in A$. For every point $p \in U$ we have $\frac{b(p)}{s(p)^d} = F\left(\frac{a(p)}{s(p)}\right) \neq 0$, hence $b(p) \neq 0$. Thus $b \in S$ and $F(x)$ is a unit in $S^{-1}A$. □

Definition 4. We call this ring $S^{-1}A$ the *ring of regular functions on U*.

If the field k is real closed and $U = V(k)$ then S is the set of divisors of the elements in $\Sigma := 1 + \Sigma A^2$, as is well known (e.g. [BCR, Cor. 4.4.5.], [KS, p.142]). Thus $S^{-1}A = \Sigma^{-1}A$, and we are back to Example 3.

Example 8. Assume that R is the total quotient ring of A, $R = \text{Quot}\, A$. The ring A is called *additively regular* [Huc, p.32], if for every $x \in R$ there exists some $a \in A$ such that $x + a$ is a "regular element", i.e. a unit in R. Of course, then condition $(*)$ is satisfied for $k := A$, and thus R is convenient over A. As Huckaba observes [Huc, p.32 f], if A is noetherian or, more generally, if the set of zero divisors of A is a union of finitely many prime ideals, then A is additively regular [Huc, p.32 f]. □

Example 9. Assume again that $R = \text{Quot}\, A$. The ring A is called a *Marot ring* [Huc, p.31], if each ideal of A, which contains a nonzero divisor, is generated by a set of non zero divisors. Marot rings form a very broad class of rings. In particular, every additively regular ring is Marot [Huc, p.33 f]. If A is Marot, $\text{Quot}\, A$ is convenient over A, cf. [Huc, Th.7.7 and Cor.7.8] and III, §3 below. But now condition $(*)$ may be violated, as we can show by examples. □

As before R denotes a ring and A a subring of R. We return to the search for Prüfer subrings of R which are not necessarily Manis in R.

Example 10. If A is Prüfer in R then R is convenient over A.

This in essence is the content of Proposition 5.1.iii   □

If R is a field then the intersection of finitely many valuation subrings of R is Prüfer in R, as is well known [Gi, p.280f]. Does the same hold if $k \subset R$ is a convenient extension and if all the valuation rings contain k? Or does this at least hold if the extension $k \subset R$ fulfills the stronger condition (∗) in 6.8? We can only prove partial results, which nevertheless seem to deserve interest.

Lemma 6.9. Let k be a subring of R, and let $v_1, \ldots, v_n$ be valuations on R with $A_{v_i} \supset k$ for every $i \in \{1, \ldots, n\}$. Given an element x of R, there exists a monic polynomial $F(T) \in k[T]$ with $F(0) = 0$ and the following property:
If $G(T) \in k[T]$ is any monic polynomial of degree ≥ 1 with absolute term $G(0) \in k^*$, then $v_i(G(F(x))) = 0$ if $v_i(x) \geq 0$, and $v_i(G(F(x))) < v_i(x)$ if $v_i(x) < 0$ $(1 \leq i \leq n)$.

Proof. For every $i \in \{1, \ldots, n\}$ we choose a monic polynomial $F_i(T) \in k[T]$ with $v_i(F_i(x)) > 0$, if such a polynomial exists. Otherwise we put $F_i(T) = 1$. We claim that the polynomial $F(T) := T^2 F_1(T) \ldots F_n(T)$ fulfills the requirements of the lemma.

Clearly $F(T)$ is monic and $F(0) = 0$. Let $G(T) \in k[T]$ be a monic polynomial of degree ≥ 1 with $G(0) \in k^*$, and let $H(T) := G(F(T))$.
Case 1. $v_i(x) < 0$. Now $v_i(H(x)) = \deg(H) \, v_i(x)$, since $H(T)$ is monic and has coefficients in A_{v_i}. We have $\deg H \geq 2$, hence $v_i(H(x)) < v_i(x)$.
Case 2. $v_i(x) \geq 0$. We have $v_i(H(x)) \geq 0$. Suppose that $v_i(H(x)) > 0$. Then $v_i(F_i(x)) > 0$, due to our choice of $F_i(T)$, hence $v_i(F(x)) > 0$. Since $G(T)$ is monic with coefficients in A_{v_i} and absolute term in $A_{v_i}^*$, we have $v_i(H(x)) = v_i(G(F(x))) = 0$. This is a contradiction. We conclude that $v_i(H(x)) = 0$.   □

Theorem 6.10 [G$_2$, Prop.22]. Assume that R has large Jacobson radical. Let $v_1, \ldots, v_n$ be special valuations on R. Then the v_i are PM and $A := \bigcap_{i=1}^{n} A_{v_i}$ is Prüfer in R.

Proof. Let $A_i := A_{v_i}$. We may assume that $A_i \neq R$ for every i. Theorem 5 tells us that the v_i are PM. Let $x \in R$ be given. We prove that there exists a monic polynomial $H(T) \in A[T]$ of degree ≥ 1 with $H(x) \in R^*$ and $1/H(x) \in A$. Then we know by Theorem 1 that $A \subset R$ is Prüfer. We apply the preceding Lemma with k the prime ring $\mathbb{Z} \cdot 1_R$ in R and $G(T) = T + 1$. Thus we have a monic polynomial $F(T) \in \mathbb{Z}[T]$ with $F(0) = 0$ and $v_i(1 + F(x)) = 0$ if $v_i(x) \geq 0$, $v_i(1 + F(x)) < v_i(x)$ if $v_i(x) < 0$.

Let $y := 1 + F(x)$. We have $v_i(y) \leq 0$ for every $i \in \{1, \ldots, n\}$. Since R has large Jacobson radical J there exists some $z \in R$ such that $yz \in J$ and $y + dz \in R^*$ for every $d \in R^*$ (cf. Prop. 4 above).

We will show that there exists an element $d \in R^*$ with $v_i(dz) > 0$ for $1 \leq i \leq n$. Then we can finish as follows: $dz \in A$, and $y + dz \in R^*$. Taking $H(T) := F(T) + 1 + dz \in A[T]$, we have $H(x) = y + dz \in R^*$ and $v_i(H(x)) = v_i(y) \leq 0$ for $1 \leq i \leq n$, hence $1/H(x) \in A$.

In order to exhibit an element $d \in R^*$ with $v_i(dz) > 0$ for $1 \leq i \leq n$ it suffices to find for every $i \in \{1, \ldots, n\}$ an element $d_i \in R^*$ with $d_i \in A$ and $v_i(d_i z) > 0$. Then $d = d_1 \ldots d_n$ has the required properties.

Case 1. $v_i(z) > 0$. Take $d_i = 1$.

Case 2. $v_i(z) \leq 0$. Now $v_i(yz) \leq 0$. We choose some $u \in R$ with $v_i(u) < 0$. This is possible since $A_i \neq R$. Applying Lemma 9 to the element uyz we obtain a monic polynomial $\Phi(T) \in k[T]$ with $\Phi(0) = 0$, such that $v_i(\Phi(uyz) + 1) < v_i(uyz) < v_i(z)$ and $v_j(\Phi(uyz) + 1) \leq 0$ for every $j \in \{1, \ldots, n\}$. The element $\Phi(uyz) + 1$ is a unit in R since it is of the form $cyz + 1$ with $c \in R$ and since $yz \in J$. We take $d_i := \frac{1}{\Phi(uyz) + 1}$. Then $d_i \in A \cap R^*$ and $v_i(d_i z) > 0$. $\qquad\square$

Starting from Theorem 10 we obtain a new class of examples of Prüfer extensions as follows.

Corollary 6.11. Assume there exists a family $(\varphi_\alpha : R_\alpha \to R \mid \alpha \in I)$ of ring homomorphisms such that R is the union of the subrings $\varphi_\alpha(R_\alpha)$, $i \in I$, and every R_α has large Jacobson radical. Assume further that W is a set of valuations on R such that for every $\alpha \in I$ the set of valuations $\{w \circ \varphi_\alpha \mid w \in W\}$ on R_α is finite. Then the intersection A of the rings A_w, $w \in W$, is Prüfer in R.

Proof. We verify condition $(5')$ in Theorem 5.2. The ring A is integrally closed in R. Let $x \in R$ be given. We choose some $\alpha \in I$ and $x_\alpha \in R_\alpha$ with $\varphi_\alpha(x_\alpha) = x$. Due to Theorem 10 the ring

$A_\alpha := \varphi_\alpha^{-1}(A)$ is Prüfer in R_α. Thus $x_\alpha \in A_\alpha[x_\alpha^2]$ by Theorem 5.2. Applying φ_α to this relation we obtain $x \in A[x^2]$. Again by Theorem 5.2 we conclude that A is Prüfer in R. $\qquad \square$

Theorem 6.12. Let k be a subring of R with the following property. (**) For every $x \in R \setminus k$ there exists a monic polynomial $\Phi_x(T) \in k[T]$, $\Phi_x \neq 1$, with $\Phi_x(x) \in R^*$ and constant term $\Phi_x(0) \in k^*$.

Let $v_1, \ldots, v_n$ be valuations on R with $A_{v_i} \supset k$ for all i. Then the intersection A of the rings A_{v_i} is Prüfer in R.

Proof. A is integrally closed in R and $k \subset A$. Let $x \in R \setminus A$ be given. We prove that there exists a monic polynomial $H(T) \in k[T]$ of degree ≥ 1 with $H(x) \in R^*$ and $1/H(x) \in A$, and then will be done by Theorem 1.

We choose a polynomial $F(T) \in k[T]$ for x as indicated in Lemma 9. Let $y := F(x)$. Certainly $y \notin A$, since $x \notin A$ and A is integrally closed in R. A fortiori $y \notin k$. The polynomial $H(T) := \Phi_y(F(T))$ fits our needs, due to Lemma 9. $\qquad \square$

Notice that, for k a subfield of a ring R, the previous condition (*) (cf. 6.8) implies (**). In particular (**) holds in Example 7 above. (**) holds also in the examples 1, 3, 4 for k the prime ring in R.

Definition 5. Let $F(T) \in R[T]$ be a nonconstant monic polynomial. Let v be a valuation on R. We call v an F-*valuation*, if $v(c) \geq 0$ for every coefficient c of F and $F(T)$ has no zero in the residue class field $\kappa(v) = \mathfrak{o}_v/\mathfrak{m}_v$. {Of course, this means that the image polynomial $\bar{F}(T) \in \kappa(v)[T]$ has no zero in $\kappa(v)$.}

Theorem 6.13. Let $(v_i \mid i \in I)$ be a family of valuations on R. Assume that A is the intersection of the valuation rings A_{v_i} $(i \in I)$. Assume also that for each $x \in R \setminus A$ there exists a monic polynomial $F_x(T) \in A[T]$ of degree $d_x \geq 1$, such that $F_x(x) \in R^*$ and every v_i is an F_x-valuation. Then A is Prüfer in R.

Proof. A is integrally closed in R. By Theorem 1 we are done if we verify that $1/F_x(x) \in A$ for each $x \in R \setminus A$, i.e. $v_i(F_x(x)) \leq 0$ for $x \in R \setminus A$ and $i \in I$. If $v_i(x) < 0$ then $v_i(F_x(x)) = d_x \cdot v_i(x) < 0$. If $v(x_i) \geq 0$ then $x \in A_{v_i}$, and $v_i(F_x(x)) = 0$ since v_i is an F_x-valuation. $\qquad \square$

Here we quote the seminal paper [R] by Peter Roquette, which in the case, that R is a field, bears close connection to Theorem 13. Roquette also obtained results on class groups which allow to conclude in important cases that A has trivial class group, hence is a Bézout domain. Our Theorem 13 generalizes the first part of [R, Theorem 1]. The second part, dealing with the class group of A, will be generalized in §7.

Example 11. Let R be any ring and let q be a power of a prime number p. Let us call a valuation v on R a *q-valuation*, if the residue class field $\kappa(v)$ is a finite field having p^r elements with $p^r \leq q$. Assume that R admits at least one q-valuation v_0. Then $x^q - x - 1 \neq 0$ for every $x \in R$, since $v_0(x^q - x - 1) \leq 0$. Assume further that $x^q - x - 1 \in R^*$ for every $x \in R$. (If R has not this property, replace R by $R\left[\frac{1}{x^q-x-1} \mid x \in R\right]$. Notice that every q-valuation on R extends to a q-valuation of this ring.) Now it follows from Theorem 13 that the intersection of the valuation rings A_v of all q-valuations on R is Prüfer in R. In particular, the q-valuations on R are PM. $\square$

This observation generalizes the well known fact, important in the theory of formally p-adic fields, that for K a formally p-adic field the so called γ-Kochen ring $\mathfrak{o}[\gamma K]$ is Prüfer in K (cf. [PR, §6]; $\mathfrak{o}[\gamma K]$ is even Bezout).

We now aim at criteria that A is Prüfer in R, which do not assume in advance that A is integrally closed in R. A prototype of the criteria to follow is a lemma by A. Dress, which states for R a field of characteristic not 2, that the subring of R generated by the elements $1/(1 + a^2)$ with $a \in F$, $a^2 \neq -1$, is Prüfer in R, cf. [D, Satz 2′], [KS, Chap III §12], [La, p.86].[*)]

Theorem 6.14. Assume that for every $x \in R \setminus A$ there exists some monic polynomical $F(T) \in A[T]$ of degree ≥ 1 with $F(x) \in R^*$, $\frac{1}{F(x)} \in A$, $\frac{x}{F(x)} \in A$. Then A is Prüfer in R.

Proof. Let B be an R-overring of A and $S := A \cap B^*$. We verify that $B = S^{-1}A$. Then we know that A is weakly surjective in every R-overring, and will be done by Theorem 5.2.

[*)] Actually Dress made the slightly stronger assumption that -1 is not a square in F.

Of course, $S^{-1}A \subset B$. Let $x \in B \setminus A$ be given. We choose a polynomial $F(T)$ as indicated in the theorem. $s := \frac{1}{F(x)} \in A \subset B$. Also $F(x) \in B$, hence $s \in S$. By assumption $a := \frac{x}{F(x)} \in A$. Thus $x = \frac{a}{s} \in S^{-1}A$. $\qquad\square$

The following remark sheds additional light both on Theorem 14 and Theorem 1.

Remark 6.15. Assume that A is integrally closed in R (e.g. A is Prüfer in R). Let $x \in R$ and let $F(T) \in A[T]$ be a monic polynomial of degree $n \geq 1$ with $F(x) \in R^*$ and $\frac{1}{F(x)} \in A$. Then $\frac{x^r}{F(x)} \in A$ for $0 \leq r \leq n$.

Proof (cf. [Gi, p.154]). We proceed by induction on r. For $r = 0$ the assertion is trivial. Assume that $1 \leq r \leq n$ and that $\frac{x^s}{F(x)} \in A$ for $0 \leq s < r$.
We write

$$F(T)^r = T^{nr} + \sum_{j=1}^{n} h_j(T)T^{(n-j)r}$$

with polynomials $h_j(T) \in A[T]$ of degree $< r$. The relation

$$\frac{1}{F(x)^{n-r}} = \frac{F(x)^r}{F(x)^n} = \left(\frac{x^r}{F(x)}\right)^n + \sum_{j=1}^{n} \frac{1}{F(x)^{j-1}} \frac{h_j(x)}{F(x)} \cdot \left(\frac{x^r}{F(x)}\right)^{n-j}$$

proves that $\frac{x^r}{F(x)}$ is integral over A, since by induction hypothesis $\frac{h_j(x)}{F(x)} \in A$ for every $j \in \{1,\ldots,n\}$. Thus $\frac{x^r}{F(x)} \in A$. $\qquad\square$

Theorem 14 and this remark imply an improvement of Example 11 as follows.

Example 12. Let again R be a ring and q a power of a prime number. Let $\varepsilon = +1$ or $\varepsilon = -1$. We assume that $x^q - x + \varepsilon \in R^*$ for every $x \in R$. Let A denote the subring of R generated by the elements $\frac{1}{x^q-x+\varepsilon}$ and B denote the subring of R generated by the elements $\frac{1}{x^q-x+\varepsilon}, \frac{x}{x^q-x+\varepsilon}$, with x running through R in both cases. B is Prüfer in R by Theorem 14. In particular B is integrally closed in R. Remark 15 tells us that B is integral over A. Thus B is the integral closure of A in R. It is evident that A and hence B is contained in the intersection H of the valuation rings A_v with v running through all q-valuations of R. $\qquad\square$

Theorem 6.16. Let k be a subring of R. (We will often choose for k the prime ring in R.) Let $F(T) \in k[T]$ be a monic polynomial of degree $d \geq 1$. Assume that $d! \in R^*$ and that $F(x) \in R^*$ for every $x \in R$ with $F(x) \notin k$. The subring A of R generated by k, the element $1/d!$ and the set $\{1/F(x)|x \in R, F(x) \notin k\}$ is Prüfer in R.

Proof. a) Let $B := \tilde{A}$, the integral closure of A in R. By Theorem 1 B is Prüfer in R. We now verify that for a given prime ideal $\mathfrak{p}$ of A we have $B \subset A_{[\mathfrak{p}]}$. Then we may conclude (Remark 5.5) that $B \subset \bigcap_{\mathfrak{p} \in \mathrm{Spec} A} A_{[\mathfrak{p}]} = A \subset B$, i.e. $B = A$, and will be done.

b) We first prove that for any $x \in B$ we have $F(x) \in A_{[\mathfrak{p}]}$. Put $y := F(x) - 1$. Suppose $F(x) \notin A_{[\mathfrak{p}]}$, hence $y \notin A_{[\mathfrak{p}]}$. Clearly $F(x) \notin k$. By hypothesis $1 + y = F(x) \in R^*$ and $\frac{1}{1+y} \in A$. Also $\frac{y}{1+y} = 1 - \frac{1}{1+y} \in A$. Since $y \notin A_{[\mathfrak{p}]}$ we conclude that $\frac{1}{1+y} \in \mathfrak{p}$, hence $\frac{y}{1+y} = 1 - \frac{1}{1+y} \notin \mathfrak{p}$. But $\frac{y}{1+y} \in (\mathfrak{p}B) \cap A = \mathfrak{p}$, since B is integral over A, contradiction! Thus indeed $F(x) \in A_{[\mathfrak{p}]}$.

c) For $\ell = 0, 1, 2, \ldots$ we successively define polynomials $\Delta^\ell F(T)$ by

$$\Delta^0 F(T) := F(T), \quad \Delta^{\ell+1} F(T) := \Delta^\ell F(T+1) - \Delta^\ell F(T).$$

For every $x \in B$ we have $F(x) \in A_{[\mathfrak{p}]}$, thus also $\Delta^\ell F(x) \in A_{[\mathfrak{p}]}$ for any $\ell \in \mathbb{N}$. But $\Delta^{d-1} F(T) = d!T + c$ with $c \in k$. Thus $(d!)x \in A_{[\mathfrak{p}]}$ for every $x \in B$. Since $1/d! \in A \subset A_{[\mathfrak{p}]}$, we conclude that $B \subset A_{[\mathfrak{p}]}$. $\qquad\square$

Example 13. We denote the prime ring in R by $\mathbb{Z} \cdot 1$. Let $d \in \mathbb{N}$. Assume that $d! \in R^*$ and $1 + x^d \in R^*$ for all $x \in R$ with $x^d \notin \mathbb{Z} \cdot 1$. The subring A of R generated by $1/d!$ and the elements $1/(1 + x^d)$ with $x \in R$, $x^d \notin \mathbb{Z} \cdot 1$ is Prüfer in R. $\qquad\square$

N.B. For $d = 2$ and R a field this example states a slight improvement of Dress's lemma cited above.

Remark. The condition $d! \in R^*$ cannot be omitted. For example, let $R := \mathbb{F}_2[T]/(1 + T^2)$ with $\mathbb{F}_2$ the field consisting of 2 elements. Let

A be the subring of R generated by the elements $1/(1+x^2)$ for all $x \in R$ with $x^2 \neq 1$. Then $A = \mathbb{F}_2$, and this not Prüfer in R, since $\mathbb{F}_2$ is not integrally closed in R.

As an illustration what has been done so far we return to Example 7. Thus let V be an affine variety over some field k which is not algebraically closed. Let U be a k-Zariski-open subset of $V(k)$, and let R be the ring of regular functions on U. We choose a monic polynomial $F(T) \in k[T]$, $F \neq 1$, which has no zeros in k.

Let B be any subring of R containing k (e.g. $B = k$). Let H_0 denote the subring $B[\frac{1}{F(x)} \mid x \in R]$ of R generated by B and the elements $\frac{1}{F(x)}$ for all $x \in R$. Let H denote the integral closure of H_0 in R.

Theorem 6.17. i) H is Prüfer in R.
ii) H is the set of all $x \in R$ such that $v(x) \geq 0$ for every Manis F-valuation v on R with $v(b) \geq 0$ for all $b \in B$.
iii) $H = B[\frac{x^i}{F(x)} \mid x \in R, 0 \leq i \leq 1]$.
iv) If the characteristic of k is zero or exceeds d, then $H = H_0$.

Proof. H is an R-Prüferring by Theorem 1. Thus H is the intersection of the valuation rings A_v with v running through the set Ω of all Manis valuations on R with $A_v \supset H$.

Let v be a Manis valuation on R. Then $v \in \Omega$ iff $A_v \supset H_0$. This means that $A_v \supset B$ and $v(\frac{1}{F(x)}) \geq 0$ for every $x \in R$. If $x \notin A_v$ then $v(F(x)) < 0$, hence $v(\frac{1}{F(x)}) > 0$ automatically. Let $x \in A_v$. Then $v(\frac{1}{F(x)}) \geq 0$ iff $v(F(x)) = 0$ iff $\bar{F}(\bar{x}) \neq 0$ for $\bar{F}(T)$ the image of $F(T)$ in $\kappa(v)[T]$ and $\bar{x}$ the image of x in $\kappa(v)$. Thus Ω is the set of all Manis F-valuations v on R with $A_v \supset B$.

The ring $H' := B[\frac{x^i}{F(x)} \mid x \in R, 0 \leq i \leq 1]$ is Prüfer in R by Theorem 14. Every valuation $v \in \Omega$ has nonnegative values on H'. Thus $H_0 \subset H' \subset H$. Since H' is integrally closed in R, we have $H' = H$. If $d! \in k^*$, then we know from Theorem 16 that H_0 is Prüfer in R and conclude that $H_0 = H$. $\qquad\square$

We finish with some examples from real algebra, using standard notions from that area, cf. [BCR], [KS]. The uninitiated reader may safely skip the following paragraphs. {For part II of the book we plan a whole section on real algebra. There we will be more explicit.}

Definition 6. We call a ring R *totally real* if, for every maximal ideal $\mathfrak{m}$ of R, the field $R/\mathfrak{m}$ is formally real. $\qquad\square$

Notice that $R/\mathfrak{m}$ is formally real iff $\mathfrak{m} \cap (1 + \Sigma R^2)$ is empty. Thus R is formally real iff $1 + \Sigma R^2 \subset R^*$.

For any ring R we denote by $H(R)$ the *real holomorphy ring* of R. This is the ring of all $f \in R$ such that there exists some natural number n with $-n \leq f \leq n$ on the real spectrum $\operatorname{Sper} R$ of R, cf. [BP].

Proposition 6.18 [BP, Th.5.13]. If R is totally real then $H(R)$ is Prüfer in R.

Proof. We know from Example 13 that the subring $H_0(R) := \mathbb{Z}\left[\frac{1}{1+f^2} \mid f \in R\right]$ of R is Prüfer in R. Clearly $H_0(R) \subset H(R)$. Thus $H(R)$ is Prüfer in R. $\qquad\square$

We give a more explicit description of the real holomorphy ring in some cases.

Examples 14. a) Let X be any topological space. Clearly the ring $R := C(X)$ of continuous $\mathbb{R}$-valued functions on X is totally real. We claim that $H(R)$ is the ring $C_b(X)$ of bounded continuous functions[*] on X. Indeed, every $x \in X$ gives us a point in the real spectrum of R, and thus $H(R) \subset C_b(X)$.[**] On the other hand, if $f \in C_b(X)$ is given, choosing some n with $|f| \leq n$ on X, we have $n^2 - f^2 = g^2$ with some $g \in C(X)$. Thus $f^2 \leq n^2$ on $\operatorname{Sper} R$, and we conclude that $f \in H(R)$. Thus $C_b(X)$ is Prüfer in $C(X)$. Of course, this can be deduced directly from Example 13 without referring to real holomorphy rings.

b) Let k be a real closed field and M a semialgebraic subset of k^N for some $N \in \mathbb{N}$. The ring $R := CS(M)$ of continuous semialgebraic k-valued functions on M is clearly totally real. By the same argument as in a) we see that $H(R)$ is the ring $CS_b(M)$ of bounded continuous

---

[*] In the classical algebraic literature (cf. [GJ]) this ring is denoted by $C^*(X)$. We have to avoid this notation due to our convention that R^* denotes the group of units of R.

[**] The reader my consult [Sch3] for information about $\operatorname{Sper} C(X)$.

semialgebraic functions on M. Thus $CS_b(M)$ is Prüfer in $CS(M)$. N. Schwartz has proved that $CS(M)$ is the Prüfer hull of $CS_b(M)$ except in the case that the semialgebraic set M "has no end points", i.e. there does not exist a point $x \in M$ which has a neighbourhood U in M which admits an semialgebraic isomorphism onto the halfopen interval $[0, 1[$ in k mapping x to 0, cf. [Sch$_2$, Example 5.13]. $\qquad\square$

c) Let A be any ring and R the ring of abstract Nash functions $N(A)$ on Sper A, i.e. $R = \mathcal{N}_A(\text{Sper } A)$ with $\mathcal{N}_A$ the Nash structure sheaf on Sper A, cf. [R]. One verifies again, as in a), that $H(R)$ is the ring $N_b(A)$ of bounded Nash functions. {Now choose $n \in \mathbb{N}$ such that $|f| < n$, and use the fact that a Nash function $h > 0$ on Sper A has a square root.} Thus $N_b(A)$ is Prüfer in $N(A)$. $\qquad\square$

d) Let U be an open subset of $\mathbb{R}^N$. As in c) one sees the following: Let R denote the ring $C^r(U)$ of real C^r-functions on U, with $r \in \mathbb{N}_0 \cup \{\infty, \omega\}$. {As usual, $C^\omega(U)$ denotes the ring of real analytic functions on U.} Then R is totally real and $H(R)$ is the subring $C_b^r(U)$ of bounded functions in $C^r(U)$. Thus $C_b^r(U)$ is Prüfer in $C^r(U)$.

e) Analogously, if U is a semialgebraic open subset of k^N for k a real closed field and $r \in \mathbb{N}_0 \cup \{\infty\}$, we can introduce the ring $R := CS^r(U)$ of semialgebraic C^r-functions on U with values in k. If $r = \infty$ this is the ring $N(U)$ of Nash functions on U, cf. [BCR, §2.9]. {In good cases, in particular if $k = \mathbb{R}$, one can also define analytic k-valued semialgebraic functions, but then observes that $C^\omega(U) = N(U)$, cf. [Br, p.265], [K$_1$, §4]}. In all these cases one sees again, as above, that R is totally real and $H(R)$ is the subring $CS_b^r(U)$ of bounded semialgebraic C^r-functions on U. $\qquad\square$

Remark 6.19. In the situation of Proposition 19 it is easily seen that $H(R)$ is totally real. Moreover, if for every $f \in R$ with $f > 0$ on Sper R there exists a square root $\sqrt{f}$ in R, then $H(R)$ coincides with $H_0(R) = \mathbb{Z}\left[\frac{1}{1+f^2} \mid f \in R\right]$. Indeed, let $a \in H(R)$ be given. Choose $n \in \mathbb{N}$ with $-n < a < n$. Then $2n + 1 > a + n + 1 > 0$ on Sper R. Set $b := \frac{a+n+1}{2n+1}$. Then $0 < b < 1$ on Sper R. By the strict Positivstellensatz (e.g. [BCR, Prop.4.4.7.ii], [KS, p.141]) it follows from $b > 0$, that b divides an element of $1 + \Sigma R^2$, hence $b \in R^*$. Now

$b^{-1} > 1$ implies

$$b = \frac{1}{1 + \left(\sqrt{b^{-1} - 1}\right)^2} \quad,$$

hence $b \in H_0(R)$ and $a \in H_0(R)$.

In particular, in all the examples $R = C(X)$, $CS(M)$, $N(A)$, ... above we have $H(R) = H_0(R)$.

§7 Principal ideal results

We start out for a generalization of the second half of Roquette's theorem 1 in [R] mentioned in §6. We will rely on techniques developed by Alan Loper in the case of subrings of fields [Lo₁], [Lo₂].

In the following we fix a ring A and a monic polynomial $F(T) \in A[T]$ of degree $d \geq 1$.

Definition 1 (cf. [Lo₁]). Let $\varphi: A \to B$ be a ring extension of A. We call the polynomial F *unit valued in B* (abbreviated: uv in B), if $F(b) \in B^*$ for every $b \in B$. {Of course, $F(b) := F^\varphi(b)$ with $F^\varphi(T)$ the image polynomial of $F(T)$ in $B[T]$.}

More precisely we then should call F "uv with respect to φ", but in the following it will be always clear which homomorphism φ from A to B is taken. We do not demand φ to be injective.

N.B. If F is uv in some extension B of A different from the null ring then certainly $d \geq 2$.

Proposition 7.1 (cf. [Lo₁, Prop.1.14]). Let $\mathfrak{m}$ be a maximal ideal of A. Then $F(T)$ is uv in $A_\mathfrak{m}$ iff $F(A) \subset A \setminus \mathfrak{m}$.

Proof. If there exists some $a \in A$ with $F(a) \in \mathfrak{m}$, then certainly $F(T)$ is not uv in $A_\mathfrak{m}$. Assume now that $F(A) \subset A \setminus \mathfrak{m}$. Suppose that $F(T)$ is not uv in $A_\mathfrak{m}$. We have some $a \in A$, $s \in A \setminus \mathfrak{m}$ with $F\left(\frac{a}{s}\right) \in \mathfrak{m}A_\mathfrak{m}$. Since the ideal $\mathfrak{m}$ is maximal there exists some $t \in A$ with $st \equiv 1 \mod \mathfrak{m}$. Then in $A_\mathfrak{m}$

$$\frac{F(at)}{1} \equiv F(\frac{a}{s}) \equiv 0 \qquad \text{mod } \mathfrak{m}A_\mathfrak{m},$$

hence $F(at) \in \mathfrak{m}$. This contradiction proves that $F(T)$ is uv in $A_\mathfrak{m}$.
$\square$

Corollary 7.2. $F(T)$ is uv in A iff $F(T)$ is uv in $A_\mathfrak{m}$ for every maximal ideal $\mathfrak{m}$ of A. $\square$

We write $F(T) = T^d + c_1 T^{d-1} + \cdots + c_d$ with $c_i \in A$, and introduce the homogenization $G(X,Y) \in A[X,Y]$ of F,

$$G(X,Y) := Y^d F(\frac{X}{Y}) = X^d + c_1 X^{d-1} Y + \cdots + c_d Y^d.$$

Proposition 7.3. Let $\mathfrak{p}$ be a prime ideal of A. The following are equivalent.

i) F is uv in $A_\mathfrak{p}$.
ii) F is uv in $k(\mathfrak{p}) = \text{Quot}(A/\mathfrak{p})$, i.e. F has no zero in $k(\mathfrak{p})$.
iii) If $x,y \in A$ and $G(x,y) \in \mathfrak{p}$, then $y \in \mathfrak{p}$.
iv) If $x,y \in A$ and $G(x,y) \in \mathfrak{p}$, then $x \in \mathfrak{p}$ and $y \in \mathfrak{p}$.

Proof. i) $\Leftrightarrow$ ii) is evident. iv) $\Rightarrow$ iii) is trivial, and iii) $\Rightarrow$ iv) is evident, since the form $G(X,Y)$ contains the term X^d.
i) $\Rightarrow$ iii): Let $x,y \in A$ and $G(x,y) \in \mathfrak{p}$. Suppose $y \notin \mathfrak{p}$. Then we have in $A_\mathfrak{p}$

$$F(\frac{x}{y}) = \frac{G(x,y)}{y^d} \in \mathfrak{p}A_\mathfrak{p}.$$

This contradicts the assumption that F is uv in $A_\mathfrak{p}$.
iii) $\Rightarrow$ i): Let $a \in A$, $s \in A \setminus \mathfrak{p}$ be given. Then $G(a,s) \in A \setminus \mathfrak{p}$. Thus

$$F(\frac{a}{s}) = \frac{G(a,s)}{s^d} \in A_\mathfrak{p}^*.$$
$\square$

Proposition 7.4 (cf. [Lo₂, Cor.2.3] for R a field). Assume that $(A,\mathfrak{p})$ is a Manis pair in some ring R. Let v denote a Manis valuation on R with $A_v = A$, $\mathfrak{p}_v = \mathfrak{p}$. The following are equivalent.

i) F is uv in $A_\mathfrak{p}$.

ii) v is an F-valuation.

iii) $v(G(x,y)) = d \min(v(x), v(y))$ for all $x, y \in R$.

Proof. The equivalence i) $\Leftrightarrow$ ii) is clear from i) $\Leftrightarrow$ ii) in Proposition 3.

i) $\Rightarrow$ iii): Let $x, y \in R$ be given. The formula is a priori valid if $v(x) < v(y)$, since $G(X, Y)$ contains the term X^d. It is also valid if $v(x) = v(y) = \infty$. Assume now that $v(x) \geq v(y) \neq \infty$. We choose some $z \in R$ with $v(yz) = 0$. This is possible since v is Manis. Then $v(xz) \geq 0$. Thus $xz \in A$ and $yz \in A \setminus \mathfrak{p}$. We know from Prop. 3 that $G(xz, yz) = z^d G(x, y) \in A \setminus \mathfrak{p}$. Thus $v(G(x, y)) = -dv(z) = dv(y)$.

iii) $\Rightarrow$ i): Let $x, y \in A$ and $G(x, y) \in \mathfrak{p}$. Then the formula in iii) tells us that $x \in \mathfrak{p}$ and $y \in \mathfrak{p}$. Thus F is uv in $A_\mathfrak{p}$ by Proposition 3. $\quad\square$

We now study finitely generated A-submodules $\mathfrak{a}$ of R with $R\mathfrak{a} = R$. These submodules should be viewed as analogues of the finitely generated fractional ideals in the classical case that A is a domain and R its quotient field. In Chapter II we will study the set of these submodules $\mathfrak{a}$ in a systematic way. Just now we are looking for criteria that some power $\mathfrak{a}^d$ is a *principal module*, i.e. $\mathfrak{a}^d = Rb$ with some $b \in R^*$.

Definition 2. Let $(a_1, \ldots, a_n)$ be a finite sequence in R. The F-*transform* of this sequence is the sequence $(b_1, \ldots, b_n)$ in R defined inductively by

$$b_1 := a_1, \quad b_i := G(b_{i-1}, a_i^{d^{i-2}}) \quad (i > 1).$$

In the following lemmas $(a_1, \ldots, a_n)$ is a sequence in R and $(b_1, \ldots, b_n)$ is its F-transform.

Lemma 7.5. Assume that all $a_i \in A$. Let $\mathfrak{p}$ be a prime ideal of A such that F is uv in $A_\mathfrak{p}$. Then $Aa_1 + \cdots + Aa_n \subset \mathfrak{p}$ iff $b_n \in \mathfrak{p}$.

Proof. If $x, y \in A$ and $t \in \mathbb{N}$, then $Ax + Ay \subset \mathfrak{p}$ iff $Ax + Ay^t \subset \mathfrak{p}$. By Proposition 3 the latter is equivalent to $G(x, y^t) \in \mathfrak{p}$. The lemma follows from this by induction on n. $\quad\square$

Lemma 7.6 (cf. [Lo₂, Cor.2.4]). Assume that A is the valuation ring A_v of a Manis valuation v on some ring R which is also an F-valuation. Then

$$v(b_n) = d^{n-1} \min\{v(a_1), \dots, v(a_n)\}.$$

The proof goes by induction on n using the formula in Proposition 4.iii. □

Lemma 7.7. Let $\mathfrak{a} := Aa_1 + \cdots + Aa_n$. Assume that F is uv in R. Then

$$R\mathfrak{a} = R \Longleftrightarrow b_n \in R^*.$$

Proof. $\Leftarrow$: This is evident since $b_n \in \mathfrak{a}$.
$\Rightarrow$: Suppose $b_n \notin R^*$. We choose a maximal ideal $\mathfrak{M}$ of R containing b_n. Our polynomial F is uv in R hence uv in $R_{\mathfrak{M}}$ by Corollary 2. Now Lemma 5, applied to F as a polynomial over R, tells us that $Ra_1 + \cdots + Ra_n \subset \mathfrak{M}$. This contradicts the assumption $R\mathfrak{a} = R$. Thus $b_n \in R^*$. □

Now we are prepared to prove a generalization of the theorem by Roquette mentioned in §6.

Theorem 7.8 (cf. [R, Th.1] for R a field). Assume that S is a set of Manis valuations on a ring R and that $A = \bigcap_{v \in S} A_v$. Assume further that there exists a monic polynomial $F(T) \in A[T]$ of degree $d \geq 1$ with the following two properties:
(i) $F(T)$ is uv in R.
(ii) Every $v \in S$ is an F-valuation.

Then A is Prüfer in R. If $\mathfrak{a}$ is any finitely generated A-submodule of R with $R\mathfrak{a} = R$ then there exists some $t \in \mathbb{N}$ such that $\mathfrak{a}^{d^t}$ is principal. More precisely, if $a_1, \dots, a_n$ is a system of generators of $\mathfrak{a}$ and $(b_1, \dots, b_n)$ is the F-transform of the sequence $(a_1, \dots, a_n)$, then

$$\mathfrak{a}^{d^{n-1}} = Ab_n.$$

Proof. Theorem 6.5 tells us that A is Prüfer in R. Let $a_1, \dots, a_n$ be a system of generators of $\mathfrak{a}$ and $(b_1, \dots, b_n)$ the F-transform of $(a_1, \dots, a_n)$. Lemma 7 tells us that $b_n \in R^*$.

It is evident that $b_n \in \mathfrak{a}^{d^{n-1}}$. The module $\mathfrak{a}^{d^{n-1}}$ is generated over A by the monomials $a_1^{e_1} \ldots a_n^{e_n}$ with $e_i \geq 0$, $e_1 + \cdots + e_n = d^{n-1}$. We now verify that

$$v(a_1^{e_1} \ldots a_n^{e_n}) \geq v(b_n) \qquad (*)$$

for every such monomial and every $v \in S$. It then follows that $a_1^{e_1} \ldots a_n^{e_n}/b_n$ is an element of A_v for every $v \in S$, hence of A, and we conclude that $\mathfrak{a}^{d^{n-1}} = Ab_n$.

The verification of $(*)$ is immediate by use of Lemma 6. Let $\gamma := \min\{v(a_1), \ldots, v(a_n)\}$. Then $v(a_1^{e_1} \ldots a_n^{e_n}) \geq (e_1 + \cdots + e_n)\gamma = d^{n-1}\gamma = v(b_n)$. $\qquad\square$

In Chapter II we will see that for A a Prüfer subring of a ring R the finitely generated A-submodules $\mathfrak{a}$ of R with $R\mathfrak{a} = R$ form an abelian group. The quotient of this group by the subgroup of principal modules should be called the *class group of A in R*. Starting with Theorem 8 it is possible to get bounds on the torsion of the class group in good cases in much the same way as Roquette has explicated for R a field [R]. Here we only quote the following theorem which is an immediate consequence of Theorem 8.

Theorem 7.9 (cf.[R, Th.2]). Assume again that $A = \bigcap_{v \in S} A_v$ for a set S of Manis valuations on some ring R. Assume further that there exist nonconstant monic polynomials $F_1(T), \ldots, F_r(T)$ with coefficients in A $(r \geq 1)$, such that for every $j \in \{1, \ldots, r\}$ the following holds

(1) F_j is uv in R.
(2) Every $v \in S$ is an F_j-valuation.

Let d denote the greatest common divisor of the degrees of $F_1, \ldots, F_r$. Then A is Prüfer in R, and for each finitely generated A-submodule $\mathfrak{a}$ of R with $R\mathfrak{a} = R$ there exists some $t \in \mathbb{N}$ such that $\mathfrak{a}^{d^t}$ is principal.
$\qquad\square$

Example 7.10. Let R be a ring such that $X^d + 1$ is uv in R for some (even) $d \in \mathbb{N}$ and $d!$ is a unit in R. Let A be a subring of R which contains $1/d!$ and the elements $1/(1+x^d)$ for all $x \in R$. Then A is Prüfer in R by Example 11 in §6. For every finitely generated A-submodule $\mathfrak{a}$ of R with $\mathfrak{a}R = R$ there exists some $t \in \mathbb{N}$ with $\mathfrak{a}^{d^t}$ principal.

Proof. A is the intersection of the rings $A_{[\mathfrak{m}]}$ with $\mathfrak{m}$ running through the maximal ideals of A (Remark 5.5). These rings are Manis in R. The polynomial $X^d + 1$ is uv in $A_{\mathfrak{m}}$ for every $\mathfrak{m}$ (Cor.2), and thus the Manis valuations giving the rings $A_{[\mathfrak{m}]}$ are $(X^d + 1)$-valuations. Theorem 8 applies.   □

In an important more special situation this result can be improved. Assume that $1 + \Sigma R^d \subset R^*$. A subring A of R containing the elements $1/(1+q)$ with $q \in \Sigma R^d$ is Prüfer in R. If $\mathfrak{a} = Ax_1 + \cdots + Ax_n$ is a finitely generated submodule of R with $R\mathfrak{a} = R$, then $\mathfrak{a}^d = A(x_1^d + \cdots + x_n^d)$. This has been proved by E. Becker and V. Powers [BP, Cor. 5.11, Cor.5.13].

A slight expansion of the techniques used so far will give us a theorem containing the result of Becker and Powers as a special case, together with a proof which is rather different from the one in [BP].

Definition 3. Let $H(X_1, \ldots, X_n) \in A[X_1, \ldots, X_n]$ be a form, i.e. a homogeneous polynomial over A in $n \geq 2$ variables. Let $\varphi \colon A \to K$ be a homomorphism into a field K. We call H *isotropic over* K, if the image form $H^\varphi(X_1, \ldots, X_n) \in K[X_1, \ldots, X_n]$ is isotropic, i.e. has a non trivial zero in K^n, and we call H *anisotropic over* K otherwise.

In the following it will be always clear which homomorphism φ is under consideration. Thus the impreciseness in this definition will do no harm.

Theorem 7.11. Let S be a set of Manis valuations on a ring R and $A := \bigcap_{v \in S} A_v$. Assume there is given a form $H(X_1, \ldots, X_n)$ over A in n variables of degree d, $n \geq 2, d \geq 1$, with the following properties:

   i) For every maximal ideal $\mathfrak{M}$ of R the form H is anisotropic over $R/\mathfrak{M}$.
   ii) For every $v \in S$ the form H is anisotropic over $\kappa(v)$.

Then A is Prüfer in R. If $\mathfrak{a}$ is an A-submodule of R generated by n elements $x_1, \ldots, x_n$ and $R\mathfrak{a} = R$ then $\mathfrak{a}^d = H(x_1, \ldots, x_n)A$.

Proof. a) We start with a proof of the second claim. Suppose that $H(x_1, \ldots, x_n)$ is not a unit in R. Then there exists a maximal ideal $\mathfrak{M}$ of R with $H(x_1, \ldots, x_n) \in \mathfrak{M}$. Since $Rx_1 + \cdots + Rx_n =$

R we conclude that H is isotropic over $R/\mathfrak{M}$, in contradiction to assumption (i) above. Thus $H(x_1, \ldots, x_n) \in R^*$.

b) Let $v \in S$ be given. We verify that

$$(*) \qquad v(H(x_1, \ldots, x_n)) = d\min\{v(x_1), \ldots, v(x_n)\}.$$

This is obvious if $v(x_i) = \infty$ for all $i \in \{1, \ldots, n\}$. Assume now that $\gamma := \min\{v(x_1), \ldots, v(x_n)\} < \infty$. We choose some $z \in R$ with $v(z) = -\gamma$, which is possible, since v is Manis. Then $v(zx_i) \geq 0$ for all $i \in \{1, \ldots, n\}$ and $v(zx_i) = 0$ for at least one i. Since H is anisotropic over $\kappa(v)$ we conclude that $v(H(zx_1, \ldots, zx_n)) = 0$, hence $v(H(x_1, \ldots, x_n)) = -dv(z) = d\gamma$, as desired.

c) Now we see, as in the proof of Theorem 8, that

$$v(x_1^{e_1} \ldots x_n^{e_n}) \geq v(H(x_1, \ldots, x_n))$$

for any integers $e_i \geq 0$ with $e_1 + \cdots + e_n = d$ and any $v \in S$, and we conclude that $x_1^{e_1} \ldots x_n^{e_n} / H(x_1, \ldots, x_n) \in A$. This proves that $\mathfrak{a}^d = H(x_1, \ldots, x_n)A$.

d) Let

$$G(X, Y) := H(X, \ldots, X, Y) = c_0 X^d + c_1 X^{d-1} Y + \cdots + c_d Y^d.$$

$c_0 = H(1, \ldots, 1, 0)$ is a unit in A, since the elements $1, \ldots, 1, 0$ generate the ideal $\mathfrak{a} = A$ and $\mathfrak{a}^d = H(1, \ldots, 1, 0)A$. We consider the monic polynomial

$$F(T) := c_0^{-1} G(T, 1) \in A[T].$$

F is uv in R, since $H(x, \ldots, x, 1) \in R^*$ for every $x \in R$. If $v(x) \geq 0$ for some $v \in S$, then $v(H(x, \ldots, x, 1)) = v(1) = 0$. Thus every $v \in S$ is an F-valuation. We conclude by Theorem 6.5 that A is Prüfer in R. $\qquad\square$

Remark. The multiplicative ideal theory in Chapter II will give a more natural proof that A is Prüfer in R.

In order to exploit Theorem 11 in the real algebraic setting, we need an easy lemma.

Lemma 7.12. Let $H(X_1, \ldots, X_n)$ be a form over a ring A of degree d in n variables with $d \geq 1, n \geq 2$. For each $i \in \{1, \ldots, n\}$ we define

$$F_i(T_1, \ldots, T_{n-1}) := H(T_1, \ldots, T_{i-1}, 1, T_i, \ldots, T_{n-1}).$$

The following are equivalent

(1) H is anisotropic over $A/\mathfrak{m}$ for every maximal ideal $\mathfrak{m}$ of A.

(2) $F_i(x_1, \ldots, x_{n-1}) \in A^*$ for all $x_1, \ldots, x_{n-1} \in A$ and $1 \leq i \leq n$.

Proof. (1) $\Longrightarrow$ (2): Let $x_1, \ldots, x_{n-1} \in A$ and $i \in \{1, \ldots, n\}$. Then $H(x_1, \ldots, x_{i-1}, 1, x_i, \ldots, x_{n-1}) \notin \mathfrak{m}$ for every maximal ideal $\mathfrak{m}$ of A. Thus $F_i(x_1, \ldots, x_{n-1}) \in A^*$.

(2) $\Longrightarrow$ (1): Suppose there exists a maximal ideal $\mathfrak{m}$ of A such that H is isotropic over $A/\mathfrak{m}$. Then there exist elements $a_1, \ldots, a_n \in A$ with $H(a_1, \ldots, a_n) \in \mathfrak{m}$ but $a_i \notin \mathfrak{m}$ for some i. We choose an element $b_i \in A$ with $a_i b_i \equiv 1 \mod \mathfrak{m}$. We have $b_i^d H(a_1, \ldots, a_n) = H(a_1 b_i, \ldots, a_n b_i) \equiv F_i(a_1 b_i, \ldots, a_{i-1} b_i, a_{i+1} b_i, \ldots, a_n b_i) \mod \mathfrak{m}$. Thus $F_i(a_1 b_i, \ldots, a_{i-1} b_i, a_{i+1} b_i, \ldots, a_n b_i) \in \mathfrak{m}$, a contradiction. $\square$

Corollary 7.13 (cf. [BP]). Let $d \in \mathbb{N}$ and let R be a ring with $1 + \Sigma R^{2d} \subset R^*$. Then the subring

$$H := H_d(R) = \mathbb{Z}\left[\frac{1}{1+q} \mid q \in \Sigma R^{2d}\right]$$

is Prüfer in R. For each finitely generated H-submodule $\mathfrak{a} = Hx_1 + \cdots + Hx_n$ of R with $\mathfrak{a}R = R$ we have $\mathfrak{a}^{2d} = (x_1^{2d} + \cdots + x_n^{2d})H$.

Proof. Applying Theorem 6.16 with $F(T) = 1 + T^{2d}$ we see that H is Prüfer in R (cf. §6, Example 13). For every maximal ideal $\mathfrak{m}$ of H we choose a Manis valuation v on R with $A_v = H_{[\mathfrak{m}]}$, $\mathfrak{p}_v = \mathfrak{m}_{[\mathfrak{m}]}$. Let S denote the set of these valuations. Then $H = \bigcap_{v \in S} A_v$ (cf. 5.5). Now, if $v \in S$, $A_v = H_{[\mathfrak{m}]}$, then $H/\mathfrak{m} = H_{[\mathfrak{m}]}/\mathfrak{m}_{[\mathfrak{m}]}$, as is easily checked, and we learn from Proposition 1.6 that $\kappa(v)$ is the quotient field of $H/\mathfrak{m}$. Since $H/\mathfrak{m}$ is already a field, we have $\kappa(v) = H/\mathfrak{m}$. Let $n \geq 2$. Using Lemma 12 we see that the form $X_1^{2d} + \cdots + X_n^{2d}$ is anisotropic in $R/\mathfrak{M}$ for every maximal ideal $\mathfrak{M}$ of R, and also anisotropic in $H/\mathfrak{m}$ for every maximal ideal $\mathfrak{m}$ of H. Now Theorem 11 gives the second claim above. $\square$

Becker and Powers have proved that $1+\Sigma R^{2d} \subset R^*$ implies $1+\Sigma R^2 \subset R^*$, and that then $H := H_d(R)$ coincides with $H_1(R)$ and the "real holomorphy ring" of R [BP, Prop.5.1 and Prop.5.7]. Thus, if $\mathfrak{a}$ is a finitely generated H-submodule of R with $R\mathfrak{a} = R$, then already $\mathfrak{a}^2$ is a principal submodule.

Chapter II: Multiplicative ideal theory

Summary:

In this chapter we study Prüfer extensions avoiding valuations as much as possible. Instead we develop and apply "multiplicative ideal theory". In contrast to the classical theory of Prüfer domains we will not use a notion of fractional ideal, but either will work with "regular modules" (see below) or with invertible modules.[1] Valuations will play a role only in §1, up to Proposition 1.8, and in the last sections §11 and §12. From Chapter I we will mainly use the sections §3 – §5.

As already said in the Introduction, Prüfer extensions are useful since they give us "families of valuations". On the other hand suitable valuations help us to understand Prüfer extensions. Thus it may be good at times to work solely in the framework of ring extensions and submodules, as we will do now, and good at other times to focus on valuations, as we will do in Chapter III.

In §1 we explain that Prüfer extensions have a pleasant multiplicative ideal theory. This will be easy. It is more demanding to *characterize* Prüfer extensions by suitable conditions from multiplicative ideal theory. Such characterizations are given in §2 and §5. We have been ambitious to make the conditions as weak as possible, which then gives strong theorems. We advise the reader to skip some proofs in §2 at first reading. The sections §3 – §9 add various facets to the axiomatics and the theory of A-submodules of a given Prüfer extension $A \subset R$, in particular of R-overrings of A, i.e. subrings of R containing A. Then in §10 we discuss a special class of Prüfer extensions, the Bezout extensions. These are the Prüfer extensions in which every invertible ideal is principal.

The final sections §11 and §12 deal with the Prüfer extensions of a noetherian ring, drawing conclusions in this case from much what has been done before in Chapters I and II.

---

[1] A useful notion of fractional ideal will be developed in part II of the book.

§1 Multiplicative properties of regular modules

Let A be a subring of a ring R. In the case that R is the total quotient ring $\operatorname{Quot} A$ of A and A is Prüfer in R it is well known that a good "multiplicative" theory exists for the ideals of A which are "regular", i.e. contain a nonzero divisor. Moreover here one has a satisfying theory of regular *fractional* ideals, similar to the classical multiplicative ideal theory of Prüfer domains [Gi]. Our goal in the present section is to establish analogous ideal theoretic results for A a Prüfer subring of an arbitrary ring R.

Definition 1. An A-submodule I of R is called *regular in R* (or *R-regular*) if $IR = R$. It is called *strongly regular in R* (or *strongly R-regular*) if $I \cap R^* \neq \emptyset$.[2)]

Example 1. Assume that $R = \operatorname{Quot} A$. Then A-module I is strongly R-regular iff I contains a nonzero divisor of A. This is the notion of regularity used traditionally in the multiplicative ideal theory of rings with zero divisors, cf. [LM], [Huc]. Moreover, now regularity and strong regularity means the same. Indeed, assume that $IR = R$. In $R = \operatorname{Quot} A$ we have an equation

$$1 = \sum_{i=1}^{r} \frac{x_i}{s} a_i$$

with $a_i \in I$, $x_i \in A$, $s \in R^*$. We conclude that $s = \sum_{1}^{r} x_i a_i \in R^* \cap I$.

$\square$

In this chapter and also the later ones regularity will play a key role, but strong regularity will only rarely be needed. Nevertheless we give one more example, where regularity and strong regularity mean the same.

[2)] Rhodes [Rh, p. 3424] uses a slightly weaker notion of R-regularity. If A is weakly surjective in R both definitions are equivalent. On the other hand, the notion of R-regularity used by Griffin [G2] means for rings with unit element strong R-regularity in our sense. (Griffin also admits certain rings without unit element.)

Example 2. If R is semilocal (i.e. the set of maximal ideals $\operatorname{Max} R$ is finite), every R-regular ideal I of A is strongly R-regular.

Proof. Let $\mathfrak{M}_1, \ldots, \mathfrak{M}_r$ denote the maximal ideals of R. Suppose that $I \cap R^* = \emptyset$. Then every $x \in I$ is contained in some $\mathfrak{M}_i$, hence $I \subset \mathfrak{M}_1 \cup \cdots \cup \mathfrak{M}_r$. By a well known lemma [Bo, II §1 Prop. 2] there exists some $i \in \{1, \ldots, r\}$ with $I \subset \mathfrak{M}_i$, hence $IR \subset \mathfrak{M}_i$. This contradicts the assumption that $IR = R$. Thus $I \cap R^* \neq \emptyset$. $\qquad\square$

Notice that the argument fails if more generally I is an R-regular A-submodule of R. We needed that I is closed under multiplication.

Definition 2. If I and J are A-submodules of R then $[I:_R J]$ denotes the A-submodule of R consisting of all $x \in R$ with $xJ \subset I$, and $(I:_A J)$ denotes the ideal of A consisting of all $x \in A$ with $xJ \subset I$, $(I:_A J) = A \cap [I:_R J]$. We usually omit the subscripts A and R here, if no ambiguity is possible.

Notice that, if I and J are ideals of A, then $(I:_A J)$ does not depend on R but is defined by I, J, A alone.

Our multiplicative theory of A-submodules of R for A Prüfer in R is based on the following three lemmas.

Lemma 1.1. Let I and J be A-submodules of R

a) $I = J$ iff $I_\mathfrak{p} = J_\mathfrak{p}$ for every maximal ideal $\mathfrak{p}$ of A.
b) $(I \cap J)_\mathfrak{p} = I_\mathfrak{p} \cap J_\mathfrak{p}$, $(I + J)_\mathfrak{p} = I_\mathfrak{p} + J_\mathfrak{p}$, and $(IJ)_\mathfrak{p} = I_\mathfrak{p} J_\mathfrak{p}$ for every $\mathfrak{p} \in \operatorname{Spec} A$.
c) If the A-module J is finitely generated, then $[I:_R J]_\mathfrak{p} = [I_\mathfrak{p}:_{R_\mathfrak{p}} J_\mathfrak{p}]$ for every $\mathfrak{p} \in \operatorname{Spec} A$.
d) If I and J are ideals of A and J is finitely generated, then $(I:_A J)_\mathfrak{p} = (I_\mathfrak{p}:_{A_\mathfrak{p}} J_\mathfrak{p})$ for every $\mathfrak{p} \in \operatorname{Spec} A$.

Here the statements a), b), d) are special cases of general facts about the localization of modules, cf. [Bo II, §3]. {Notice that $(I:_A J)$ is the annulator ideal of $I + J/I$.} Statement c) can be verified in a straightforward way. $\qquad\square$

Lemma 1.2. Assume that A is the valuation ring A_v of a local Manis valuation v on R. Let I be an A-submodule of R. The following are equivalent.

i) I is strongly regular in R

ii) I is regular in R

iii) $I \not\subset \operatorname{supp} v$

iv) $\operatorname{supp} v \subsetneqq I$

v) I is v-convex, $I \neq \operatorname{supp} v$.

Proof. Let $\mathfrak{q} := \operatorname{supp} v$, $\mathfrak{p} := \mathfrak{p}_v$. A is a local ring with maximal ideal $\mathfrak{p}$, and R is a local ring with maximal ideal $\mathfrak{q}$ (Prop.I.1.3.), hence $A^* = A \setminus \mathfrak{p}$, $R^* = R \setminus \mathfrak{q}$. We have $I \cap R^* \neq \emptyset$ iff $I \not\subset \mathfrak{q}$ iff $IR \not\subset \mathfrak{q}$ iff $IR = R$. This proves the equivalence of (i), (ii), (iii). The implication (iv) $\Rightarrow$ (iii) is trivial.

(iii) $\Rightarrow$ (iv): It suffices to verify that $\mathfrak{q} \subset Ax$ for every $x \in R \setminus \mathfrak{q}$. But this is obvious: If $y \in \mathfrak{q}$ then $y = (yx^{-1})x$ and $yx^{-1} \in \mathfrak{q} \subset A$.

(iv) $\Rightarrow$ (v) holds by Corollary I.1.9. The reverse implication is trivial. $\square$

Lemma 1.3. Assume again that v is a local Manis valuation on R and $A = A_v$. Let I be a *finitely generated* A-submodule of R, $I = Ax_1 + \cdots + Ax_r$ with $v(x_1) \leq v(x_2) \leq \cdots \leq v(x_r)$.

i) For any two elements $x, y \in R$ with $v(x) \neq \infty$ we have $Ay \subset Ax$ iff $v(x) \leq v(y)$.

ii) The following are equivalent

  a) $v(x_1) \neq \infty$,

  b) $x_1 \in R^*$,

  c) I is R-regular.

If these properties hold then $I = Ax_1$.

Proof. i) Since $v(x) \neq \infty$ and v is local and Manis, the element x is a unit in R (Prop.I.1.3.). Let $z := yx^{-1}$. Clearly $Ay \subset Ax$ iff $Az \subset A$. This means that $z \in A$, i.e. $v(z) \geq 0$, and is equivalent to $v(y) \geq v(x)$.

ii) The equivalence a) $\Longleftrightarrow$ b) is clear from Prop.I.1.3. The implications b) $\Longrightarrow$ c) and c) $\Longrightarrow$ a) are trivial. If a) holds then we know from part i) that $I = Ax_1$. $\square$

Theorem 1.4. Assume that A is a Prüfer subring of R. Let I, J, K be A-submodules of R.

(1) If at least one of the modules I, J, K is R-regular, then
$$I \cap (J + K) = (I \cap J) + (I \cap K).$$

(2) If either I is R-regular, or both J and K are R-regular then $(I + J) \cap (I + K) = I + (J \cap K)$.

(3) If I is R-regular and K is finitely generated then $(I + J : K) = (I : K) + (J : K)$.

(4) If I or J is R-regular then $I(J \cap K) = (IJ) \cap (IK)$.

(5) If I is R-regular then $(I + J)(I \cap J) = IJ$.

(6) If $I \subset J$, and if J is finitely generated and R-regular, then $J(I : J) = I$.

(7) If I is finitely generated and R-regular, and if $IJ = IK$, then $J = K$.

(8) Assume that J is R-regular and both J and K are finitely generated. Then $(I : J \cap K) = (I : J) + (I : K)$.

Proof. We first deal with the assertions (1) – (7). By Lemma 1 it suffices to prove these statements for $R_{\mathfrak{p}}$, $A_{\mathfrak{p}}$, $I_{\mathfrak{p}}$, $J_{\mathfrak{p}}$, $K_{\mathfrak{p}}$ with $\mathfrak{p}$ running through the prime ideals of A. (The maximal ideals would suffice.) Each such ring $A_{\mathfrak{p}}$ is a local Manis subring of $R_{\mathfrak{p}}$ with maximal ideal $\mathfrak{p}_{\mathfrak{p}}$. If one of the modules I, J, K is regular or finitely generated then the same holds for the localized module. Thus it suffices to prove the assertions in the case that $A = A_v$ with v a local Manis valuation on R. Let $\mathfrak{q} = \operatorname{supp} v$. By Lemma 2 an A-submodule I of R is either contained in $\mathfrak{q}$ or it properly contains $\mathfrak{q}$, and the modules $I \underset{\neq}{\supset} \mathfrak{q}$ are precisely the R-regular A-modules of R. They form a chain, since they are v-convex. Observing also that a finitely generated A-submodule J of R with $J \not\subset \mathfrak{q}$ is generated by a unit of R (Lemma 3), it is an easy exercise to verify all the statements (1) – (7) except (4) by case distinctions.

Concerning (4) there is a small problem in the case that neither J nor K is R-regular. Now I is R-regular, hence v-convex and different from $\mathfrak{q}$. If I is finitely generated, then Lemma 3 tells us that $I = Ax_1$ with $x_1 \in R^*$. In this case (4) is again evident. In general we argue as follows. In order to prove the nontrivial inclusion $(IJ) \cap (IK) \subset I(J \cap K)$, we pick an element $x \in (IJ) \cap (IK)$. There exists a finitely generated R-regular A-module $I_0 \subset I$ such that $x \in (I_0 J) \cap (I_0 K)$. Since, as observed, $(I_0 J) \cap (I_0 K) = I_0(J \cap K)$, we conclude that $x \in I_0(J \cap K) \subset I(J \cap K)$.

We turn to the proof of (8). Here we have to be more careful, since we do not know whether $J \cap K$ is finitely generated. For every prime ideal $\mathfrak{p}$ of A,

$$(I : J \cap K)_\mathfrak{p} \subset (I_\mathfrak{p} : (J \cap K)_\mathfrak{p}) = (I_\mathfrak{p} : J_\mathfrak{p} \cap K_\mathfrak{p}).$$

Since $K_\mathfrak{p} \subset J_\mathfrak{p}$ or $J_\mathfrak{p} \subset K_\mathfrak{p}$ we further have

$$(I_\mathfrak{p} : J_\mathfrak{p} \cap K_\mathfrak{p}) = (I_\mathfrak{p} : J_\mathfrak{p}) + (I_\mathfrak{p} : K_\mathfrak{p}) = [(I : J) + (I : K)]_\mathfrak{p}.$$

Thus $(I : J \cap K) \subset (I : J) + (I : K)$. The reverse inclusion is trivial. $\square$

We discuss a consequence of assertion (5) in this theorem.

Lemma 1.5. Assume that A is Prüfer in R. Let I and J be A-submodules of R with $I \cap J = A$. Then $IJ = I + J$.

Proof. $IJ = (I + J)(I \cap J) = (I + J)A = I + J$. $\square$

Proposition 1.6. Assume again that A is Prüfer in R. Let B_1 and B_2 be R-overrings of A. Then $B_1 B_2 = B_1 + B_2$.

Proof. One applies the preceding lemma to the Prüfer extension $B_1 \cap B_2 \subset R$ and the $(B_1 \cap B_2)$-modules B_1 and B_2. $\square$

We expand this proposition to a new characterization of Prüfer extensions.

Theorem 1.7 (cf. [Rh, Th.2.1]). Let $A \subset R$ be a ring extension with A integrally closed in R. The following are equivalent.
(1) A is Prüfer in R.
(2) For any two R-overrings B and C of A the sum $B + C$ is a subring of R, i.e. $B + C = BC$.
(3) $xy \in A[x] + A[y]$ for any two elements x, y of R.

Proof. The implication (1) $\Rightarrow$ (2) is covered by the preceding proposition, and (2) $\Rightarrow$ (3) is trivial.
(3) $\Rightarrow$ (1): Let $x \in R$ be given. Then $x^5 = x^2 \cdot x^3 \in A[x^2] + A[x^3]$. Condition (6) in Theorem I.5.2 is fulfilled, and we conclude by that theorem that A is Prüfer in R. $\square$

Proposition 1.8. Assume that A is Prüfer in R, and that $I_1, \ldots, I_r$ are A-submodules of R with $I_1 + \cdots + I_r$ being R-regular. Then, for every $n \in \mathbb{N}$,

$$(I_1 + \cdots + I_r)^n = I_1^n + \cdots + I_r^n.$$

Proof. By Lemma 1 it suffices to prove this in the case $A = A_v$ with v a local Manis valuation on R. Let $\mathfrak{q} := \operatorname{supp} v$. By Lemma 2 for every $j \in \{1, \ldots, r\}$ either $\mathfrak{q} \nsubseteq I_j$ or $I_j \subset \mathfrak{q}$, and the first case happens at least once, since $\mathfrak{q} \nsubseteq I_1 + \cdots + I_r$. Moreover, if $\mathfrak{q} \nsubseteq I_j$, $\mathfrak{q} \nsubseteq I_k$ then $I_j \supset I_k$ or $I_k \supset I_j$, since I_j and I_k both are v-convex. After a change of numeration we may assume that $I_1 \supset I_j$ for $j = 2, \ldots, r$. Now the claim of the proposition is evident. $\square$

Again we can expand this proposition to a characterization of Prüfer extensions.

Proposition 1.9. Let $A \subset R$ be a ring extension with A integrally closed in R. Let an integer $n \geq 2$ be given. The following are equivalent.

(1) A is Prüfer in R.
(2) $xy \in A + Ax^n + Ay^n$ for every $x, y \in R$.
(3) $x \in A + Ax^n$ for every $x \in R$.

Proof. The implication (2) $\Rightarrow$ (3) is obvious. (Take $y = 1$.) The implication (1) $\Rightarrow$ (2) follows by applying the preceding Proposition 8 with $r = 3$, $I_1 = A$, $I_2 = Ax$, $I_3 = Ay$. Finally condition (3) implies condition (6) in Theorem I.5.2, hence (1). $\square$

Remark. The implication (1) $\Rightarrow$ (3) can also be proved as follows. If A is Prüfer in R and $x \in R$, then $A = (A : x)(A + Ax)$ by Theorem I.5.2. From this we deduce for any $m \in \mathbb{N}$

$$A = (A : x)^m (A + Ax)^m \subset (A : x)^m \sum_{k=0}^{m} Ax^k \subset (A : x) + (A : x)^m x^m,$$

since $(A : x)^m x^k \subset (A : x)$ for $0 \leq k < m$.
Multiplying by x we obtain

$$Ax \subset A + (A : x)^m x^{m+1},$$

which is slightly sharper than condition (3) for $n = m + 1$. $\qquad\qquad$ □

Definition 3. An A-submodule I of R is called R-*invertible* (or *invertible in* R), if there exists an A-submodule J of R with $IJ = A$.

Remarks 1.10. a) In this situation the A-module I is finitely generated and $J = [A : I]$. Indeed, we have an equation $1 = \sum_{i=1}^{r} x_i y_i$ with $x_i \in I$, $y_i \in J$, from which one concludes easily that $I = \sum_{i=1}^{r} A x_i$. Let $K := [A : I]$. Then $J \subset K$, thus $IK = A$. We get $J = (KI)J = K(IJ) = K$. We also have $J = \sum_{i=1}^{r} A y_i$.

b) The argument in a) also shows the following: If I is R-invertible and is generated by r elements, then $[A : I]$ is generated by r elements.

c) Let $x \in R$. The A-module Ax is R-invertible iff $x \in R^*$. Then $[A : Ax] = Ax^{-1}$.

d) If I is R-invertible then I is R-regular. Indeed, from $IJ = A$ we get $IR = R$ by multiplying with the A-module R.

e) If A is the valuation ring A_v of a local Manis valuation v on R, then I is R-invertible iff I is finitely generated and R-regular, as is evident from Lemma 3 above and the preceding remarks a), c), d). In this case $I = Ax$ with some $x \in R^*$. $\qquad\qquad$ □

Definition 4. If an A-submodule I of R is R-invertible then we call $[A : I]$ *the inverse of* I *(in* R*)*.

For later use we quote a general lemma on invertible modules.

Lemma 1.11. Let $A \subset R$ be any ring extension. Assume that $\mathfrak{a}$ and $\mathfrak{b}$ are A-submodules of R and that $\mathfrak{a}$ is R-invertible.

a) $[A : \mathfrak{a}]\mathfrak{b} = [\mathfrak{b} : \mathfrak{a}]$, and $\mathfrak{a}[\mathfrak{b} : \mathfrak{a}] = \mathfrak{b}$.

b) If $\mathfrak{b} \subset \mathfrak{a}$ then $[\mathfrak{b} : \mathfrak{a}] = (\mathfrak{b} : \mathfrak{a})$.

Proof. a): Let $x \in [A : \mathfrak{a}]$ and $b \in \mathfrak{b}$. For any $a \in \mathfrak{a}$ we have $a(xb) = (ax)b \in Ab \subset \mathfrak{b}$. Thus $xb \in [\mathfrak{b} : \mathfrak{a}]$. This proves that $[A : \mathfrak{a}]\mathfrak{b} \subset [\mathfrak{b} : \mathfrak{a}]$. In order to verify the reverse inclusion we start with a relation $1 = \sum_{i=1}^{n} a_i r_i$, with $a_i \in \mathfrak{a}$, $r_i \in [A : \mathfrak{a}]$. Let $x \in [\mathfrak{b} : \mathfrak{a}]$. Then $b_i := x a_i \in$

$\mathfrak{b}$ $(1 \le i \le n)$. Thus $x = \sum_{i=1}^{n} b_i r_i \in \mathfrak{b}[A\colon \mathfrak{a}]$.

Multiplying the relation $[A\colon \mathfrak{a}]\mathfrak{b} = [\mathfrak{b}\colon \mathfrak{a}]$ by $\mathfrak{a}$ we obtain $\mathfrak{b} = \mathfrak{a}[\mathfrak{b}\colon \mathfrak{a}]$.
b): Let $x \in [\mathfrak{b}\colon \mathfrak{a}]$. Then $Ax = [A\colon \mathfrak{a}]\mathfrak{a}x \subset [A\colon \mathfrak{a}]\mathfrak{b} \subset [A\colon \mathfrak{a}]\mathfrak{a} = A$, hence
$x \in A$. Thus $[\mathfrak{b}\colon \mathfrak{a}] = (\mathfrak{b}\colon \mathfrak{a})$.   □

We turn back to Prüfer extensions.

Example 1.12. If A is Prüfer in R then, for every $x \in R$, the ideal
$(A\colon x)$ of A is the inverse of the A-module $A + Ax$. Thus $(A\colon x)$ is
R-invertible. It is generated by 2 elements.

Proof. Clearly $(A\colon x) = [A\colon A + Ax]$. The implication $(1) \Rightarrow (8)$ in
Theorem I.5.2 gives us the first claim. Then Remark 10.b tells us
that $(A\colon x)$ is generated by two elements.   □

Theorem 1.13. If A is Prüfer in R then every finitely generated
R-regular A-submodule of R is R-invertible.

Proof. Let I be a finitely generated R-regular A-submodule of R,
and let $J := [A\colon I]$. By Remark 10.e above and Lemma 1 we have
$I_{\mathfrak{p}} J_{\mathfrak{p}} = A_{\mathfrak{p}}$ for every $\mathfrak{p} \in \operatorname{Spec} A$. Thus $IJ = A$.   □

Corollary 1.14. Assume that A is Prüfer in R. Let I and J be finitely generated R-regular A-submodules of R. Then the A-modules
$I \cap J$, $[I\colon J]$ and $(I\colon J)$ are R-invertible.

Proof. By Th. 13 the modules I, J, and $I + J$ are R-invertible.
The claim follows from the identities $IJ = (I + J)(I \cap J)$ (Th. 4),
$J[I\colon J] = I$ (Lemma 11), and $(I\colon J) = A \cap [I\colon J]$.   □

For R-regularity instead of R-invertibility similar statements hold in
a more general setting.

Proposition 1.15. Assume that A is ws in R.

a) If I and J are R-regular A-submodules of R then $I \cap J$ is R-
   regular.
b) If, in addition, J is finitely generated then also $[I\colon J]$ and $(I\colon J)$
   are R-regular.

Proof. a): There exist finitely generated A-submodules $I_0 \subset I$, $J_0 \subset J$, such that $RI_0 = R$ and $RJ_0 = R$. If I_0 is generated by elements $x_1, \ldots, x_r$, then $\mathfrak{a} := \prod_{i=1}^{r} (A : x_i)$ is an R-regular ideal of A with $\mathfrak{a}I_0 \subset A$. For the same reason there exists an R-regular ideal $\mathfrak{b}$ of A with $\mathfrak{b}J_0 \subset A$. We have $R(\mathfrak{a}\mathfrak{b}I_0J_0) = R$. Now $\mathfrak{a}\mathfrak{b}I_0J_0 = (\mathfrak{a}I_0)(\mathfrak{b}J_0) \subset (\mathfrak{a}I_0) \cap (\mathfrak{b}J_0) \subset I_0 \cap J_0 \subset I \cap J$. Thus $R(I \cap J) = R$.
b): We choose an R-regular ideal $\mathfrak{a}$ with $\mathfrak{a}J \subset A$. Then $\mathfrak{a}IJ \subset I$, hence $\mathfrak{a}I \subset [I : J]$. The module $\mathfrak{a}I$ is R-regular. Thus $[I : J]$ is R-regular. As already proved, it follows that the intersection $A \cap [I : J] = (I : J)$ is R-regular. $\square$

Corollary 1.16. If A is ws in R then an A-submodule I of R is R-regular iff the ideal $A \cap I$ of A is R-regular. $\square$

We have seen in Chapter I that for $A \subset R$ a Prüfer extension every R-overring B of A is integrally closed in R (Th.I.5.2). We now expand this result considerably.

Let $A \subset R$ be any ring extension.

Definition 5. Let L be an A-submodule of R. We call an element $x \in R$ *integral over* L, if there exists a relation

$$(*) \qquad x^n + l_1 x^{n-1} + \cdots + l_{n-1} x + l_n = 0$$

with $l_i \in L^i$ $(1 \leq i \leq n)$, and we call the set $\tilde{L}$ consisting of these elements x the *integral closure of* L in R. If $\tilde{L} = L$, then we say that L is *integrally closed in* R. $\square$

Remark. The set $\tilde{L}$ is an $\tilde{A}$-submodule of R, with $\tilde{A}$ denoting the integral closure of A in R. This can be derived from the following. Let t be an indeterminate over R. In $R[t]$ we consider the subring $B := \sum_{i \in \mathbb{N}_0} L^i t^i$, reading $L^\circ := A$. As is easily checked, a given element x of R is integral over L iff tx is integral over B. Thus $t\tilde{L} = \tilde{B} \cap (tR)$, with $\tilde{B}$ the integral closure of B in $R[t]$. $\square$

Theorem 1.17. If $A \subset R$ is Prüfer and L is an R-regular A-submodule of R then L is integrally closed in R.

For the proof we need two lemmas which are valid for an arbitrary ring extension $A \subset R$.

Lemma 1.18. Let L be an A-submodule of R, and assume that $x \in R$ is integral over L. Then, for every $y \in R$, the element xy is integral over Ly.

Proof. We multiply an integrality relation $(*)$, as given in Definition 5, by y^n and obtain the relation

$$(xy)^n + (l_1 y)(xy)^{n-1} + \cdots + (l_{n-1} y^{n-1})(xy) + \cdots + l_n y^n = 0,$$

which proves the claim. $\qquad\square$

Lemma 1.19. Let $(L_\alpha \mid \alpha \in \Lambda)$ be a direct system of A-submodules L_α of R and $L := \varinjlim_{\alpha \in \Lambda} L_\alpha = \bigcup_\alpha L_\alpha$. Then the integral closure $\tilde{L}$ of L is the union of the integral closures $\tilde{L}_\alpha$.

We omit the easy proof. $\qquad\square$

Proof of Theorem 17. Using Lemma 19 we easily retreat to the case that the A-module L is finitely generated. Now L is R-invertible. Let $x \in R$ be integral over L. Given some $y \in L^{-1}$, the element xy is integral over Ly by Lemma 18. A fortiori, xy is integral over $LL^{-1} = A$. Since A is Prüfer in R this implies $xy \in A$ (cf. Th.I.5.2). We now have proved that $xL^{-1} \subset A$. Multiplying by L we obtain $xA \subset L$, i.e. $x \in L$. $\qquad\square$

§2 Characterisation of Prüfer extensions by the behavior of their regular ideals

The definition of Prüfer extensions in I, §5 involves Manis valuations. But, as explained in the Introduction, we are very much interested in criteria for a given ring extension $A \subset R$ to be Prüfer, which do not mention valuations either explicitly or implicitly. Such criteria have been given in §5 and §6 of Chapter I and then in §1 of the present chapter (cf. Th.1.7, Prop.1.9). The criteria in §1 are characterizations of Prüfer extensions within the class of extensions $A \subset R$

with A integrally closed in R. We now strive for characterizations in the class of ws (= weakly surjective) extensions. Sometimes we will be forced to replace this class by the more narrow class of "tight" extensions to be defined now.

As before R is a ring and A a subring of R.

Definition 1. We say that A is *tight in* R, or that R is a *tight extension of* A, if for every $x \in R \setminus A$ there exists an R-invertible ideal I of A with $Ix \subset A$.*)

Notice that this implies $(A:x)R = R$ for every $x \in R$. Thus a tight extension is a weakly surjective extension (cf. Th.I.3.13). Notice also that in the case $R = \text{Quot}\,A$ the ring A is always tight in R since then $(A:x) \cap R^* \neq \emptyset$ for every $x \in R$. If A is Prüfer in R then, of course, A is tight in R.

We postpone a study of tight extensions to §4, but already now use this class of extensions.

Theorem 2.1. The following are equivalent for the subring A of R.

(1) A is Prüfer in R.
(2) (a) A is weakly surjective in R. (b) Every finitely generated R-regular ideal I of A is R-invertible.
(3) (a) A is tight in R. (b) Every finitely generated ideal of A which contains an R-invertible ideal of A is itself R-invertible.
(4) (a) A is tight in R. (b) For any R-invertible ideal I of A and any $a \in A$ the ideal $I + Aa$ is again R-invertible.

Proof. The implications $(1) \Longrightarrow (2)$ and $(1) \Longrightarrow (3)$ are evident from the above and Theorem 1.13.
(2) or $(3) \Longrightarrow (1)$: Given a prime ideal $\mathfrak{p}$ of A and an element x in $R \setminus A_{[\mathfrak{p}]}$ we verify that there exists an element $x' \in \mathfrak{p}$ with $xx' \in A \setminus \mathfrak{p}$. This will prove that $(A_{[\mathfrak{p}]}, \mathfrak{p}_{[\mathfrak{p}]})$ is a Manis pair in R (Th.I.2.4), and we are done. First assume that (2) holds. We have an equation $\sum_{i=1}^{n} a_i r_i = 1$ with $a_i \in (A:x)$, $r_i \in R$. Let I denote the

---

*) In the literature (e.g. [AB], [Eg], [Rh]) these extensions run under the name "invertible transforms". This sounds misleading to us.

ideal $(a_1, a_2, \ldots, a_n, a_1 x, a_2 x, \ldots, a_n x)$ of A. It is R-regular, hence $I[A\colon I] = A$ by (2) (b). Thus we have an equation

$$1 = \sum_{i=1}^{n} a_i b_i + x\left(\sum_{i=1}^{n} a_i c_i\right)$$

with $b_i, c_i \in [A\colon I]$. If $b \in [A\colon I]$ then $ba_i \in A$, $ba_i x \in A$, hence $ba_i \in \mathfrak{p}$ for every $i \in \{1, \ldots, n\}$, since $x \notin A_{[\mathfrak{p}]}$. We put $x' = \sum_{i=1}^{n} a_i c_i$, $y = \sum_{i=1}^{n} a_i b_i$. Then $x' \in \mathfrak{p}$, $y \in \mathfrak{p}$, $xx' = 1 - y \in A \setminus \mathfrak{p}$, as desired.

Now assume that (3) holds. Then $(A\colon x)$ contains an R-invertible ideal $(a_1, \ldots, a_n)$. Let $I := (a_1, \ldots, a_n, a_1 x, \ldots, a_n x)$. By (3) (b) we again have $I[A\colon I] = A$, and we can finish as before.

It is obvious that (3) implies (4). Now assume that (4) holds. Let $I = (a_1, \ldots, a_n)$ be a finitely generated ideal of A containing an R-invertible ideal I_0 of A. Then we may write $I = I_0 + Aa_1 + \cdots + Aa_n$. Applying (4) (b) n times we see that I is R-invertible. Thus (4) implies (3). $\qquad\square$

Theorem 2.2. Assume that A is tight in R. The following are equivalent.

(1) A is Prüfer in R.
(2) A is integrally closed in R. For any finitely many elements $a_1, \ldots, a_n$ of A which generate an R-regular ideal $(a_1, \ldots, a_n)$ of A,
$$(a_1, \ldots, a_n)^2 = (a_1^2, \ldots, a_n^2)$$

(3) A is integrally closed in R. For any finitely many elements $a_1, \ldots, a_n$ of A which generate an R-regular ideal $(a_1, \ldots, a_n)$ of A, there exists some integer $m \geq 2$ with
$$(a_1, \ldots, a_n)^m = (a_1^m, \ldots, a_n^m).$$

Proof. We know from §5 that an R-Prüfer ring is integrally closed in R. Thus the implication (1) $\Longrightarrow$ (2) {and (1) $\Longrightarrow$ (3) as well} is covered by Proposition 1.8. The implication (2) $\Longrightarrow$ (3) is trivial. (3) $\Longrightarrow$ (1): In order to prove that A is Prüfer in R we verify condition (6) in Theorem I.5.2. Let $x \in R \setminus A$ be given. Since A is tight in

R there exists an invertible ideal $I \subset (A\!:x)$. Write $I = (a_1, \ldots, a_n)$ with finitely many $a_i \in A$. We have $1 = \sum_{i=1}^{n} a_i r_i$ with some $r_i \in [A\!:I]$. Let $J := (a_1, \ldots, a_n, a_1 x, \ldots, a_n x)$. This ideal of A is regular in R. By assumption (3) there exists some integer $m \geq 2$ with

$$J^m = (a_1^m, \ldots, a_n^m, a_1^m x^m, \ldots, a_n^m x^m).$$

Let Λ denote the set of multiindices $\alpha = (\alpha_1, \ldots, \alpha_n) \in \mathbb{N}_0^n$ with $\alpha_1 + \cdots + \alpha_n = m$. For any $\alpha = (\alpha_1, \ldots, \alpha_n) \in \Lambda$ we put $a^\alpha := \prod_{i=1}^{n} a_i^{\alpha_i}$, $r^\alpha = \prod_{i=1}^{n} r_i^{\alpha_i}$.

Now observe that $a^\alpha x \in J^m$ for every $\alpha \in \Lambda$. Thus we have an equation

(1) $$a^\alpha x = \sum_{j=1}^{n} b_{\alpha j} a_j^m + x^m \sum_{j=1}^{n} c_{\alpha j} a_j^m$$

for every $\alpha \in \Lambda$ with elements $b_{\alpha j}, c_{\alpha j} \in A$. Raising the relation $1 = \sum_{i=1}^{n} a_i r_i$ to the m-th power we have

(2) $$1 = \sum_{\alpha \in \Lambda} \binom{m}{\alpha} a^\alpha r^\alpha.$$

Combining (1) and (2) we obtain

$$x = \sum_{\alpha \in \Lambda} \binom{m}{\alpha} r^\alpha a^\alpha x = b + x^m c$$

with $b := \sum_{\alpha \in \Lambda} \sum_{j=1}^{n} \binom{m}{\alpha} b_{\alpha j} r^\alpha a_j^m$, $c := \sum_{\alpha \in \Lambda} \sum_{j=1}^{m} \binom{m}{\alpha} c_{\alpha j} r^\alpha a_j^m$.

Since $r_i \in [A\!:I]$ for $i = 1, \ldots, n$, we have $r^\alpha a_j^m \in A$ for every $\alpha \in \Lambda$, and we conclude that $b \in A$, $c \in A$. Thus $x \in A + Ax^m$, and condition (6) in Theorem I.5.2 is fulfilled. □

The now proved theorems 1 and 2 leave something to be desired. For a given ring extension $A \subset R$ it will be often much more difficult to verify that A is tight in R than that A is ws in R. Thus it will pay to understand the notion of R-invertibility of an ideal I of A in terms of R-regularity and A-intrinsic properties of I. Fortunately this is possible, even in various ways.

Definition 2. Let A be any ring (commutative, with 1, as always) and M an A-module.

a) M is called a *multiplication module*, if for every submodule N of M there exists an ideal I of A with $N = IM$, cf. [Ba$_1$]. Notice that then we may choose $I = (N:M):= \{a \in A \mid aM \subset N\}$.

b) M is called *locally principal*, if for every $\mathfrak{p} \in \operatorname{Spec} A$ the $A_{\mathfrak{p}}$-module $M_{\mathfrak{p}}$ is generated by one element. {It suffices to know this for the maximal ideals $\mathfrak{p}$ of A.}

In the following we will apply this terminology to A-submodules $\mathfrak{a}$ of R for $A \subset R$ a ring extension. (The word "locally principal" alludes to the idea, that we regard such modules as something like fractional ideals.)

Proposition 2.3. (cf. [Bo, II §5, Th.4] in the case $R = S^{-1}A$.) Let $\mathfrak{a}$ be an R-regular A-submodule of R. The following are equivalent.

(1) $\mathfrak{a}$ is R-invertible.
(2) $\mathfrak{a}$ is a multiplication module.
(3) $\mathfrak{a}$ is finitely generated and locally principal.

Assume either that R is a ring of quotients of A (cf. I §3, Def.4) or that $\mathfrak{a} \subset A$. Then (1), (2), (3) are equivalent to:

(4) $\mathfrak{a}$ is a projective A-module of rank 1.
(5) $\mathfrak{a}$ is a projective A-module.

Proof. The implication (1) $\implies$ (2) is covered by Lemma 1.11. (2) $\implies$ (3): We first verify that $\mathfrak{a}$ is finitely generated, using an argument from [LS, p.4358f]. Since $\mathfrak{a}$ is multiplicative, we have $A a = (Aa:\mathfrak{a})\mathfrak{a}$ for every $a \in \mathfrak{a}$. We introduce the ideal $\mathfrak{r} := \sum_{a \in \mathfrak{a}} (Aa:\mathfrak{a})$ of A. We infer that $\mathfrak{a}\mathfrak{r} = \mathfrak{a}$. From this we want to conclude that $\mathfrak{r} = A$. We choose elements $b_1, \ldots, b_n$ in $\mathfrak{a}$ such that $\mathfrak{b} = \sum_{i=1}^{n} Ab_i$ is R-regular. For every $i \in \{1, \ldots, n\}$ we have $Ab_i = (Ab_i:\mathfrak{a})\mathfrak{a} = (Ab_i:\mathfrak{a})\mathfrak{r}\mathfrak{a} = \mathfrak{r}b_i$. We choose elements $c_i \in \mathfrak{r}$ with $b_i = c_i b_i$ $(1 \le i \le n)$. Then

$\mathfrak{b} \cdot \prod_{i=1}^{n} (1 - c_i) = 0$. Since $\mathfrak{b}$ is R-regular, this implies $\prod_{i=1}^{n} (1 - c_i) = 0$. Thus $1 \in \mathfrak{r}$.

We now choose elements $a_1, \ldots, a_m \in \mathfrak{a}$ and $x_j \in (Aa_j : \mathfrak{a})$ with $1 = \sum_{j=1}^{m} x_j$. We have $\mathfrak{a} \subset \sum_{j=1}^{m} x_j \mathfrak{a} \subset \sum_{j=1}^{m} Aa_j$, hence $\mathfrak{a} = \sum_{j=1}^{m} Aa_j$. This proves that $\mathfrak{a}$ is finitely generated.

Given a prime ideal $\mathfrak{p}$ of A, it remains to verify that the $A_\mathfrak{p}$-module $\mathfrak{a}_\mathfrak{p}$ is principal. From the equation $1 = \sum_{j=1}^{m} x_j$ above we conclude that there is some index $j \in \{1, \ldots, m\}$ such that $\frac{x_j}{1} \in A_\mathfrak{p}$ is a unit in $A_\mathfrak{p}$, hence $(Aa_j : \mathfrak{a})_\mathfrak{p} = A_\mathfrak{p}$. Let $d_j := \frac{a_j}{1} \in \mathfrak{a}_\mathfrak{p}$. We have (cf. Lemma 1.1.d) $(A_\mathfrak{p} d_j : \mathfrak{a}_\mathfrak{p}) = (Aa_j : \mathfrak{a})_\mathfrak{p} = A_\mathfrak{p}$, hence $\mathfrak{a}_\mathfrak{p} = 1 \cdot \mathfrak{a}_\mathfrak{p} \subset A_\mathfrak{p} d_j$. Thus $\mathfrak{a}_\mathfrak{p} = A_\mathfrak{p} d_j$.

(3) $\Longrightarrow$ (1): For every $\mathfrak{p} \in \operatorname{Spec} A$ we have $\mathfrak{a}_\mathfrak{p} = A_\mathfrak{p} u_\mathfrak{p}$ with some $u_\mathfrak{p} \in A_\mathfrak{p}$. Since $\mathfrak{a}$ is finitely generated, we have $[A : \mathfrak{a}]_\mathfrak{p} = [A_\mathfrak{p} : \mathfrak{a}_\mathfrak{p}]$. Thus

$$(\mathfrak{a}[A : \mathfrak{a}])_\mathfrak{p} = \mathfrak{a}_\mathfrak{p}[A_\mathfrak{p} : \mathfrak{a}_\mathfrak{p}] = A_\mathfrak{p} u_\mathfrak{p} \cdot A_\mathfrak{p}(u_\mathfrak{p})^{-1} = A_\mathfrak{p}$$

for every $\mathfrak{p} \in \operatorname{Spec} A$. We conclude that $\mathfrak{a}[A : \mathfrak{a}] = A$.

(1) $\Longrightarrow$ (4): This is clear by first principles of the theory of projective modules, cf. [Bo, II §5 Th.1] and also Lemma 4.1 below. For this implication we do not need the additional assumption made in the middle of the proposition.

(4) $\Longrightarrow$ (5) is trivial. (5) $\Longrightarrow$ (1): Since $\mathfrak{a}$ is projective there exists a family $(\varphi_i \mid i \in I)$ of A-linear maps from $\mathfrak{a}$ to A and a family $(a_i \mid i \in I)$ of elements of $\mathfrak{a}$ such that, for every $x \in \mathfrak{a}$, $\varphi_i(x) = 0$ for almost all $i \in I$ and $x = \sum_{i \in I} \varphi_i(x) a_i$, cf. e.g. [CE, Chap.VII, Prop.3.1]. Since $\mathfrak{a}$ is R-regular there further exists a finite subset K of I and a family $(r_k \mid k \in K)$ in R such that $1 = \sum_{k \in K} a_k r_k$. We now choose a finite subset J of I such that $\varphi_i(a_k) = 0$ for every $i \in I \setminus J$ and $k \in K$. Then

$$1 = \sum_{k \in K} \left(\sum_{j \in J} \varphi_j(a_k) a_j \right) r_k.$$

Introducing the elements $b_j := \sum_{k \in K} \varphi_j(a_k) r_k$ for $j \in J$ we have $1 = \sum_{j \in J} a_j b_j$. We are done, if we verify that the b_j are elements of $[A : \mathfrak{a}]$. We need the following fact, which we will prove below.

(*) If $\varphi\colon \mathfrak{a} \to A$ is an A-linear map then, $\varphi(x)y = x\varphi(y)$ for any two elements x, y of $\mathfrak{a}$.

For each $i \in I$ we have

$$a_i b_j = \sum_{k \in K} a_i \varphi_j(a_k) r_k = \sum_{k \in K} \varphi_j(a_i) a_k r_k = \varphi_j(a_i) \in A.$$

Thus indeed $b_j \in [A\colon \mathfrak{a}]$ for every $j \in J$.

Proof of (*): If $\mathfrak{a} \subset A$, then clearly $\varphi(x)y = \varphi(xy) = x\varphi(y)$. Assume now that R is a ring of quotients of A. Then the ideals $(A\colon x)R$ and $(A\colon y)R$ are dense in R. Thus also $\mathfrak{c}R$ is dense in R for $\mathfrak{c} := (A\colon x) \cap (A\colon y)$. If $c \in \mathfrak{c}$ we have $c[\varphi(x)y - x\varphi(y)] = \varphi(x)(cy) - (cx)\varphi(y) = \varphi(xcy) - \varphi(cxy) = 0$. We conclude that $\varphi(x)y = x\varphi(y)$.   □

In the following we will apply Proposition 3 in the case that $\mathfrak{a}$ is an ideal of A. Later we will need Proposition 3 for more general A-modules. Then it will be important to know that R is a ring of quotients of A.

As before (I, §3) we denote the complete ring of quotients of a ring A by $Q(A)$.

Definition 3. We call an ideal $\mathfrak{a}$ of A *invertible*, if $\mathfrak{a}$ is $Q(A)$-invertible.[*]

Proposition 2.4. An ideal $\mathfrak{a}$ of A is invertible iff $\mathfrak{a}$ is a finitely generated projective A-module and a dense ideal. Then $\mathfrak{a}$ is projective of rank 1.

Proof. We may assume in advance that $\mathfrak{a}$ is finitely generated. If $\mathfrak{a}$ is $Q(A)$-invertible then $\mathfrak{a}$ is projective of rank 1 by the preceding proposition, and certainly $(0\colon \mathfrak{a}) = 0$. Assume now that $\mathfrak{a}$ is dense in A and projective. Let $a_1, \ldots, a_n$ be generators of the ideal $\mathfrak{a}$. We have A-linear forms $\varphi_i\colon \mathfrak{a} \to A$ $(i = 1, \ldots, n)$ with $x = \sum_{i=1}^{n} \varphi_i(x) a_i$ for all $x \in A$. The φ_i can be interpreted as elements of $Q(A)$ [Lb,

---

[*] This definition differs from the one given in Huckaba's book [Huc, p. 29].

§2.3]. Then $1 = \sum_{i=1}^{n} a_i \varphi_i$, and $\varphi_i \in [A : \mathfrak{a}]$ for every i. Thus $\mathfrak{a}$ is invertible in $Q(A)$.   □

From the last two propositions we infer

Scholium 2.5. Let $A \subset R$ be a ring extension and $\mathfrak{a}$ an ideal of A. Then $\mathfrak{a}$ is R-invertible iff $\mathfrak{a}$ is invertible and $\mathfrak{a}R = R$.   □

We now can improve a part of Theorem 1 as follows.

Theorem 2.6. Given a ring extension $A \subset R$ the following are equivalent.
(1) A is Prüfer in R.
(2) A is weakly surjective in R. Every finitely generated R-regular ideal I of A is locally principal.
(3) A is weakly surjective in R. Every finitely generated R-regular ideal I of A is a multiplication ideal.

Proof. This is evident from Proposition 3 and the equivalence (1) $\Leftrightarrow$ (2) in Theorem 1.   □

We will give two more characterizations of Prüfer extensions within the class of ws extensions. For this we need the following easy lemma.

Lemma 2.7. Let I be a finitely generated ideal of a local ring C. Assume that there exists some $a \in I$ such that $I = Ca + I^2$. Then I is principal.

Proof. If $I = C$ this is evident. We now assume that $I \neq C$. Then I is contained in the maximal ideal $\mathfrak{m}$ of C. We have $I = Ca + I\mathfrak{m}$. By Nakayama's lemma this implies that $I = Ca$.   □

Definition 4 [Fu]. A ring C is called *arithmetical*, if the lattice of its ideals is distributive, i.e. $I \cap (J + K) = (I \cap J) + (I \cap K)$ for any three ideals I, J, K of C. {Equivalently, $I + (J \cap K) = (I + J) \cap (I + K)$ for all I, J, K.}

We will use basic results about arithmetical rings, which in full generality are due to C.U. Jensen [J_1]. For the convenience of the reader we reproduce the proofs of these results in Appendix B.

Theorem 2.8. (cf. [AP, Th.2] for $R = \operatorname{Quot} A$). Assume that A is ws in R. The following are equivalent.

(1) A is R-Prüfer.
(2) The lattice of R-regular ideals of A is distributive, i.e.
$I \cap (J + K) = (I \cap J) + (I \cap K)$ for any three R-regular ideals I, J, K of A.
(3) For every finitely generated R-regular ideal I of A the ring A/I is arithmetical.

Proof. The implication (1) $\Longrightarrow$ (2) is covered by Theorem 1.4, and the implication (2) $\Longrightarrow$ (3) is trivial.

(3) $\Rightarrow$ (1): We verify condition (2) in Theorem 6. Let I be a finitely generated R-regular ideal of A and $\mathfrak{m}$ a maximal ideal of A. We have to prove that the ideal $J := I_{\mathfrak{m}}$ of the local ring $C := A_{\mathfrak{m}}$ is principal. We may assume that $J \neq C$. By hypothesis (3) the ring A/I^2 is arithmetical. Thus also C/J^2 is arithmetical, and is local. According to [J_1, Th.1] (cf. App.B, Th.1) this implies that the set of ideals of C/J^2 is totally ordered by inclusion. It follows that the ideal J/J^2 is principal. Lemma 7 now tells us that J is principal. $\square$

Here is a variant of Theorem 8 for tight extensions.

Theorem 2.9. Assume that A is tight in R and that for every R-invertible ideal I of A the ring A/I is arithmetical. Then A is Prüfer in R.

Proof. We verify condition (3) in Theorem 1. Let I be a finitely generated ideal of A containing an R-invertible ideal $\mathfrak{a}$ of A. We have to prove that I itself is R-invertible. By hypothesis the ring $\bar{A} := A/\mathfrak{a}^2$ is arithmetical. The ideal $\bar{I} := I/\mathfrak{a}^2$ of $\bar{A}$ is finitely generated. It contains the ideal $\bar{\mathfrak{a}} := \mathfrak{a}/\mathfrak{a}^2$ of $\bar{A}$. By a well known theorem about arithmetical rings ([J_1, Th.2], cf. App.B, Th.4) there exists an ideal J of A containing $\mathfrak{a}$, such that $\bar{\mathfrak{a}} = \bar{I} \cdot \bar{J}$, with $\bar{J} := J/\mathfrak{a}^2$. Since $\mathfrak{a}^2 \subset IJ$, this implies $\mathfrak{a} = IJ$. Since $\mathfrak{a}$ is R-invertible, also I is R-invertible. $\square$

Theorem 2.10 (cf. [AP, Th.4] for $R = \operatorname{Quot} A$, [Rh, Th.2.1] for A tight in R). Assume that A is ws in R. Let $\mathcal{F}$ denote the set of R-regular ideals of A. The following are equivalent.

(1) A is Prüfer in R.
(2) If $I, J, K \in \mathcal{F}$ and K is finitely generated, then $(I + J): K = (I: K) + (J: K)$.
(3) If $I, J, K \in \mathcal{F}$ and both J and K are finitely generated, then $I: (J \cap K) = (I: J) + (I: K)$.
(4) For ideals $I, J \in \mathcal{F}$ with $I \subset J$, J finitely generated, there exists an ideal K of A with $I = JK$.

Proof. The implications $(1) \Rightarrow (2) - (4)$ are covered by Theorem 1.4. (2) or $(3) \Rightarrow (1)$: Let $\mathfrak{a}$ be an R-regular ideal of A. From (2) or (3) one obtains analogous statements for the ideals of $A/\mathfrak{a}$, with the word "R-regular" cancelled. It is well known that these statements imply that $A/\mathfrak{a}$ is arithmetical ([J_1, Th.3], cf. App.B, Th.5). It follows from Theorem 8 that A is Prüfer in R.
$(4) \Rightarrow (1)$: Let I be a finitely generated R-regular ideal of A. We prove that the ring A/I is arithmetical and then again will be done by Theorem 9. According to Theorem 2 in [J_1], (cf. App.B, Th.4) it suffices to verify for a given finitely generated ideal $J \supset I$ of A that J/I is a multiplication ideal of A/I. This is clear from hypothesis (4). □

Remark. We mention another proof of the implication $(4) \Rightarrow (1)$: Let again I be a finitely generated R-regular ideal of A. Then I/I^2 is a multiplication ideal of A/I^2 by hypothesis (4). By a well known theorem on multiplication ideals [An][*] it follows that I/I^2 is locally principal. Now Lemma 7 tells us that the ideal I of A is locally principal, and Theorem 6 tells us that A is Prüfer in R. □

Under the stronger condition that A is tight in R we find more multiplicativity conditions on ideals of A which are equivalent to the conditions $(1) - (4)$ in the just proved theorem. The proof of the new equivalences will be somewhat harder than the preceding proofs, but the new conditions are cute.

Theorem 2.11 (cf. [Rh, Th.2.1]). Let $A \subset R$ be tight extension. As before, let $\mathcal{F}$ denote the set of R-regular ideals of A. The following are equivalent.

---

[*] Every finitely generated multiplication ideal $\mathfrak{a}$ of a ring A is locally principal, cf. [An, Th.3].

(1) A is Prüfer in R.

(5) If $I, J, K \in \mathcal{F}, IJ = IK$, and I is finitely generated, then $J = K$.

(6) $I(J \cap K) = (IJ) \cap (IK)$ for all $I, J, K \in \mathcal{F}$.

(7) $(I + J)(I \cap J) = IJ$ for all $I, J \in \mathcal{F}$.

Proof. The implications $(1) \Rightarrow (5) - (7)$ are covered by Theorem 1.4.

$(5) \Longrightarrow (1)$: We verify condition (2) of Theorem 2. We first prove that A is integrally closed in R. Let $x \in R$ be integral over A. We have

$$x^{m+1} + c_1 x^m + c_2 x^{m-1} + \cdots + c_m = 0$$

with some $m \in \mathbb{N}$ and $c_i \in A$. Let $J := A + Ax + \cdots + Ax^m$. Then $J^2 = J$. We choose an R-invertible ideal $I \subset (A : x)$. This is possible since A is tight in R. Let $K := I^m J$. Then $I^m \subset K \subset A$. Thus K is an R-regular ideal of A. We have $K^2 = I^{2m} J^2 = I^{2m} J = I^m K$. By assumption (5) this implies $K = I^m$, i.e. $I^m J = I^m$. Since I is R-invertible we conclude that $J = A$, i.e. $x \in A$.

Let $(b_1, \ldots, b_n)$ be an R-regular ideal of A. We have

$$(b_1, \ldots, b_n)^{n+1} = (b_1, \ldots, b_n)^{n-1}(b_1^2, \ldots, b_n^2).$$

By (5) this implies $(b_1, \ldots, b_n)^2 = (b_1^2, \ldots, b_n^2)$. Condition (2) of Theorem 2 is fulfilled.

$(6) \Longrightarrow (7)$: A priori we have $(I + J)(I \cap J) \subset IJ$. From (6) we obtain

$$(I + J)(I \cap J) = [(I + J)I] \cap [(I + J)J] \supset IJ.$$

$(7) \Longrightarrow (1)$: We verify condition (4) in Theorem 1. Let $\mathfrak{a}$ be an R-invertible ideal of A and $a \in A$. We have to prove that $\mathfrak{a} + Aa$ is again R-invertible. We use the following complicated lemma, the proof of which will be given below.

Lemma 2.12. Let $A \subset R$ be a ring extension. Assume that I is an R-invertible ideal and J is a finitely generated ideal of A. Assume further that

$$(I + J)(I \cap (I^2 + J)) = I(I^2 + J).$$

Then, for every $\mathfrak{p} \in \operatorname{Spec} A$, either $I_{\mathfrak{p}} \subset J_{\mathfrak{p}}$ or $J_{\mathfrak{p}} \subset I_{\mathfrak{p}}$. $\qquad\square$

We apply this lemma to $I := \mathfrak{a}$ and $J := A a$. This is possible by (7).
Let $\mathfrak{p} \in \operatorname{Spec} A$ be given.

Case 1. $J_\mathfrak{p} \subset I_\mathfrak{p}$. Now $(I + J)_\mathfrak{p} = I_\mathfrak{p}$, which is invertible in $R_\mathfrak{p}$.

Case 2. $I_\mathfrak{p} \subset J_\mathfrak{p} = A_\mathfrak{p} \cdot \frac{a}{1}$. The $A_\mathfrak{p}$-module $I_\mathfrak{p}$ is $R_\mathfrak{p}$-regular, since it
is invertible in $R_\mathfrak{p}$. Thus also $J_\mathfrak{p}$ is $R_\mathfrak{p}$-regular. This means $\frac{a}{1} \in R_\mathfrak{p}^*$.
We conclude that $(I + J)_\mathfrak{p} = J_\mathfrak{p}$ is again invertible in $R_\mathfrak{p}$. Thus
$(I + J)_\mathfrak{p}$ is invertible in $R_\mathfrak{p}$ for every $\mathfrak{p} \in \operatorname{Spec} A$. It follows that
$I + J = \mathfrak{a} + A a$ is invertible in R. $\square$

Proof of the lemma. Since $I \cap (I^2 + J) = I^2 + I \cap J$, it follows from
our assumption that $(I + J)(I^2 + I \cap J) = I(I^2 + J)$. Multiplying
both sides with $[A : I]$ we obtain

$$I^2 + J = (I + J)(I + (I \cap J)[A : I]).$$

We expand the right hand side:

$$I^2 + J = I^2 + JI + I \cap J + J(I \cap J)[A : I].$$

By Lemma 6 we have $(I \cap J)[A : I] = (I \cap J : I)$.
We obtain

$$I^2 + J = I^2 + I \cap J + J(I \cap J : I).$$

Since $I^2(I \cap J : I) \subset I(I \cap J) \subset I \cap J$, we may also write

$(*)$ $$I^2 + J = I^2 + I \cap J + (I^2 + J)(I \cap J : I).$$

Let now $\mathfrak{p} \in \operatorname{Spec} A$ be given. Assume that $I_\mathfrak{p} \not\subset J_\mathfrak{p}$. Then $I_\mathfrak{p} \not\subset$
$(I \cap J)_\mathfrak{p} = I_\mathfrak{p} \cap J_\mathfrak{p}$. Thus the ideal $(I_\mathfrak{p} \cap J_\mathfrak{p} : I_\mathfrak{p})$ of $A_\mathfrak{p}$ is different from
$A_\mathfrak{p}$, hence contained in $\mathfrak{p}_\mathfrak{p}$.

From $(*)$ we obtain by Lemma 1.1

$$(I^2 + J)_\mathfrak{p} = (I_\mathfrak{p}^2 + I_\mathfrak{p} \cap J_\mathfrak{p}) + (I^2 + J)_\mathfrak{p}(I_\mathfrak{p} \cap J_\mathfrak{p} : I_\mathfrak{p}).$$

Since the $A_\mathfrak{p}$-module $(I^2 + J)_\mathfrak{p}$ is finitely generated, Nakayama's
lemma gives us $I_\mathfrak{p}^2 + I_\mathfrak{p} \cap J_\mathfrak{p} = (I^2 + J)_\mathfrak{p}$, so

$$J_\mathfrak{p} \subset I_\mathfrak{p}^2 + I_\mathfrak{p} \cap J_\mathfrak{p} \subset I_\mathfrak{p}.$$ $\square$

§3 Describing a Prüfer extension by its lattice of regular ideals

Let $A \subset B$ be any ring extension. We denote the set of ideals of A by $\mathcal{J}(A)$ and the subset of B-regular ideals of A by $\mathcal{F}(B/A)$. The ordering by the inclusion relation makes $\mathcal{J}(A)$ a lattice with meet $\mathfrak{a} \wedge \mathfrak{b} = \mathfrak{a} \cap \mathfrak{b}$ and join $\mathfrak{a} \vee \mathfrak{b} = \mathfrak{a} + \mathfrak{b}$. As a further operation on $\mathcal{J}(A)$ we have the multiplication $(\mathfrak{a}, \mathfrak{b}) \mapsto \mathfrak{a}\mathfrak{b}$. It is of interest to study the relations between multiplication and the lattice operations.

Clearly the subset $\mathcal{F} := \mathcal{F}(B/A)$ of $\mathcal{J}(A)$ has the following four properties:

R0. $A \in \mathcal{F}$.
R1. If $I \in \mathcal{F}$ and $I \subset J$ then $J \in \mathcal{F}$.
R2. If $I \in \mathcal{F}$ and $J \in \mathcal{F}$ then $IJ \in \mathcal{F}$.
R3. For every $I \in \mathcal{F}$ there exists a finitely generated ideal $I_0 \subset I$ with $I_0 \in \mathcal{F}$.

One verifies easily that, if $\mathcal{F}$ is any subset of $\mathcal{J}(A)$ fulfilling R0 – R3, then also the following properties hold:

R4. If $I \in \mathcal{F}$ and $J \in \mathcal{F}$ then $I \cap J \in \mathcal{F}$.
R5. If $I \in \mathcal{J}(A)$ and $\sqrt{I} \in \mathcal{F}$ then $I \in \mathcal{F}$.[*]

In particular $\mathcal{F}$ is a filter on $\mathcal{J}(A)$. It is also true that, if $\mathcal{F}$ is a filter on $\mathcal{J}(A)$ with the properties R3 and R5 then R2 holds. But let us stop here looking in general at the axiomatics of filters on $\mathcal{J}(A)$ with such properties. {We will continue with this business in volume II.}

Let $Y(B/A)$ denote the set $\operatorname{Spec} A \cap \mathcal{F}(B/A)$ of B-regular *prime* ideals of A. As before (I, §4) let $X(B/A)$ denote the image of the restriction map from $\operatorname{Spec} B$ to $\operatorname{Spec} A$. Clearly $X(B/A)$ and $Y(B/A)$ are disjoint.

Lemma 3.1. Let $\mathfrak{a}$ be an ideal of A. The following are equivalent.

i) $\mathfrak{a} \in \mathcal{F}(B/A)$.
ii) If $\mathfrak{p}$ is a prime ideal of A containing $\mathfrak{a}$ then $\mathfrak{p} \in Y(B/A)$.
iii) If $\mathfrak{p}$ is a prime ideal of A containing $\mathfrak{a}$ then $\mathfrak{p} \notin X(B/A)$.

---

[*] We use the notion $\sqrt{I} = \{x \in A \mid x^n \in I \text{ for some } n\}$.

Proof. The implications i) $\Rightarrow$ ii) $\Rightarrow$ iii) are trivial. iii) $\Rightarrow$ i): Suppose that $\mathfrak{a}B \neq B$. Then there exists a prime ideal $\mathfrak{q}$ of B containing $\mathfrak{a}B$. The prime ideal $\mathfrak{q} \cap A$ of A is an element of $X(B/A)$ and contains $\mathfrak{a}$, a contradiction. Thus $\mathfrak{a}B = B$, i.e. $\mathfrak{a} \in \mathcal{F}(B/A)$. $\qquad\square$

According to Lemma 1 the lattice $\mathcal{F}(B/A)$ is determined both by the sets $X(B/A)$ and $Y(B/A)$. If A is weakly surjective in B then we know in addition from Theorem I.4.8 that

$$Y(B/A) = \operatorname{Spec} A \setminus X(B/A). \qquad (*)$$

Lemma 3.2. Assume that A is weakly surjective in B and that $X(B/A) = \operatorname{Spec} A$. Then $B = A$.

Proof. Suppose there exists some $x \in B$ with $x \notin A$. Then $(A:x) \neq A$. We choose a maximal ideal $\mathfrak{m}$ of A containing $(A:x)$. Since A is weakly surjective in B we have $(A:x)B = B$, hence $\mathfrak{m}B = B$, i.e. $\mathfrak{m} \in Y(B/A)$. This contradicts our hypothesis that $X(B/A) = \operatorname{Spec} A$. $\qquad\square$

Theorem 3.3. Let $A \subset R$ be a ring extension. Let B and C be R-overrings of A, and assume that A is weakly surjective both in B and C. The following are equivalent.
- i) $B \subset C$.
- ii) $X(B/A) \supset X(C/A)$.
- iii) $\mathcal{F}(B/A) \subset \mathcal{F}(C/A)$.

Proof. The implication i) $\Rightarrow$ ii) is trivial. The implications ii) $\Leftrightarrow$ iii) are evident from Lemma 1 and statement $(*)$ above. We now prove that ii) implies i). Let $D := B \cdot C$. We will verify that $D = C$ using Lemma 2, and then will be done.

Let $\mathfrak{q} \in \operatorname{Spec} C$ be given, and $\mathfrak{p} := \mathfrak{q} \cap A$. Then $\mathfrak{p} \in X(C/A) \subset X(B/A)$. Thus there exists a prime ideal $\mathfrak{r}$ of B with $\mathfrak{r} \cap A = \mathfrak{p}$. {In fact $\mathfrak{r} = \mathfrak{p}B$ and $\mathfrak{q} = \mathfrak{p}C$, as we know from Th.I.4.8.} Now $D = B \otimes_A C$ by Prop.I.4.2. The prime ideals $\mathfrak{r}$ and $\mathfrak{q}$ of B and C both lie over the same prime ideal $\mathfrak{p}$ of A. Thus there exists a prime ideal $\mathfrak{P}$ of D lying over $\mathfrak{r}$ and $\mathfrak{q}$. This implies $\mathfrak{q} \in X(D/C)$. We have proved that $X(D/C) = \operatorname{Spec} C$, and we conclude by Lemma 2 that $D = C$. $\qquad\square$

In particular, choosing $R = Q(A)$ or $R = M(A)$ in the theorem, we see that a weakly surjective extension B of a ring A is determined up to isomorphism by the lattice of ideals $\mathcal{F}(B/A)$. Which sublattices of $\mathcal{J}(A)$ occur here? How can a weakly surjective extension B of A be described starting from the corresponding lattice $\mathcal{F} = \mathcal{F}(B/A)$? We will answer these questions for Prüfer extensions.

It has already been proved by Lazard that a weakly surjective (= flat epimorphic) extension B of A is classified by the subset $X(B/A)$ of $\operatorname{Spec} A$ [L, Prop. IV.2.5].

Lemma 3.4 (cf. [Eg, Cor. 4]). Let $A \subset R$ be a ring extension and I an R-invertible ideal of A. Pick a relation $1 = \sum\limits_{i=1}^{n} a_i x_i$ with $a_i \in I$, $x_i \in [A:I]$. Then $I = \bigcap\limits_{i=1}^{n} (A:x_i)$.

Proof. We have $Ix_i \subset A$, hence $I \subset (A:x_i)$ for every i. Thus $I \subset \bigcap\limits_{i}(A:x_i)$. If $a \in \bigcap\limits_{i}(A:x_i)$ is given then $a = \sum\limits_{i=1}^{n} a_i(ax_i) \in I$. This proves the claim. $\qquad\square$

Theorem 3.5. Let $A \subset R$ be a ring extension. Assume that $\mathcal{F}$ is a subset of $\mathcal{J}(A)$ with the properties R0, R1, R2 from above. Assume that also the following holds:

P1. For every $I \in \mathcal{F}$ there exists an R-invertible ideal $I_0 \subset I$ with $I_0 \in \mathcal{F}$.

P2. $\mathcal{F}$ satisfies one of the conditions (2) – (7) listed in Theorems 2.10 and 2.11, or one of the following ones:

    (i) A is integrally closed in R. If $a_1, \ldots, a_n \in A$ and $(a_1, \ldots, a_n) \in \mathcal{F}$ then there exists some natural number $d \geq 2$ such that $(a_1, \ldots, a_n)^d = (a_1^d, \ldots, a_n^d)$.

    (ii) The lattice $\mathcal{F}$ is distributive.

    (iii) For every finitely generated $I \in \mathcal{F}$ the ring A/I is arithmetical.

    (iv) Every finitely generated ideal $I \in \mathcal{F}$ is R-invertible.

Then $A_{[\mathcal{F}]} := \{x \in R \mid (A:x) \in \mathcal{F}\}$ is an R-overring of A. The ring A is Prüfer in $A_{[\mathcal{F}]}$, and $\mathcal{F}$ is the set of $A_{[\mathcal{F}]}$-regular ideals of A.

Remark 3.6. Conversely, if B is an R-overring of A, in which A is Prüfer, then $\mathcal{F} := \mathcal{F}(B/A)$ fulfills R0 – R2, the condition P1, and *all* the conditions listed under P2, as we know from §1. Thus $B = A_{[\mathcal{F}]}$ by Theorem 3 and the present theorem.

Proof of Theorem 5. a) It follows from R1 that

$$A_{[\mathcal{F}]} = \{x \in R \mid \exists I \in \mathcal{F} \ \text{ with } \ Ix \subset A\}.$$

Using R0 and R2 it is easily verified that $A_{[\mathcal{F}]}$ is a subring of R containing A.

b) In the sequel we denote the R-overring $A_{[\mathcal{F}]}$ of A by B. We now prove: If $I \in \mathcal{F}$ and I is R-invertible, then I is B-invertible. In order to do this we pick a relation $1 = \sum\limits_{i=1}^{n} a_i x_i$ with $a_i \in I$ and $x_i \in [A :_R I]$. By Lemma 4 we know that $I = \bigcap\limits_{i}(A : x_i)$. Thus $(A : x_i) \in \mathcal{F}$ for $i = 1, \ldots, n$, which means that $x_i \in B$ for $i = 1, \ldots, n$. Our relation $1 = \sum\limits_{i} a_i x_i$ now shows that I is B-invertible.

c) We prove that A is tight in B. Let $x \in B$ be given, thus $(A : x) \in \mathcal{F}$. By hypothesis P1 there exists an R-invertible ideal $I_0 \subset I$ with $I_0 \in \mathcal{F}$. As proved above, I_0 is B-invertible.

d) We prove that $\mathcal{F} = \mathcal{F}(B/A)$. Let an ideal $I \in \mathcal{F}$ be given. Again by hypothesis P1 there exists an R-invertible ideal $I_0 \subset I$ with $I_0 \in \mathcal{F}$. As proved in step b), I_0 is B-invertible. This implies that I is B-regular, i.e. $I \in \mathcal{F}(B/A)$.

Let now an ideal $I \in \mathcal{F}(B/A)$ be given. Since I is B-regular, we have a relation

$$1 = \sum_{i=1}^{n} a_i x_i \tag{$*$}$$

with $a_i \in I$, $x_i \in B$. All the ideals $(A : x_i)$ are elements of $\mathcal{F}$. By R4 also $\bigcap\limits_{i}(A : x_i) \in \mathcal{F}$. For any $a \in \bigcap\limits_{i}(A : x_i)$ we have

$$a = \sum_{i=1}^{n} a_i(a x_i) \in I$$

by relation $(*)$ above. Thus $\bigcap\limits_{i}(A : x_i) \subset I$, and $I \in \mathcal{F}$ by R1.

e) We finally prove that A is Prüfer in B.

If $\mathcal{F}$ has one of the properties (2) – (7) of Theorems 2.10 and 2.11, this is clear by those theorems. If $\mathcal{F}$ has property P2(i) or P2(ii) or P2(iii), A is Prüfer in B by Theorem 2.2 or Theorem 2.8 respectively. Assume finally that $\mathcal{F}$ has property P2(iv). We now verify for $\mathcal{F} = \mathcal{F}(B/A)$ the condition (4) in Theorem 2.10 and then will be done. Let I and J be elements of $\mathcal{F}$ with $I \subset J$ and J finitely generated. J is R-invertible by hypothesis P2(iv). We infer from Lemma 1.11 that $I = JK$ with $K = (I:J)$. $\square$

§4 Tight extensions

In §2 we obtained various characterizations of Prüfer extensions by multiplicative ideal theory within the class of tight extensions. In order to make these criteria more useful we now develop a somewhat constructive view of tight extensions.

In this section A is a ring and $A \subset R$ is a ring extension. We start with some observations on R-invertible A-submodules of R, more or less all well known.

Lemma 4.1. Every R-invertible A-submodule I of R is a projective A-module of rank 1. {If $I \subset A$ or $I \supset A$, we know this already from Prop. 2.3.}

Proof. We choose elements $a_1, \ldots, a_r$ in I, $x_1, \ldots, x_r$ in $[A:_R I]$ with

$$a_1 x_1 + \cdots + a_r x_r = 1. \tag{$*$}$$

Every x_i gives us a homomorphism $\varphi_i: I \to A$, $\varphi_i(z) := x_i z$ and, for every $x \in A$, $x = \sum_{i=1}^{r} a_i \varphi_i(x)$. Thus I is projective [CE, Chap. VII, Prop. 3.1].

We consider the special case that A is local with maximal ideal $\mathfrak{m}$. At least one of the summands $a_i x_i$ in $(*)$ is not an element of $\mathfrak{m}$, say $a_1 x_1 \notin \mathfrak{m}$. Then $u := a_1 x_1$ is a unit of A, and $y := u^{-1} x_1 \in [A:_R I]$, $a_1 y = 1$. This implies that I is a free A-module generated by a_1. In general we conclude for every $\mathfrak{p} \in \operatorname{Spec} A$ that $I_\mathfrak{p}$ is a free $A_\mathfrak{p}$-module of rank 1.   □

Starting from Lemma 1 it is easy to verify the following remarks.

Remarks 4.2. Assume that I and J are invertible A-submodules of R.

a) The A-module homomorphism $I \otimes_A J \to IJ$, sending a tensor $x \otimes_A y$ to xy, is an isomorphism.

b) The natural A-module homomorphism $\Phi: [A:_R I] \longrightarrow \operatorname{Hom}_A(I, A)$, $\Phi(x)(z):= xz$, is an isomorphism.

c) More generally, the natural A-module homomorphism
   $\Psi: [A:_R I] \longrightarrow \operatorname{Hom}_A(IJ, J)$, $\Psi(x)(z):= xz$, is an isomorphism.
                                                                                       □

If $A \subset R$ is a ring extension and J is an A-submodule of R containing A, then the union of the sets J^n, $n \geq 0$, is the R-overring of A generated by J,

$$A[J] = \bigcup_{n \geq 0} J^n.$$

(We put $J^0 = A$.) We are interested in the special case that $J = [A:_R I]$ for some R-invertible ideal I of A. The following proposition tells us, that the extension $A \subset A[J]$ is, up to canonical isomorphism, determined by A and I alone.

Proposition 4.3. Assume that I is an ideal of A and that $A \subset R$, $A \subset T$ are ring extensions in both of which I is invertible. Let $J:= [A:_R I]$, $K:= [A:_T I]$. There exists a unique ring homomorphism $\varphi: A[J] \to T$ over A. It is injective and has the image $A[K]$. It maps J onto K.

Proof. For any $n \in \mathbb{N}_0$ we define $\varphi_n: J^n \xrightarrow{\sim} K^n$ as the unique A-module isomorphisms which makes the triangle

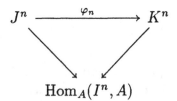

commuting. Here the oblique arrows are canonical isomorphisms, cf. Remark 2.b above. For $x \in J^n$, $a \in I^n$, we have $\varphi_n(x) \cdot a = x \cdot a$. We claim that $\varphi_n(x) = \varphi_m(x)$, if $x \in J^n$ and $n < m$.

Let $a \in I^m$. Then $a \in I^n$ and $\varphi_n(x)a = xa = \varphi_m(x)a$. Since the ideal I^m is dense in A (i.e. has zero annullator), this implies that $\varphi_n(x) = \varphi_m(x)$.

Let $\varphi \colon A[J] \to A[K]$ denote the A-module homomorphism with $\varphi(x) = \varphi_n(x)$ for $x \in J^n$, $n \in \mathbb{N}_0$. It is evident that φ is bijective, since φ maps J^n bijectively onto K^n for every n. We verifiy that φ is a ring homomorphism. Let $x, y \in A[J]$ be given. We choose $n, m \in \mathbb{N}_0$ with $x \in J^n$, $y \in J^m$. For $a \in I^{n+m}$ we have $\varphi(y)a = ya \in I^m$, and then

$$\varphi(x)\varphi(y)a = xya = \varphi(xy)a.$$

Since I^{n+m} is dense in A we conclude that $\varphi(x)\varphi(y) = \varphi(xy)$.

If $\psi \colon A[J] \to T$ is any ring homomorphism over A and $x \in J^n$, then we have for $a \in I^n$

$$\psi(x)a = \psi(x)\psi(a) = \psi(xa) = xa = \varphi(x)a,$$

and we conclude that $\psi = \varphi$. $\square$

Definition 1. Let I be an invertible ideal of A, i.e. I is invertible in the complete quotient ring $Q(A)$ of A (cf. §2, Def. 3). We denote the inverse $[A:_{Q(A)} I]$ of I in $Q(A)$ by I^{-1}, and then have the $Q(A)$-overring $A[I^{-1}]$ of A.

Convention 4.4. If $A \subset R$ is a ring extension in which I is invertible and $J := [A:_R I]$, then we usually identify $J = I^{-1}$, $A[J] = A[I^{-1}]$. This is justified by Proposition 3.

Using this convention we may state: If the ideal I is invertible in R then $A[I^{-1}] \subset R$. In other words, $A[I^{-1}]$ is the universal ring extension in which I becomes invertible.

Proposition 4.5. Assume that I is invertible in the extension $A \subset R$, hence $A[I^{-1}] \subset R$.

i) $A[I^{-1}]$ is the set of all $x \in R$ with $I^n x \subset A$ for some $n \in \mathbb{N}$.

ii) The extension $A \subset A[I^{-1}]$ is tight.

Proof. We have $A[I^{-1}] = \bigcup_n J^n$ with $J := I^{-1} = [A :_R I]$. If $x \in J^n$ then $I^n x \subset A$, and I^n is invertible in $A[I^{-1}]$. This proves that A is tight in $A[I^{-1}]$ (Recall §2, Def. 1). If $x \in R$ and $I^n x \subset A$ for some n, then $x \in [A :_R I^n] = (I^n)^{-1} = J^n$.   □

Proposition 4.6. Assume that I and J are R-invertible ideals of A. Then the R-overring $A[I^{-1}, J^{-1}]$ of A generated by I^{-1} and J^{-1} coincides with both $A[(IJ)^{-1}]$ and $A[I^{-1}] \otimes_A A[J^{-1}]$.

Proof. Clearly $A[I^{-1}, J^{-1}]$ is the smallest R-overring B of A in which both I and J become invertible. But this is also the smallest R-overring in which IJ becomes invertible. Thus $A[I^{-1}, J^{-1}] = A[(IJ)^{-1}]$. Since A is weakly surjective in $A[I^{-1}]$ and $A[J^{-1}]$ we have $A[I^{-1}] \otimes_A A[J^{-1}] = A[I^{-1}, J^{-1}]$, cf. Prop. I.4.2.   □

Definitions 2. a) We denote the set of invertible ideals of A by $\operatorname{Inv} A$ and the set of R-invertible ideals of A by $\operatorname{Inv}(A, R)$. {Thus $\operatorname{Inv} A = \operatorname{Inv}(A, Q(A))$ and $\operatorname{Inv}(A, R) \subset \operatorname{Inv} A$.}

b) The *tight hull* $T(A, R)$ of A in the ring extension $A \subset R$ is defined as the union of the R-overrings $A[I^{-1}]$ with I running through $\operatorname{Inv}(A, R)$.

c) The *tight hull* $T(A)$ of A is the union of the $Q(A)$-overrings $A[I^{-1}]$ of A with I running through $\operatorname{Inv}(A)$, i.e. $T(A) = T(A, Q(A))$.[*]

This terminology is justified by the following theorem.

Theorem 4.7. i) A is tight in $T(A, R)$. If $A \subset B$ is any tight subextension of $A \subset R$ then $B \subset T(A, R)$.

ii) If $A \subset B$ is any tight ring extension then there exists a unique homomorphism $\varphi : B \to T(A)$ over A. Moreover, φ is injective.

Proof. a) If $I, J \in \operatorname{Inv}(A, R)$ then both $A[I^{-1}]$ and $A[J^{-1}]$ are contained in $A[(IJ)^{-1}]$. This proves that the set $T(A, R)$ is a subring of R. If $x \in T(A, R)$ then there exists some $I \in \operatorname{Inv}(A, R)$ with

---

[*] Caution: In the literature $T(A)$ often stands for the total quotient ring of A. We denote this ring by $\operatorname{Quot}(A)$.

$x \in A[I^{-1}]$, hence $I^n x \subset A$ for some $n \in \mathbb{N}$, i.e. $I^n \subset (A:x)$. The ideal I^n is invertible in $A[I^{-1}]$, hence invertible in $T(A, R)$. This proves that $A \subset T(A, R)$ is tight.

b) Let now B be an R-overring of A in which A is tight. Let $x \in B$. There exists a B-invertible ideal I of A with $Ix \subset A$. Of course, I is also invertible in R, and $x \in A[I^{-1}]$ by Prop. 5. Thus $x \in T(A, R)$. This proves that $B \subset T(A, R)$.

c) Let $A \subset B$ be any tight extension of A. Since this extension is weakly surjective, there exists a unique homomorphism $\varphi: B \to Q(A)$ over A, and φ is injective (cf. I, §3). A is tight in $\varphi(B)$ and thus $\varphi(B) \subset T(A, Q(A)) = T(A)$. □

Remarks 4.8. i) In I, §3 we introduced the weakly surjective hull $M(A, R)$, and in I, §5 we introduced the Prüfer hull $P(A, R)$ of A in R. It is now evident that

$$P(A, R) \subset T(A, R) \subset M(A, R).$$

Taking $R = Q(A)$ we obtain $P(A) \subset T(A) \subset M(A)$.
ii) We observed in §1 that the extension $A \subset \text{Quot} A$ is tight. Thus $\text{Quot}(A) \subset T(A)$. □

Example 4.9. If A is semilocal, i.e. has only finitely many maximal ideals, then $T(A) = \text{Quot} A$.

Proof. Let $x \in T(A)$ be given. There exists an invertible ideal I of A with $Ix \subset A$. We know that the A-module I is projective of rank 1 (Lemma 4.1 or Prop.2.3). Since A is semilocal, this implies $I = As$ with s a non-zero divisor of A. We have $sx \in A$ and $x \in \text{Quot} A$. □

We now generalize our construction of $T(A, R)$.

Definition 3. a) We call a set $\mathfrak{G}$ of ideals of A *multiplicative* if it is not empty and closed under multiplication ($I \in \mathfrak{G}, J \in \mathfrak{G} \Rightarrow IJ \in \mathfrak{G}$).

b) Let $\mathfrak{G}$ be a multiplicative subset of $\text{Inv}(A, R)$. Then we denote the union of the subsets I^{-1} of R, with I running through $\mathfrak{G}$, by $A_{[\mathfrak{G}]}^R$. In the case $R = Q(A)$ we denote this set more briefly by $A_{\mathfrak{G}}$.

c) The *saturation* $\hat{\mathfrak{G}}$ of a multiplicative subset $\mathfrak{G}$ of Inv A is defined as the set of all $I \in$ Inv A such that $I \supset J$ for some $J \in \mathfrak{G}$. We call $\mathfrak{G}$ *saturated* if $\mathfrak{G} = \hat{\mathfrak{G}}$. □

Notice that $\hat{\mathfrak{G}}$ is again multiplicative and that $A^R_{[\mathfrak{G}]} = A^R_{[\hat{\mathfrak{G}}]}$. Notice also that $\text{Inv}(A, R)$ is a saturated multiplicative subset of Inv A, and that for $\mathfrak{G} = \text{Inv}(A, R)$ we have $A^R_{[\mathfrak{G}]} = T(A, R)$. In particular, $A_{\text{Inv}A} = T(A)$. If I is an invertible ideal of A then, taking $\mathfrak{G} := \{I^n | n \geq 0\}$, we obtain $A_{\mathfrak{G}} = A[I^{-1}]$.

Proposition 4.10. Let $\mathfrak{G}$ be a multiplicative subset of $\text{Inv}(A, R)$.

a) $A^R_{[\mathfrak{G}]}$ is a subring of R containing A, and A is tight in $A^R_{[\mathfrak{G}]}$. This subring is the set of all $x \in R$ with $Ix \subset A$ for some $I \in \mathfrak{G}$.

b) If $A \subset T$ is another ring extension of A with $\mathfrak{G} \subset \text{Inv}(A, T)$ then there exists a unique ringhomomorphism $\varphi: A^R_{[\mathfrak{G}]} \to T$ over A. It is injective and has the image $A^T_{[\mathfrak{G}]}$.

The proof of these facts goes by straightforward arguments, most of which have been used above (cf. the proofs of Prop. 5 and Th. 7). □

We now can state a classification theorem for tight extensions as follows.

Theorem 4.11. i) If $A \subset B$ is a tight extension then B is canonically isomorphic to $A_{\text{Inv}(A,B)}$ over A.

ii) If $\mathfrak{G}$ is a multiplicative subset of $\text{Inv}(A)$ then $\text{Inv}(A, A_{\mathfrak{G}}) = \hat{\mathfrak{G}}$. Thus the isomorphism classes of tight extensions of A correspond uniquely with the saturated multiplicative subsets of Inv A.

Proof. i) Let $\mathfrak{G} := \text{Inv}(A, B)$. Let $x \in B$ be given. Since A is tight in B there exists some $I \in \mathfrak{G}$ with $Ix \subset A$. Thus $x \in A^B_{[\mathfrak{G}]}$, and we conclude that $B = A^B_{[\mathfrak{G}]}$. Proposition 10 tells us that B is canonically isomorphic to $A_{\mathfrak{G}}$ over A.

ii) Let now $\mathfrak{G}$ be any multiplicative subset of Inv A and $B := A_{\mathfrak{G}}$. Of course, $\hat{\mathfrak{G}} \subset \text{Inv}(A, B)$. We have to prove equality. Let $I \in \text{Inv}(A, B)$ be given. The A-module $I^{-1} = [A:_B I]$ is finitely generated. We choose generators $x_1, \ldots, x_n$ of I^{-1}. Then we choose ideals $J_k \in \mathfrak{G}$ with $x_k \in J_k^{-1}$ $(1 \leq k \leq n)$. Let $J := J_1 \ldots J_n \in \mathfrak{G}$. We have

$x_k \in J^{-1}$ for every k, hence $I^{-1} \subset J^{-1}$, hence $J \subset I$. This proves $\hat{\mathfrak{G}} = \mathrm{Inv}(A, B)$. □

Corollary 4.12. Let $A \subset B$ and $A \subset C$ be tight subextensions of $A \subset R$. Then $B \subset C$ iff $\mathrm{Inv}(A, B) \supset \mathrm{Inv}(A, C)$. □

For finitely generated tight extensions we obtain the following more precise classification theorem.

Theorem 4.13. i) The finitely generated tight extensions of A are (up to isomorphism) the extensions $A[I^{-1}]$ with I running through $\mathrm{Inv}\, A$.

ii) Let I and J be invertible ideals of A. Then $A[J^{-1}] \subset A[I^{-1}]$ (as subrings of $Q(A)$) iff $\sqrt{I} \subset \sqrt{J}$. {We use the notation $\sqrt{I} := \{x \in A \mid x^n \in I \text{ for some } n \in \mathbb{N}\}$.}

Proof. a) If I is an invertible ideal of A then the ideal I^{-1} is finitely generated, say by elements $x_1, \ldots, x_r$. These elements also generate the ring $A[I^{-1}]$ over A.
b) Assume now that $A \subset R$ is a finitely generated tight extension. Let $x_1, \ldots, x_r$ be generators of the ring R over A. For each x_k we choose an ideal $I_k \in \mathrm{Inv}(A, R)$ with $x_k \in I_k^{-1}$. (Recall Th. 11.i). Let $I := I_1 I_2 \ldots I_r$. Then $I \in \mathrm{Inv}(A, R)$ and $x_k \in I^{-1}$ for $k = 1, \ldots, r$. Thus $A[I^{-1}]$ contains all the x_k. We conclude that $A[I^{-1}] = R$.
c) Let I and J be invertible ideals of A. We have $A[I^{-1}] = A_{\mathfrak{G}}$ with $\mathfrak{G} := \{I^n \mid n \geq 0\}$. Now $A[J^{-1}] \subset A[I^{-1}]$ iff $J \in \mathrm{Inv}(A, A_{\mathfrak{G}})$ iff $J \in \hat{\mathfrak{G}}$ (by Theorem 11). This means that $J \supset I^n$ for some $n \in \mathbb{N}$, and that is equivalent to $\sqrt{J} \supset \sqrt{I}$, since I is finitely generated. □

Corollary 4.14. Let I be an invertible ideal of A. The following are equivalent.
 a) $\sqrt{I} = \sqrt{As}$ for some $s \in A$.
 b) There exists a non-zero-divisor s of A with $A[I^{-1}] = A[s^{-1}]$.
 c) There exists a multiplicative subset S of A, consisting of non-zero-divisors, with $A[I^{-1}] = S^{-1}A$.

Proof. $\sqrt{I} = \sqrt{As}$ means that $I^n \subset As \subset I^m$ for some natural numbers $m \leq n$. Then s is certainly a non-zero-divisor, hence As is an invertible ideal. Theorem 13.ii now gives the equivalence of a) and b). It remains to verify that c) implies b). If $A[I^{-1}] =$

$S^{-1}A$ then there exists some $s \in S$ with $I^{-1} \subset As^{-1}$, since I^{-1} is a finitely generated A-module. This implies that $A[I^{-1}] \subset A[s^{-1}]$. But trivially $A[s^{-1}] \subset S^{-1}A = A[I^{-1}]$. Thus b) holds. $\qquad\square$

From Theorem 13 and its corollary we immediately obtain a characterization of the ring extensions $A \subset R$ for which every R-overring of A is of the form $S^{-1}A$. In the case that R is a field this criterion is well known and runs under the name "Pendleton's criterion", cf. [Pe].

Proposition 4.15 (Rhodes, [Rh, p. 3439]). Let $A \subset R$ be a ring extension. The following are equivalent.
(1) Every R-overring B of A has the form $B = S^{-1}A$ with some multiplicative subset S of A (consisting of non-zero-divisors).
(2) A is Prüfer in R. For every finitely generated R-regular ideal I of A there exists some element s of A with $\sqrt{I} = \sqrt{As}$.

Proof. (1) $\Rightarrow$ (2): A is Prüfer in R since condition (2) in Theorem I.5.2 holds. If I is a finitely generated R-regular ideal then I is R-invertible, hence $A[I^{-1}] \subset R$. We conclude from our assumption (1) and Corollary 14 that $I = \sqrt{As}$ for some $s \in A$.
(2) $\Rightarrow$ (1): It suffices to consider the R-overrings of A which are finitely generated over A. If B is one of these then $A \subset B$ is Prüfer, hence tight. By Theorem 13 we have $B = A[I^{-1}]$ with I an R-invertible ideal of A. We conclude by our assumption (2) and Corollary 14 that $B = A[s^{-1}]$ with some non-zero-divisor s of A. $\qquad\square$

We mention an important special case.

Proposition 4.16. Let $A \subset R$ be a ring extension, and assume that every projective A-module of rank 1 is free (e.g. A is semilocal).
i) A is tight in R iff there exists a multiplicative subset S of A with $R = S^{-1}A$.
ii) A is Prüfer in R iff for every R-overring B of A there exists a multiplicative subset S_1 of A with $B = S_1^{-1}A$.

N.B. Of course, the multiplicative sets S and S_1 are forced to consist of nonzero divisors.

Proof. If I is an invertible ideal of A then we learn from Proposition 2.3 that $I = As$ with some nonzero divisor S of A. {We used this argument already in 4.9.} Now Theorem 11 gives us the first claim i). If we already know that $A \subset R$ is tight then it follows from Theorem I.5.2 that $A \subset R$ is Prüfer iff every subextension $A \subset B$ of $A \subset R$ is tight. This gives the second claim. $\square$

We learned in §3 that a tight extension $A \subset R$ is characterized by the set $\mathcal{F}(R/A)$ of R-regular ideals in A. (More generally this holds for $A \subset R$ weakly surjective, cf. Th. 3.3.) We learned now that $A \subset R$ is characterized by the set $\mathrm{Inv}(A, R)$ of R-invertible ideals. How is $\mathcal{F}(R/A)$ related to $\mathrm{Inv}(A, R)$?

Proposition 4.17. Assume that $A \subset R$ is tight. Then $\mathcal{F}(R/A)$ is the set of all ideals $\mathfrak{a}$ of A which contain some ideal $I \in \mathrm{Inv}(A, R)$.

Proof. Of course, if I is R-invertible and $\mathfrak{a} \supset I$ then $\mathfrak{a}$ is R-regular. Let now an R-regular ideal $\mathfrak{a}$ be given. We choose elements $a_1, \ldots, a_r$ in $\mathfrak{a}$, $x_1, \ldots, x_r$ in R with $1 = \sum_{i=1}^{r} a_i x_i$. Then we choose an ideal $I \in \mathrm{Inv}(A, R)$ such that $x_i \in I^{-1}$ for $1 \leq i \leq r$. This is possible since A is tight in R. Let $\mathfrak{a}_0 := \sum_{i=1}^{r} A a_i$. We have $A \subset \mathfrak{a}_0 I^{-1}$, hence $I \subset \mathfrak{a}_0$. Since $\mathfrak{a}_0 \subset \mathfrak{a}$, we conclude that $I \subset \mathfrak{a}$. $\square$

Conversely we know for *any* ring extension $A \subset R$ from Scholium 2.5 that
$$\mathrm{Inv}(A, R) = \mathrm{Inv}(A) \cap \mathcal{F}(R/A).$$

How can we describe the Prüfer hull $P(A, R)$ of A in R using theorem 11?

Definition 4. We call an ideal I of A *Prüfer* if I is invertible and A is Prüfer in $A[I^{-1}]$. Given an extension $A \subset R$ we say that I is *R-Prüfer* if in addition $I^{-1} \subset R$, i.e. I is R-invertible. We denote the set of all Prüfer ideals by $\Pi(A)$ and the set of R-Prüfer ideals by $\Pi(A, R)$. We have $\Pi(A, R) = \Pi(A) \cap \mathrm{Inv}(A, R)$. $\square$

It is evident from the theory of Prüfer hulls (I, §5) that $\Pi(A) = \mathrm{Inv}(A, P(A))$ and $\Pi(A, R) = \mathrm{Inv}(A, P(A, R))$. It is clear that $\Pi(A, R)$ is multiplicative and saturated.

Theorem 11 tells us that

$$P(A) = A_{\Pi(A)}, \ P(A, R) = A^R_{[\Pi(A,R)]}.$$

Thus a good description of the Prüfer hulls hinges on a good insight into Prüfer ideals.

Just now we can state two criteria for an ideal I to be Prüfer.

Theorem 4.18. Let I be an ideal of A. I is Prüfer iff every finitely generated ideal of A containing some power I^n is invertible.

Proof. We may assume in advance that I is invertible. Let $\mathfrak{a}$ be any ideal of A. Applying Theorem 11.ii to $\mathfrak{G} := (I^n | n \geq 0)$ we learn that $\mathfrak{a}$ is invertible in $A[I^{-1}]$ iff $\mathfrak{a}$ is invertible and $\mathfrak{a} \supset I^n$ for some $n \in \mathbb{N}$. Now Theorem 2.1, applied to the extension $A \subset A[I^{-1}]$, tells us, that I is Prüfer iff every finitely generated ideal of A containing some power I^n $(n \geq 1)$ is invertible.   □

Theorem 4.19. An ideal I of A is Prüfer iff the following two conditions hold:

a) Every finitely generated ideal $\mathfrak{a} \supset I$ of A is invertible.
b) $(\mathfrak{a} + I)(\mathfrak{a} \cap I) = \mathfrak{a}I$ for every (finitely generated) ideal $\mathfrak{a}$ of A with $\mathfrak{a} \supset I^n$ for some $n \in \mathbb{N}$.

Proof. If I is Prüfer then certainly a) holds, and $(\mathfrak{a}+I)(\mathfrak{a}\cap I) = \mathfrak{a}I$ for *every* ideal $\mathfrak{a}$ of I, as follows from Theorem 1.4 (5). Assume now that both conditions a) and b) are fulfilled. Let $\mathfrak{a}$ be a finitely generated ideal of A with $\mathfrak{a} \supset I^n$ for some $n \in \mathbb{N}$. We verify by induction on n that $\mathfrak{a}$ is invertible. Then we will know by the preceding theorem 18 that I is Prüfer.

Nothing has to be done for $n = 1$. Assume that $n > 1$. The ideal $\mathfrak{a} + I$ is finitely generated, hence is invertible by condition a). Also I is invertible by condition a). We now work in the ring $A[I^{-1}]$. From $I \supset \mathfrak{a} \cap I \supset I^n$ we conclude that $A \supset (\mathfrak{a} \cap I)I^{-1} \supset I^{n-1}$. Thus $(\mathfrak{a} \cap I)I^{-1}$ is an ideal of A. It is finitely generated. Using the induction hypothesis we see that $(\mathfrak{a} \cap I)I^{-1}$ is invertible. By condition b) we have $\mathfrak{a} = (\mathfrak{a} + I)(\mathfrak{a} \cap I)I^{-1}$, and we conclude that $\mathfrak{a}$ is invertible.   □

Remark 4.20. If I is a Prüfer ideal of A and $A \subset C$ a weakly surjective ring extension, then IC is a Prüfer ideal of C.

Indeed, the ideal IC of C is invertible with $(IC)^{-1} = I^{-1}C$. We have $C[I^{-1}C] = C \cdot A[I^{-1}]$, and C is Prüfer in $C \cdot A[I^{-1}]$ by Theorem I.5.10.

In particular, for every multiplicative subset S of A the ideal $S^{-1}I$ of $S^{-1}A$ is again Prüfer. $\square$

§5 Distributive submodules

In this whole section A is an arbitrary ring (as always, commutative with 1).

One of our major problems is to obtain a good understanding of the Prüfer hull $P(A)$ of A, or – what is nearly the same – a good understanding of the Prüfer ideals (cf. §4, Def. 4) of A. In this section we will approach such an understanding by the notion of "distributivity".

Definition 1. a) Let M be an A-module. An A-submodule N of M is called *distributive in* M (or a *distributive submodule of* M) if

$$N \cap (N' + N'') = (N \cap N') + (N \cap N'')$$

for any two submodules N', N'' in M.

b) An *ideal* I of A is called *distributive* if I is distributive in the A-module A.

Example 5.1. If $A \subset R$ is a Prüfer extension then every A-submodule I of R with $IR = R$ is distributive in R, as has been shown in §1 (Th.1.4). In particular, every R-regular ideal of A is distributive. Taking $R = P(A)$ we conclude that every Prüfer ideal of A is distributive.

C.U. Jensen has studied distributive ideals under the name of "D-ideals" [J$_2$]. He obtained rather complete results in the case that A is noetherian.

Proposition 5.2 ([J$_2$], [Gr$_3$] for $M = A$). Let M be an A-module and N an A-submodule of M. The following are equivalent.

(1) N is distributive in M.

(2) If N' and N'' are submodules of M, then $N + (N' \cap N'') = (N + N') \cap (N + N'')$.

(3) If N' and N'' are submodules of M, then $N' \cap (N + N'') = (N' \cap N) + (N' \cap N'')$.

(4) If N' and N'' are submodules of M, then $N' + (N \cap N'') = (N' + N) \cap (N' + N'')$.

(5) For every maximal ideal $\mathfrak{m}$ of A and every submodule N' of M the $A_{\mathfrak{m}}$-modules $N_{\mathfrak{m}}$ and $N'_{\mathfrak{m}}$ are comparable, i.e. $N_{\mathfrak{m}} \subset N'_{\mathfrak{m}}$ or $N'_{\mathfrak{m}} \subset N_{\mathfrak{m}}$.

Proof. $(5) \Rightarrow (1)$: Let N' and N'' be submodules of M. It suffices to verify for any maximal ideal $\mathfrak{m}$ of A that

$$N_{\mathfrak{m}} \cap (N'_{\mathfrak{m}} + N''_{\mathfrak{m}}) = (N_{\mathfrak{m}} \cap N'_{\mathfrak{m}}) + (N_{\mathfrak{m}} \cap N''_{\mathfrak{m}}).$$

This can be easily done by case distinctions, since both $N'_{\mathfrak{m}}$ and $N''_{\mathfrak{m}}$ are comparable with $N_{\mathfrak{m}}$.

$(1) \Rightarrow (5)$: We may assume that A is local with maximal ideal $\mathfrak{m}$. In order to prove comparability of N with any submodule N' of M it suffices to consider the case that N' is monogenic, $N' = Am$ with some $m \in M$. Assume that $m \notin N$. We have to verify that $N \subset N'$. Let $n \in N$ be given. Let $N'' := A(m + n)$. Then

$$m \in (N + N') \cap (N + N'') = N + (N' \cap N'').$$

Thus $m = n_0 + am$, $am = b(m + n)$ with $n_0 \in N$, $a \in A$, $b \in A$. From the equation $(1 - a)m = n_0$ we conclude that $1 - a \in \mathfrak{m}$, since $m \notin N$. Thus $a \in A^*$. From the equation $(a - b)m = bn$ we then conclude that $a - b \in \mathfrak{m}$ for the same reason. Thus $b \in A^*$ and $n = b^{-1}(a - b)m \in N'$. This proves that $N \subset N'$.

The implications $(2) \Leftrightarrow (5)$, $(3) \Leftrightarrow (5)$, $(4) \Leftrightarrow (5)$ can be proved in the same way. {We will not need the conditions (2), (3), (4) in the sequel.} $\qquad\square$

If N_1 and N_2 are submodules of an A-module M then we define

$$(N_1 \colon N_2) = (N_1 \colon_A N_2) := \{x \in A \mid xN_2 \subset N_1\}.$$

Proposition 5.3 ([J$_2$] for $M = A$).
Let M be an A-module and N a finitely generated submodule of M.
The following are equivalent.

(1) N is distributive in M.
(2) $(N:N') + (N':N) = A$ for every finitely generated submodule N' of M.
(3) $(N:Ax) + (Ax:N) = A$ for every $x \in M$.

Proof. We may retreat to the case that A is local. (1) $\Rightarrow$ (2): By Proposition 2 we have $N \subset N'$ or $N' \subset N$. Thus $(N':N) = A$ or $(N:N') = A$, which implies in both cases that $(N':N)+(N:N') = A$. (2) $\Rightarrow$ (3) is trivial. (3) $\Rightarrow$ (1): We prove condition (5) in Proposition 2. It suffices to verify that $N \subset Ax$ for a given $x \in M$ with $x \notin N$. By hypothesis we have elements $a \in (N:Ax)$ and $b \in (Ax:N)$ with $a + b = 1$. Now a is not a unit of A, since $x \notin N$. Thus $b \in A^*$ and $N \subset Ax$.   □

Theorem 5.4. A ring extension $A \subset R$ is Prüfer iff A is distributive in the A-module R.

Proof. We know from Theorem I.5.2 that A is Prüfer in R iff $(A:x) + x(A:x) = A$ for every $x \in R$. Now $x(A:x) = Ax \cap A = (Ax:A)$. The claim follows from the preceding Proposition 3.   □

In the special case $R = \text{Quot}\,A$ this theorem has already been proved by T.M.K. Davison [Dvs, Prop.4.1].

The theorem implies a characterization of Prüfer ideals by distributivity. We need two easy lemmas.

Lemma 5.5. Assume that $(M_\lambda \mid \lambda \in \Lambda)$ is a directed family of submodules of an A-module M with $\bigcup_{\lambda \in \Lambda} M_\lambda = M$. Let N be a finitely generated submodule of M and let $\gamma \in \Lambda$ be chosen with $N \subset M_\gamma$. Then N is distributive in M iff N is distributive in M_λ for every $\lambda \geq \gamma$.

This is evident from property (3) in Proposition 3. {We could equally well use other properties from above which characterize distributivity.}

Lemma 5.6. Let I and J be distributive ideals of A. Assume that I is also invertible. Then the ideal IJ is distributive.

Proof. We may assume that A is local. Let K be any ideal of A. We verify that K is comparable with IJ. (Recall $(1) \Leftrightarrow (5)$ in Prop. 2.) If $K \supset I$ then, of course, $K \supset IJ$. Otherwise $K \subset I$. The ideals KI^{-1} and J are comparable, and thus the ideals K and IJ are comparable.   □

Theorem 5.7. An ideal I of A is Prüfer iff I is invertible and distributive.

Proof. We may assume in advance that I is invertible. That I is Prüfer means by definition that A is Prüfer in $A[I^{-1}] = \bigcup_{n \geq 1} I^{-n}$. By Theorem 4 this holds iff A is distributive in $A[I^{-1}]$, and by Lemma 5 this means that A is distributive in I^{-n} for every $n \in \mathbb{N}$. Now $M \mapsto I^n M$ is a bijection from the set of submodules M of I^{-n} onto the set of ideals of A, and this bijection preserves the inclusion relation. Thus A is distributive in I^{-n} iff $I^n = I^n A$ is distributive in $A = I^n I^{-n}$. We conclude that I is Prüfer iff I^n is distributive in A for every $n \in \mathbb{N}$. By Lemma 6 this holds if I is distributive in A.
□

This theorem, together with Theorems 1.4 and 1.13 implies the following remarkable fact.

Corollary 5.8. If I is an invertible distributive ideal of A, then every finitely generated ideal $J \supset I$ is again invertible and distributive. □

Second proof of Corollary 8 (more direct). It suffices to consider the case $J = I + Aa$ with one element $a \in A$. We first verify that J is locally principal. Let $\mathfrak{m}$ be a maximal ideal of A. Then the ideals $I_\mathfrak{m}$ and $aA_\mathfrak{m}$ of $A_\mathfrak{m}$ are comparable. Thus $J_\mathfrak{m} = I_\mathfrak{m}$ or $J_\mathfrak{m} = aA_\mathfrak{m}$. In both cases $J_\mathfrak{m}$ is generated by one element.

We now know that J is invertible (cf. Prop.2.3). Let $K := J^{-1}I \subset A$. The A-module (= ideal) I is distributive in A. A fortiori $I = JK$ is distributive in K. By a similar argument as in the proof of Theorem 7 we conclude that J is distributive in A, since K is invertible.   □

It is clear from Lemma 5.6 that the invertible distributive ideals of A form a multiplicative set. We have just proved in a direct way that this multiplicative set of invertible ideals is saturated. Thus we have obtained a second proof that the set $\Pi(A)$ of Prüfer ideals is multiplicative and saturated, a fact already stated and proved in §4. Recall from §4 that $P(A) = A_{\Pi(A)}$.

Definition 2 [Dvs, p.31]. An element t of A is called a *strong divisor in A*, if t is not a zero divisor in A and for every $b \in A$ either $At \subset Ab$ or $Ab \subset At$. We denote the set of strong divisors in A by $sd(A)$. □

Remark 5.9. It has been proved by T.M.K. Davison [Dvs, p.31] that the set $sd(A)$ is multiplicatively closed and saturated in A. If $t \in sd(A)$, the ideal At is clearly invertible and distributive, hence Prüfer. Thus A is Prüfer in $sd(A)^{-1}A$. □

Proposition 5.10. If the ring A is local then $P(A) = sd(A)^{-1}A$. More generally, every Prüfer extension of A has the form $R = S^{-1}A$ with S a multiplicative subset of $sd(A)$.

Proof. Every Prüfer ideal I of A is principal, since I is a projective A-module of rank one (cf.Prop.2.3), and now such modules are free. Thus $I = At$ with some nonzero divisor t of A. Condition (5) in Proposition 2 tells us that t is a strong divisor in A. Now the claims of the proposition are obvious. □

§6 Transfer theorems

If E and L are subfields of some field with $E \cap L$ Galois in E then a well known theorem, running in the classical (at least German) literature under the label "transfer theorem" ("Translationssatz"), tells us that the extension $L \subset EL$ is again Galois and the restriction homomorphism $\mathrm{Gal}(EL/L) \to \mathrm{Gal}(E/L \cap E)$ between the Galois groups is an isomorphism, hence the fields F between $E \cap L$ and E correspond uniquely with the field F' between L and EL via $F' = LF$ and $F = F' \cap E$.

We will prove in this section theorems on Prüfer extensions of similar flavour, which we again will call transfer theorems.

Definition 1. If $A \subset R$ is a ring extension then $D(A,R)$ denotes the set of R-invertible A-submodules of R. We regard $D(A,R)$ as an ordered abelian group with multiplication $(I,J) \mapsto IJ$, and ordering $I \leq J$ iff $I \subset J$. The submonoid of "negative" elements is the set $\text{Inv}(A,R)$ of R-invertible ideals of A, since the neutral element is the A-module A.

Remarks 6.1. If A is Prüfer in R then $D(A,R)$ is a distributive lattice ordered group (cf. e.g. [BKW]), with the lattice operations $I \wedge J = I \cap J$ and $I \vee J = I + J$. Indeed, we know from §1 that if I and J are R-invertible A-modules then $I + J$ and $I \cap J$ are again R-invertible and, as in any lattice ordered group, $IJ = (I \vee J)(I \wedge J)$ (cf.Th.1.4). In particular, the group $D(A,R)$ is generated by $\text{Inv}(A,R)$. In explicit terms, if $I \in D(A,R)$ then $I = K_1 K_2^{-1}$ with $K_1 = I \cap A$, $K_2 = (I + A)^{-1}$. It is also evident that, for any $I, J \in D(A,R)$ we have $(I + J)^{-1} = I^{-1} \cap J^{-1}$ and $(I \cap J)^{-1} = I^{-1} + J^{-1}$. Distributivity is clear by Theorem 1.4.   □

We will exploit the fact that $D(A,R)$ is lattice ordered more thoroughly only later.

We now look at the following situation: R and C are subrings of some ring T and A is a subring of both R and C. {Later we will assume that $A \subset R$ is Prüfer and $A \subset C$ is ws.}

We want to compare the ring extensions $A \subset R$ and $C \subset RC$. We have an evident homomorphism

$$\varphi: D(A,R) \longrightarrow D(C,RC), \quad \varphi(I) := CI,$$

of abelian groups which is compatible with the orderings. Indeed, if I is an R-invertible A-module in R, then $(IC) \cdot (I^{-1}C) = C$. Thus IC is an RC-invertible C-module with $(IC)^{-1} = I^{-1}C$.

Lemma 6.2. If $R \cap C = A$ then φ is injective.

Proof. Let $I \in D(A,R)$ be given with $IC = C$. Then $I \subset C \cap R = A$. Also $I^{-1}C = C$, hence $I^{-1} \subset A$, hence $A \subset I$. Thus $I = A$.   □

Lemma 6.3. Assume that $A \subset R$ is Prüfer and $R \cap C = A$. Let J be an RC-invertible C-submodule of RC and I an R-regular A-submodule of R with $IC = J$. Then I is R-invertible and $I = J \cap R$.

Proof. J is a finitely generated C-module. Thus we can choose an R-regular finitely generated A-module $I_0 \subset I$ with $I_0 C = J$. Since A is Prüfer in R, the module I_0 is R-invertible. (Recall Th.1.13.) We have $I_0 \subset I \subset J \cap R$. We claim that $I_0 = J \cap R$, and then will be done. Let $x \in J \cap R$ be given. The module $I_1 := I_0 + Ax$ is again finitely generated and R-regular, hence R-invertible. $I_1 C = J = I_0 C$. Lemma 2 tells us that $I_1 = I_0$, i.e. $x \in I_0$. Thus $J \cap R = I_0$. $\square$

Starting from now we nearly always assume in this section that $A \subset R$ is Prüfer and $A \subset C$ is ws. Notice that then $C \subset RC$ is again Prüfer (Th.I.5.10). Recall also that, if $A \subset R$ and $A \subset C$ are any ws extensions, we may imbed both R and C over A in a unique way into the weakly surjective hull $M(A)$. Thus RC has a completely unambiguous meaning, $RC = R \otimes_A C$, cf. I, §3 and §4.

Theorem 6.4 (Transfer theorem for invertible modules).
Assume that A is Prüfer in R and ws in C. Assume in addition that $R \cap C = A$. Then the homomorphism $\varphi : D(A, R) \to D(C, RC)$ is an isomorphism of lattice ordered groups. The preimage of an element $J \in D(C, RC)$ is $\varphi^{-1}(J) = J \cap R$.

Proof. It is clear that the homomorphism φ respects the lattice operation $I_1 \vee I_2 = I_1 + I_2$. Since $(I_1 \cap I_2)(I_1 + I_2) = I_1 I_2$, it also respects the lattice operation $I_1 \wedge I_2 = I_1 \cap I_2$. By Lemma 2 φ is injective. Moreover, if $I \in D(A, R)$, $J \in D(C, RC)$, and $\varphi(I) = J$, then Lemma 3 tells us that $I = J \cap R$. Thus it only remains to prove that φ is surjective. Since $D(C, RC)$ is generated by $\mathrm{Inv}(C, RC)$, it suffices to verify for a given RC-regular ideal J of C that there exists an R-regular ideal I of A with $IC = J$.

Let $I := J \cap C$. Since $A \subset C$ is ws, we have $IC = J$ (cf. Prop.I.4.6). We now verify that I is R-regular. Then it will follow from Lemma 3 that I is R-invertible, and we will be done.

We choose finitely many elements $x_1, \ldots, x_r$ in R such that $J^{-1} \subset Cx_1 + \cdots + Cx_r$, which is possible since $J^{-1} \subset RC$. Then $A + Ax_1 + \cdots + Ax_r$ is an R-invertible A-submodule of R, since $A \subset R$ is Prüfer.

Let $K := (A + Ax_1 + \cdots + Ax_r)^{-1}$. We have $J^{-1} \subset K^{-1}C = (KC)^{-1}$, hence $KC \subset J$. It follows that $K \subset J \cap A = I$. This implies that I is R-regular. $\qquad\square$

Theorem 6.5 (Transfer theorem for regular modules).
Assume that $R \cap C = A$, and that A is Prüfer in R and ws in C. $I \mapsto IC$ is a bijection from the set of R-regular A-submodules of R to the set of RC-regular C-submodules of RC. The inverse mapping is $J \mapsto J \cap R$.

Proof. a) If I is an R-regular A-submodule of R then, of course, IC is an RC-regular C-submodule of RC. Let now J be an RC-regular C-submodule of RC and $I := J \cap R$. We prove that I is the unique R-regular A-submodule K of R with $KC = J$, and then will be done.
b) Of course, $IC \subset J$. Let $x \in J$ be given. We choose a finitely generated RC-regular C-submodule J_0 of J with $x \in J_0$. Then J_0 is RC-invertible since the extension $C \subset RC$ is Prüfer. By Theorem 4 we know that $J_0 = (J_0 \cap R)C$, and $J_0 \cap R$ is R-invertible. Since $J_0 \cap R \subset I$, we conclude that I is R-regular. Moreover $x \in (J_0 \cap R)C \subset IC$. Thus $IC = J$.
c) Let finally K be an R-regular A-submodule of R with $KC = J$. Then $K \subset J \cap C = I$. Let $x \in I$ be given. We have $x \in J = KC$. We choose a finitely generated R-regular A-submodule K_0 of K with $x \in K_0C$. Now K_0 is R-invertible since $A \subset R$ is Prüfer. Applying again Theorem 4 we obtain $x \in (K_0C) \cap R = K_0 \subset K$. This proves that $K = I$. $\qquad\square$

Corollary 6.6 (Transfer theorem for overrings).
As before, assume that A is Prüfer in R and ws in C. The R-overrings B of $R \cap C$ correspond uniquely with the RC-overrings E of C via $E = BC$ and $B = E \cap R$.

Proof. $R \cap C$ is Prüfer in R and ws in C. Thus, replacing A by $R \cap C$, we may assume that $R \cap C = A$. If B is an R-overring of A then BC is an RC-overring of C, and if E is an RC-overring of C then $E \cap C$ is an R-overring of A. The claim follows from Theorem 5, since overrings are regular modules. $\qquad\square$

Remark. It is possible to deduce this corollary directly from Theorem 4 instead of Theorem 5, since an R-overring B of A is classified by the submonoid $\text{Inv}(A, B)$ of $\text{Inv}(A, R)$ and an R-overring E of C is classified by the submonoid $\text{Inv}(C, E)$ of $\text{Inv}(C, RC)$, cf. Theorem 3.11. We do not give the – rather obvious – details but mention that such a proof is more parallel to Galois theory than our present one. The analogy becomes even closer if we work with the subgroups $D(A, B)$ of $D(A, R)$ and $D(C, E)$ of $D(C, RC)$, which is possible as well. $\qquad\qquad\square$

Corollary 6.7. Assume that $R \cap C = A$ and, as before, that A is Prüfer in R and ws in C. The R-regular prime ideals $\mathfrak{p}$ of A correspond uniquely with the RC-regular prime ideals $\mathfrak{q}$ of C via $\mathfrak{q} = \mathfrak{p}C$, $\mathfrak{p} = \mathfrak{q} \cap A$.

Proof. If $\mathfrak{p}$ is an R-regular prime ideal of A then Theorem 5 tells us that $\mathfrak{p}C$ is an RC-regular ideal of C and $\mathfrak{p}C \neq C$. Since A is ws in C it follows that $\mathfrak{p}C$ is a prime ideal of C and $(\mathfrak{p}C) \cap A = \mathfrak{p}$, cf. Th.I.4.8. If $\mathfrak{q}$ is an RC-regular prime ideal of C then the prime ideal $\mathfrak{q} \cap A$ of A is R-regular by Theorem 5, and $(\mathfrak{q} \cap A)C = \mathfrak{q}$, as is already clear from Theorem I.4.8. $\qquad\qquad\square$

Corollary 6.8. Assume that A is Prüfer in R and ws in C.

a) If I is an R-regular A-submodule of R, then $R \cap (IC) = I(R \cap C)$.
b) The homomorphism $\varphi \colon D(A, R) \to D(C, RC)$, $\varphi(I) := IC$, is surjective and has the kernel $D(A, R \cap C)$.

Proof. a) Let $J := I(R \cap C)$. Then $IC = JC$, hence $R \cap (IC) = J$ by Theorem 5.

b) Let $J \in D(C, RC)$ be given, and $K := J \cap R$. Then $K \in D(R \cap C, R)$ and $KC = J$ by Theorem 5. It is easy to find an R-regular finitely generated A-submodule I of R with $I(R \cap C) = K$. Since $A \subset R$ is Prüfer, the module I is R-invertible. We have $\varphi(I) = J$. This proves the surjectivity of φ.

If $I \in D(A, R)$ is given then, again by Theorem 5 (or Theorem 4), $IC = C$ if and only if $R \cap (IC) = R \cap C$. But $R \cap (IC) = I(R \cap C)$. Thus $IC = C$ iff $I(R \cap C) = R \cap C$. $\qquad\qquad\square$

We give some formulas on regular modules which will be useful here and then.

Proposition 6.9. Assume that A is Prüfer in R and ws in C, and that $R \cap C = A$. If I_1 and I_2 are R-regular A-submodules of R then $(I_1 \cap I_2)C = (I_1 C) \cap (I_2 C)$.

Proof. This holds if I_1 and I_2 are R-invertible A-modules, since then our map $\varphi: D(A, R) \to D(C, RC)$, $\varphi(I) = IC$, is an isomorphism of lattice ordered groups (Th.4). In general, it is trivial that $(I_1 \cap I_2)C \subset (I_1 C) \cap (I_2 C)$. Let $x \in (I_1 C) \cap (I_2 C)$ be given. We choose finitely generated R-regular A-modules $K_1 \subset I_1$ and $K_2 \subset I_2$ with $x \in K_1 C$ and $x \in K_2 C$. Since $A \subset R$ is Prüfer, both K_1 and K_2 are R-invertible. Thus

$$x \in (K_1 C) \cap (K_2 C) = (K_1 \cap K_2)C \subset (I_1 \cap I_2)C. \qquad \square$$

Corollary 6.10. Under the same assumptions on A, R and C, if I_1 and I_2 are R-regular A-submodules of R and J is any C-submodule of RC, then $(I_1 \cap I_2)J = (I_1 J) \cap (I_2 J)$.

Proof. $(I_1 \cap I_2)J = (I_1 \cap I_2)CJ = [(I_1 C) \cap (I_2 C)] \cdot J = (I_1 J) \cap (I_2 J)$ by the preceding proposition and Theorem 1.4 (4). $\qquad \square$

Proposition 6.11. Assume again that A is Prüfer in R and ws in C, and that $R \cap C = A$. Let I be an A-submodule of R and J an RC-regular C-submodule RC. Then
(1) $[J + (IC)] \cap R = (J \cap R) + I$.
If in addition I is R-regular, then
(2) $J \cap (IC) = (J \cap I)C$,
(3) $(JI) \cap R = (J \cap R)I$.

Proof. We will use Theorem 5 constantly. Let $K := J \cap R$. Then $J = KC$. We have $J + (IC) = (KC) + (IC) = (K + I)C$. Since $K + I$ is R-regular, it follows that $K + I = [J + (IC)] \cap R$, i.e. (1).

Assume now that I is R-regular. Since also K is R-regular it follows from Proposition 9 that $(K \cap I)C = J \cap (IC)$, i.e. (2). Finally $JI = (KI)C$, and KI is R-regular. Thus $KI = (JI) \cap R$, i.e. (3). $\square$

We now state a proposition which in some sense gives converses to the transfer theorem for overrings (Corollary 6).

Proposition 6.12. Let $A \subset T$ be any ring extension, and let R and C be T-overrings of A. We consider the map ψ from the set

of R-overrings B of A to the set of RC-overrings of C defined by
$\psi(B) := BC$.

a) If $A \subset R$ is Prüfer and ψ is surjective, then $C \subset RC$ is Prüfer.

b) If $C \subset RC$ is Prüfer and ψ is injective, then $A \subset R$ is Prüfer and
$C \cap R = A$.

Proof. a): We verify for a given RC-overring D of C that C is ws
in D, and then will be done by Theorem I.5.2. By assumption there
exists an R-overring B of A with $BC = D$. Since $A \subset B$ is ws also
$C \subset BC$ is ws, cf. Prop.I.3.10.

b): From $AC = (R \cap C)C = C$ we conclude that $A = R \cap C$. We now
verify for a given R-overring B of A that B is integrally closed in
R, and then will be done, again by Theorem I.5.2 (using a different
part of that theorem). Let $\tilde{B}$ denote the integral closure of B in R.
Then $\tilde{B}C$ is integral over BC. Since $C \subset RC$ is Prüfer we conclude
that $\tilde{B}C = BC$. Since ψ is injective it follows that $\tilde{B} = B$. $\qquad\square$

§7 Polars and factors in a Prüfer extension

In the whole section $A \subset R$ *is a Prüfer extension.* We start with a
lemma of general nature.

Lemma 7.1. Let $(I_\lambda \mid \lambda \in \Lambda)$ be a family of A-modules in R and
also J an A-module in R. Assume that either J or at least one of
the I_λ is R-regular. Then $(\sum_{\lambda \in \Lambda} I_\lambda) \cap J = \sum_{\lambda \in \Lambda} (I_\lambda \cap J)$.

Proof. Given an element x in $(\sum_\lambda I_\lambda) \cap J$ we have to verify that $x \in$
$\sum_\lambda (I_\lambda \cap J)$. We choose a finite set $S \subset \Lambda$ such that $x \in (\sum_{\lambda \in S} I_\lambda) \cap J$.
If J is R-regular then we obtain from Theorem 1.4.(1) by an easy
induction that $(\sum_{\lambda \in S} I_\lambda) \cap J = \sum_{\lambda \in S} (I_\lambda \cap J)$, and we are done. Assume
now that there exists some $\lambda_0 \in \Lambda$ such that I_{λ_0} is R-regular. Then
we may add λ_0 to the set S, hence assume that $\lambda_0 \in S$. Now we
obtain again from Theorem 1.4.(1) that $(\sum_{\lambda \in S} I_\lambda) \cap J = \sum_{\lambda \in S} (I_\lambda \cap J)$
by an induction, which runs over sets $S' \subset S$ containing λ_0. $\qquad\square$

Proposition 7.2. Let I be an A-module in R with $A \subset I$. Then there exists an A-module J in R, such that $I \cap J = A$ and every A-module $K \subset R$ with $I \cap K \subset A$ is contained in J.

Proof. Let J denote the sum of all A-modules K in R with $I \cap K \subset A$. Then $I \cap J \subset A$ by the preceding lemma. But $A \subset J$, hence $I \cap J = A$. □

Definition 1. We call the maximal module J occuring in Proposition 2 the *polar of I (in R)*, and denote it by I°. If necessary, we will use the more precise notation $I^{\circ,R}$ instead. □

Proposition 7.3. Let again I be an A-submodule of R containing A. Then I° is the set of all $x \in R$ with $I \cap (Ax) \subset A$.

Proof. Let $x \in R$ be given. Then $x \in I^\circ$ iff $Ax \subset I^\circ$ iff $A + Ax \subset I^\circ$ iff $I \cap (A + Ax) = A$. Since $I \cap (A + Ax) = (I \cap A) + [I \cap (Ax)]$ and $I \cap A = A$, the claim follows. □

Remark 7.4. If B is an R-overring of A and I is an A-submodule of B containing A then $I^{\circ,B} = B \cap I^{\circ,R}$, as follows immediately from Proposition 2 or Proposition 3. □

Remarks 7.5. It is obvious that for two A-modules $J_1 \subset J_2$ in R with $A \subset J_1$ we have $J_1^\circ \supset J_2^\circ$. If I is any A-module in R containing A then $I \subset I^{\circ\circ}$. Here we have written $I^{\circ\circ} := (I^\circ)^\circ$. From this it follows in the usual way that $I^{\circ\circ\circ} = I^\circ$. □

Proposition 7.6. Let $(I_\lambda \mid \lambda \in \Lambda)$ be a family of A-modules in R, all containing A. Then $(\sum_{\lambda \in \Lambda} I_\lambda)^\circ = \bigcap_{\lambda \in \Lambda} I_\lambda^\circ$.

Proof. Let $J := \bigcap_{\lambda \in \Lambda} I_\lambda^\circ$. It is obvious that $(\sum_{\lambda \in \Lambda} I_\lambda)^\circ \subset J$. On the other hand, Lemma 1 tells us that $(\sum_{\lambda \in \Lambda} I_\lambda) \cap J = A$. Thus $J \subset (\sum_{\lambda \in \Lambda} I_\lambda)^\circ$. □

Theorem 7.7. The polar I° of any A-submodule I of R with $A \subset I$ is a subring of R, hence an R-overring of A.

Proof. It suffices to verify the following: Let J_1 and J_2 be finitely generated A-modules in R with $I \cap J_1 = I \cap J_2 = A$. Then $I \cap (J_1 J_2) = A$.

Since A is Prüfer in R, the module J_1 is R-invertible (cf. Th.1.13). Using Theorem 1.4.(3) we obtain $J_1^{-1}(I \cap (J_1 J_2)) = (J_1^{-1} I) \cap J_2$. Now $J_1^{-1} \subset A$, hence $J_1^{-1} I \subset I$. Thus $J_1^{-1}(I \cap (J_1 J_2)) \subset I \cap J_2 = A$. Multiplying by J_1 we obtain $I \cap (J_1 J_2) \subset J_1$. Intersecting with I, we obtain $I \cap (J_1 J_2) \subset I \cap J_1 = A$. We have $A \subset J_1 J_2$, hence $I \cap (J_1 J_2) = A$. $\qquad\square$

If again I is an A-module in R containing A then, as usual, let $A[I]$ denote the ring generated by I over A in R. Notice that $A[I]$ is the subring of R generated by I alone since $A \subset I$, and that $A[I]$ is the union of the modules I^n with n running through $\mathbb{N}$.

Corollary 7.8. Let I be an A-module in R containing A. Then $I^\circ = A[I]^\circ = (I^n)^\circ$ for every $n \in \mathbb{N}$.

Proof. A priori we have $I^\circ \supset (I^n)^\circ \supset A[I]^\circ$. But $I^{\circ\circ}$ is a subring of R by the preceding theorem. It contains I, hence $A[I]$. From $A[I] \subset I^{\circ\circ}$ we obtain $A[I]^\circ \supset I^{\circ\circ\circ} = I^\circ$. $\qquad\square$

Proposition 7.9. Let I be an R-invertible ideal of A, and $C := A[I^{-1}]^\circ$. Then $\mathrm{Inv}(A, C)$ is the set of all R-invertible ideals J of A with $I + J = A$, and C is the union of the rings $A[J^{-1}]$ with J running through these ideals.

Proof. Let $J \in \mathrm{Inv}(A, R)$ be given. Then J is C-invertible iff $J^{-1} \subset C$, which means that $A[I^{-1}] \cap J^{-1} = A$. Since $A[I^{-1}]$ is the union of the modules $I^{-n}(n \in \mathbb{N})$, the latter condition means $I^{-n} \cap J^{-1} = A$ for all $n \in \mathbb{N}$. Taking inverses we see that this is equivalent to $I^n + J = A$ for all $n \in \mathbb{N}$. But $I + J = A$ implies $I^n + J = A$ for every $n \in \mathbb{N}$. Thus $\mathrm{Inv}(A, C)$ is the set of all $J \in \mathrm{Inv}(A, R)$ with $I + J = A$. The last claim is now obvious (cf. Theorem 4.11.i). $\qquad\square$

If B is an R-overring of A then $BB^\circ = B + B^\circ$, cf. Proposition 1.6. It may happen that $BB^\circ = R$. We strive for a good understanding of this situation.

Lemma 7.10. Let I, J_1, J_2 be A-modules in R. Assume that at least one of these modules is R-regular, and that $I \cap J_1 \subset J_2$, $I + J_1 \subset I + J_2$. Then $J_1 \subset J_2$.

Proof. $J_1 = (I + J_1) \cap J_1 \subset (I + J_2) \cap J_1 = (I \cap J_1) + (J_2 \cap J_1) \subset J_2$. Here we have used Lemma 1 (or Th. 1.4.(1)). $\square$

Definition 2. Let I and J be A-submodules of R, and assume that $A \subset I$. We call J a *complement of I in R (over A)* if $I \cap J = A$ and $I + J = R$.

Theorem 7.11. Let I be an A-module in R with $A \subset R$.
  i) Then I has at most one complement J in R.
  ii) If I has a complement J then $J = I^\circ$ and $I = J^\circ$. Both I and J are R-overrings of A.

Proof. The first assertion is evident by Lemma 10. Assume now that J is a compelement of I. Then $I \cap J = I \cap I^\circ = A$, and $J \subset I^\circ$, hence also $I + J = I + I^\circ = R$. Thus I° is a complement of A, and we conclude that $J = I^\circ$. Also I is a complement of J, and we conclude that $I = J^\circ$. By Theorem 7 both I and J are subrings of R. $\square$

Definition 3. a) We call an R-overring B of A a *factor of R (over A)*, if the A-module B has a complement, i.e. $BB^\circ = B + B^\circ = R$. We then also say that the extension $A \subset B$ is a factor of the Prüfer extension $A \subset R$.
b) If B is a factor of R over A with complement C then we write $R = B \times_A C$. We call such a decomposition of R a *factorisation of R over A*.
c) A factor B of R over A is called *trivial* if $B = A$ or $B = R$. The Prüfer extension $A \subset R$ is called *reducible* if it has a nontrivial factor. It is called *irreducible* if it is nontrivial but does not have a nontrivial factor. We then also say that R is *reducible* (resp. *irreducible over A*). $\square$

Remarks 7.12. If B is any R-overring of A then $D := BB^\circ = B + B^\circ$ is the largest R-overring of B such that B is a factor of D over A. Also, if C is a second R-overring of A, then $BC = B + C = B \times_{B \cap C} C$. $\square$

We want to prove that, if B_1 and B_2 are factors of R over A then $B_1 B_2$ and $B_1 \cap B_2$ are again factors of R over A.

Lemma 7.13. Let B be a factor of R over A and let C be an R-overring. Then $B \cap C$ is a factor of C over A with the complement $B° \cap C$.

Proof. We have $(B \cap C) \cap (B° \cap C) = A$. From $B + B° = R$ we deduce that $C = C \cap (B + B°) = (B \cap C) + (B° \cap C)$. $\square$

Proposition 7.14. If B_1 and B_2 are factors of R over A then also $B_1 \cap B_2$ and $B_1 B_2$ are factors over A. The complements are $(B_1 B_2)° = B_1° \cap B_2°$ and

$$(B_1 \cap B_2)° = B_1° + B_2° = (B_1° \cap B_2) + B_2° = (B_2° \cap B_1) + B_1°.$$

Proof. Lemma 13 tells us that $B_1 \cap B_2 + B_1 \cap B_2° = B_1$ and $B_1° \cap B_2 + B_1° \cap B_2° = B_1°$. Inserting this in $B_1 + B_1°$ we obtain that $(B_1 \cap B_2) + J = R$ with

$$J := (B_1 \cap B_2°) + (B_1° \cap B_2) + (B_1° \cap B_2°).$$

One readily verifies that $(B_1 \cap B_2) \cap J = A$. Thus $B_1 \cap B_2$ is a factor with complement J. Clearly $J = (B_1° \cap B_2) + B_2° = (B_2° \cap B_1) + B_1°$. Now $J \subset B_1° + B_2°$ and $(B_1 \cap B_2) \cap (B_1° + B_2°) = A$. Thus also $B_1° + B_2°$ is a complement of $B_1 \cap B_2$, which forces $J = B_1° + B_2°$. It follows that $(B_1° + B_2°)° = J° = B_1 \cap B_2$. Applying all this to $B_1°$, $B_2°$ instead of B_1, B_2 we learn that $B_1 + B_2$ is a factor of R over A with complement $B_1° \cap B_2°$. Notice that $B_1 + B_2 = B_1 B_2$. $\square$

Proposition 7.15. Let D be an R-overring of A. Assume that $R = B \times_A C$. Then $D = (B \cap D) \times_A (C \cap D)$ and $R = (BD) \times_D (CD)$.

Proof. $(B \cap D) \cap (C \cap D) = A \cap D = A$, and $(BD) \cap (CD) = (B \cap C)D = AD = D$. On the other hand, $R = B + C$, hence $D = D \cap (B + C) = (D \cap B) + (D \cap C)$, and $R = BD + CD$. Here we have used parts (1) and (4) of Theorem 1.4. $\square$

Proposition 7.16 (Transfer principle for polars and factors). Let $A \subset C$ be a ws ring extension. We regard R and C as subrings of $RC = R \otimes_A C$. Assume that $R \cap C = A$.

(i) Let I be an A-submodule of R containing A. Let $I°$ denote the polar of I in R and $(IC)°$ denote the polar of the C-module IC in RC. Then $(IC)° = (I°)C$.

(ii) Let B be an R-overring of A. Then B is a factor of R over A iff BC is a factor of RC over C. In this case $RC = (BC) \times_C (B°C)$.

Proof. Statement (i) is obvious from the fact that we have an isomorphism $J \mapsto JC$ from the lattice of R-regular A-submodules J of R to the lattice of RC-regular C-submodules of RC, as follows from Theorem 6.5. If B is an R-overring of R over A then $(BC)° = B°C$, as just proved, and $BC + B°C = (B + B°)C = RC$. In view of Theorem 6.5 this gives us the second claim (ii). $\square$

We will continue our study of polars and factors in part II of the book.

§8 Decomposition of regular modules

In this section we assume again that $A \subset R$ is a Prüfer extension. We *further assume that B and C are R-overrings of A with $B \cap C = A$.*

Theorem 8.1. a) Let I be a BC-regular A-submodule of BC. Then there exists a unique pair (J, K), consisting of a B-regular A-submodule J of B and a C-regular A-submodule K of C, such that $I = (JC) \cap (BK)$. We have $J = (IC) \cap B$ and $K = (IB) \cap C$.

b) Conversely, if J is a B-regular A-submodule of B and K is a C-regular A-submodule of C then $I := (JC) \cap (BK)$ is a BC-regular A-submodule of BC.

Proof. Assertion b) of the theorem is trivial. Assume now that I is a BC-regular A-submodule of BC. Then IC is a BC-regular C-submodule of BC. We define $J := B \cap (IC)$. This is an A-submodule of B. According to the transfer theorem 6.5 the module J is B-regular and $JC = IC$. By the same theorem $K := C \cap (IB)$ is an A-regular submodule of C and $IB = KB$. We have $(JC) \cap (KB) = (IC) \cap (IB) = I(C \cap B)$, by using Theorem 1.4.(4).

Let now J' be any B-regular A-submodule of B and K' any C-regular A-submodule of C such that $I = (J'C) \cap (BK')$. Then, again by Theorem 1.4.(4), $IB = [(J'C) \cap (K'B)]B = (J'CB) \cap (K'B) = (BC) \cap (K'B) = K'B = KB$, and Theorem 6.5 tells us that $K' = K$. By the same argument $J' = J$. □

Definition 1. In the situation of Theorem 1.a we call J the B-*component* and K the C-*component* of the given BC-regular A-module J. Notice that J is the unique B-regular A-module in B with $JC = IC$, and K is the unique C-regular A-module in C with $KB = IB$.

Example 8.2. Let I be an A-submodule of BC with $A \subset I$. Then I has the B-component $I \cap B$ and the C-component $I \cap C$, and $I = (I \cap B)(I \cap C) = (I \cap B) + (I \cap C)$.

Proof. Applying Theorem 1.4.(4) we learn that $(I \cap B)C = (IC) \cap (BC) = IC$. The A-module $I \cap B$ is certainly B-regular, since it contains A. Thus $I \cap B$ is the B-component of I. For the same reason $I \cap C$ is the C-component of I. We have

$$I = I \cap (B + C) = (I \cap B) + (I \cap C)$$

by Theorems 1.4.(1) and (1.7), and then $I = (I \cap B)(I \cap C)$, since $I \cap B \cap C = A$, cf. Th.1.4.(5). □

Remark 8.3. Assume that J is the B-component and K is the C-component of a BC-regular A-submodule I of BC. Then $J \cap K = A \cap I$.

Proof. $J \cap K = (IC) \cap B \cap (IB) \cap C = (IB) \cap (IC) \cap A = [I(B \cap C)] \cap A = (IA) \cap A = I \cap A$, where we again have used Theorem 1.4.(4). □

For BC-regular ideals of A we have the following simpler decomposition theorem.

Theorem 8.4.
a) Let I be a BC-regular ideal of A. Then there exists a unique pair (J, K) consisting of a B-regular ideal J of A and a C-regular ideal K of A such that $I = J \cap K$. The A-module J is the B-component of I and K is the C-component of A.

b) Conversely, if J is a B-regular ideal of A and K is a C-regular ideal of A then $J \cap K$ is a BC-regular ideal of A. Also $J + K = A$ and $J \cap K = JK$.

Proof. a): Assume that I is a (BC)-regular ideal of A. We *define* the A-modules J and K as the B-component and the C-component of I respectively. Remark 3 tells us that $I = J \cap K$. We have $JC = IC \subset C$, hence $J = (JC) \cap B \subset A$, and also $K \subset A$, and we know that J is B-regular and K is C-regular.

Let now J' be any B-regular ideal of A and K' be any C-regular ideal of A with $J' \cap K' = I$. Then $(J'C) \cap (K'B) = (J'B) \cap (J'C) \cap (K'B) \cap (K'C) = [J'(B \cap C)] \cap [K'(B \cap C)] = J' \cap K' = I$, and Theorem 1 tells us that $J' = J$, $K' = K$.

b): Let J be a B-regular and K a C-regular ideal of A. Then both J and K are BC-regular, hence JK and $J \cap K$ are BC-regular. We have $(JK)B = (JB)(KB) = KB$ and $(J \cap K)B = (JB) \cap KB = B \cap KB = KB$. Thus $(JK)B = (J \cap K)B$. Also $(JK)C = (J \cap K)C$. As just proved, this implies $JK = J \cap K$. Also $(J + K)B = B + KB = B = AB$ and $(J + K)C = AC$, hence $J + K = A$. □

Remark 8.5. Notice that $C \subset B^\circ$. Assume that I is a BC-regular A-submodule of BC. Then I is also BB°-regular. Thus I has a B-component $J = (IC) \cap B$ and a C-component $K = (IB) \cap C$. But I has also a B-component $J' = (IB^\circ) \cap B$ and a B°-component $K' = (IB) \cap B^\circ$. Fortunately $J = J'$ and $K = K'$.

Proof. Both J and J' are B-regular and $JB^\circ = (JC)B^\circ = (IC)B^\circ = IB^\circ = J'B^\circ$. Theorem 6.5 tells us that $J = J'$. Also $KB = IB = K'B$, and both K, K' are B°-regular. Thus, by the same theorem, $K = K'$. □

Thus we can say, that the B-component and the C-component of I only depend on A, B, and I, as long as I is BC-regular.

Remarks 8.6. Assume that I_1 and I_2 are BC-regular A-submodules of BC. Let J_1, J_2 denote the B-components and K_1, K_2 denote the C-components of I_1 and I_2 respectively.

a) $I_1 \subset I_2$ iff $J_1 \subset J_2$ and $K_1 \subset K_2$.

b) $I_1 + I_2$ has the B-component $J_1 + J_2$ and the C-component $K_1 + K_2$.

c) $I_1 \cap I_2$ has the B-component $J_1 \cap J_2$ and the C-component $K_1 \cap K_2$.

d) $I_1 I_2$ has the B-component $J_1 J_2$ and the C-component $K_1 K_2$.

Proof. It suffices to prove the claims about the B-components. We have $I_1 C = J_1 C$, $I_2 C = J_2 C$, hence $(I_1 + I_2)C = (J_1 + J_2)C$,

$$(I_1 \cap I_2)C = (I_1 C) \cap (I_2 C) = (J_1 C) \cap (J_2 C) = (J_1 \cap J_2)C,$$

and $(I_1 I_2)C = (J_1 J_2)C$. In view of the tranfer theorem 6.5, applied to the extensions $A \subset B$ and $A \subset C$, now all claims are evident. $\square$

Proposition 8.7. Let I be a (BC)-regular ideal of A. The B-component J of I is the smallest B-regular ideal of A containing I.

Proof. Let K denote the C-component of I. Then $I = J \cap K \subset J$. Let now $\mathfrak{a}$ be B-regular ideal of A with $I \subset \mathfrak{a}$. Then $\mathfrak{a}$ is BC-regular. The trivial equation $\mathfrak{a} = \mathfrak{a} \cap A$ tells us that $\mathfrak{a}$ has the B-component $\mathfrak{a}$ and the C-component A. It follows by Remark 8.6.a that $J \subset \mathfrak{a}$.
$\square$

Proposition 8.8. Let I be a BC-regular A-submodule of BC, and let J, K denote the B- and C-component of I respectively. Then I is BC-invertible (i.e. finitely generated) iff J is B-invertible and K is C-invertible. In this case I^{-1} has the B-component J^{-1} and the C-component K^{-1}.

Proof. The A-module A has the B-component A and the C-component A. The claim now follows easily from Remark 6.d. $\square$

Recall from §6, that $D(A, R)$ denotes the group of R-invertible A-modules in R, and from §4, that $\mathrm{Inv}(A, R)$ denotes the submonoid of R-invertible ideals of A.

Theorem 8.9. The map $\varphi \colon D(A, B) \times D(A, C) \to D(A, BC)$, $\varphi(J, K) := (JC) \cap (BK)$, is an isomorphism of lattice ordered abelian groups. Here the ordering on $D(A, B) \times D(A, C)$ is given by $(J, K) \leq (J', K')$ iff $J \subset J'$ and $K \subset K'$.

Proof. This is evident from Theorem 1, Remarks 6, and Proposition 8. □

Scholium 8.10. Notice that, by this theorem, the map $(J, K) \mapsto J \cap K = JK$,

$$\mathrm{Inv}(A, B) \times \mathrm{Inv}(A, C) \longrightarrow \mathrm{Inv}(A, BC),$$

is an isomorphism of lattice ordered abelian semigroups. The inverse map sends a BC-ideal I of A to the pair (J, K) consisting of the B-component J and the C-component K of I. We have $I^{-1} = (J^{-1}C) \cap (BK^{-1}) = J^{-1}K^{-1}$, but $J^{-1} \cap K^{-1} = A$. Also $I^{-1} = J^{-1} + K^{-1}$, as is evident from $I = J \cap K$, and also from Lemma 1.5. □

Theorem 8.11. Let $\mathfrak{p}$ be a BC-regular prime ideal of A. Then $\mathfrak{p}$ is either B-regular or C-regular. $\mathfrak{p}$ has the B-component $\mathfrak{p}$ and C-component A in the first case, and the B-component A and C-component $\mathfrak{p}$ in the second case.

Proof. Let J be the B-component and K the C-component of $\mathfrak{p}$. We have $J = (\mathfrak{p}C) \cap B$, $K = (\mathfrak{p}B) \cap C$, and $\mathfrak{p} = J \cap K$. Since $\mathfrak{p}$ is prime, either $J \subset \mathfrak{p}$ or $K \subset \mathfrak{p}$. If $J \subset \mathfrak{p}$ then $\mathfrak{p}$ is B-regular, and $K = A$, hence $\mathfrak{p} = J \cap A = J$. If $K \subset \mathfrak{p}$ then $\mathfrak{p}$ is C-regular, and $\mathfrak{p} = K$. □

Corollary 8.12. If $R = BC$, i.e. $R = B \times_A C$, the set $Y(R/A)$ of R-regular prime ideals of A is the disjoint union of $Y(B/A)$ and $Y(C/A)$. □

We give an application of this corollary which will be very useful later on. Recall that for any R-overring D of A the restriction map $\mathrm{Spec}\, D \to \mathrm{Spec}\, A$ is a bijection onto its image $X(D/A)$, and that $X(D/A)$ is the complement of $Y(D/A)$ in $\mathrm{Spec}\, A$. {This holds more generally if A is ws in R, cf. §3 and I, §4.}

Theorem 8.13. As before we assume that A is Prüfer in R. Let $(B_i \mid 1 \leq i \leq r)$ be a finite family of R-overrings of A. Then

a) $X(B_1 B_2 \ldots B_r / A) = \bigcap_{i=1}^{r} X(B_i/A)$,

b) $X(\bigcap_{i=1}^{r} B_i/A) = \bigcup_{i=1}^{r} X(B_i/A)$.

Proof. It suffices to verify this for $r = 2$. We first work in the special case that $B_1 \cap B_2 = A$. Then Corollary 12 tells us that $Y(B_1/A) \cup Y(B_2/A) = Y(B_1 B_2/A)$ and $Y(B_1/A) \cap Y(B_2/A) = \emptyset$. Taking complements in $\operatorname{Spec} A$ we obtain a) and b). If $A' := B_1 \cap B_2$ is different from A, then we obtain a) and b) by applying the restriction map $\operatorname{Spec} A' \to \operatorname{Spec} A$ to the equations $X(B_1 B_2/A') = X(B_1/A') \cap X(B_2/A')$ and $\operatorname{Spec} A' = X(B_1/A') \cup X(B_2/A')$. $\qquad\square$

We return to the situation that $A \subset R$ is Prüfer and B, C are R-overrings of A with $B \cap C = A$.

Proposition 8.14. Let I be a BC-regular A-submodule of BC and let I_1 be an A-submodule of B. Let J denote the B-component and K denote the C-component of I. The BC-regular module $I + I_1$ has the B-component $J + I_1$ and the C-component $K + (I_1 \cap A)$. In particular, if I_1 is an ideal of A, then $I + I_1$ has the B-component $J + I_1$ and the C-component $K + I_1$.

Proof. $(I + I_1)C = IC + I_1C = JC + I_1C = (J + I_1)C$, and $J + I_1$ is a B-regular A-submodule of A. Thus $J + I_1$ is the B-component of $I + I_1$. The C-component is $[(I + I_1)B] \cap C = (IB + I_1B) \cap C = [(IB) \cap C] + [(I_1B) \cap C] = J + (I_1 \cap C) = J + (I_1 \cap A)$, since $I_1 \cap C \subset B \cap C = A$. $\qquad\square$

If both I and I_1 are BC-regular ideals of A, we have the following funny consequence. (Recall Remark 6.b.)

Corollary 8.15. Let I and I_1 be BC-regular ideals of A. Let J and J_1 denote the B-components of I and I_1 respectively. Then $J + I_1 = I + J_1 = J + J_1$. $\qquad\square$

Corollary 8.16. Let J be the B-component of a (BC)-regular ideal I of A. Let $\mathfrak{a}$ be any B-regular ideal of A with $\mathfrak{a} \subset J$. Then $J = \mathfrak{a} + I$.

Proof. Take $I_1 = \mathfrak{a}$ in the preceding corollary. $\qquad\square$

§9 Prüfer overmodules

We enlarge and deepen some results of §5 by working with "overmodules" of rings instead of ring extensions. This means some sort of linearisation of the basic theory of Prüfer extensions.

In the following A is a ring (commutative, with 1, as always).

Definition 1. An *overmodule of A* is an A-module M which contains the ring A as an A-submodule. Such an overmodule M is called *Prüfer* if $(A\!:\!m) + (A\!:\!m)m = A$ for every $m \in M$. Alternatively we then say that *A is Prüfer in M*. Here $(A\!:\!m)$ denotes the ideal $\{a \in A \mid am \in A\}$ of A.

Notice that if $A \subset R$ is a ring extension then R may be also regarded as an overmodule of A. The A-overmodule R is Prüfer iff the ring extension $A \subset R$ is Prüfer, as we know from the implication (1) $\Leftrightarrow$ (8) in Theorem I.5.2.

Theorem 9.1. Let M be an overmodule of A. Then A is Prüfer in M iff A is a distributive submodule of M.

This can be proved in exactly the same way as Theorem 5.4 (which is a special case of the present theorem). $\square$

It turns out that every Prüfer overmodule of A can be embedded over A in the complete ring of quotients $Q(A)$ (cf. I, §4) in a unique way. In order to prove this we need a lemma.

Definition 2. Let N be an overmodule of A. We call an ideal I of A *dense in N*, if $In \neq 0$ for every element $n \neq 0$ of N.

Lemma 9.2. If M is a Prüfer overmodule of A then, for every $m \in M$, the ideal $(A\!:\!m)$ is dense in M (hence also dense in A).

Proof. We may assume that $m \in M \backslash A$, since otherwise $(A\!:\!m) = A$. From $(A\!:\!m) + (A\!:\!m)m = A$ we conclude that $(A\!:\!m)$ is dense in A and also that $(A\!:\!m)m \neq 0$, since otherwise $(A\!:\!m) = A$, which would imply $m \in A$.

Let n be a nonzero element of M. We want to prove that $(A\!:\!m)n \neq 0$. We know already that $(A\!:\!n)n \neq 0$. Since $(A\!:\!n)n \subset A$

and $(A:m)$ is dense in A, it follows that $(A:m)(A:n)n \neq 0$. This implies $(A:m)n \neq 0$ $\square$

Lemma 9.3. Let M be an overmodule of A. Assume that for every $m \in M$ the ideal $(A:m)$ is dense in A, and that $(A:m)m \neq 0$ if $m \neq 0$. Then there exists a unique A-module homomorphism $\varphi: M \to Q(A)$ with $\varphi(a) = a$ for every $a \in A$, and φ is injective.

Proof. $Q(A)$ is the inductive limit of the A-modules $\mathrm{Hom}_A(\mathfrak{a}, A)$ with $\mathfrak{a}$ running through the dense ideals of A. Since the natural maps $\mathrm{Hom}_A(\mathfrak{a}, A) \to Q(A)$ are injective, we feel free to regard any A-linear form $h: \mathfrak{a} \to A$, with $\mathfrak{a}$ a dense ideal of A, as an element of $Q(A)$.

For every $m \in M$ we introduce the A-linear form $\widehat{m}: (A:m) \to A$, $\widehat{m}(a) := am$, and we define $\varphi: M \to Q(A)$ by $\varphi(m) := \widehat{m}$.

If $m, n \in M$ we have $(\widehat{m} + \widehat{n})(x) = (m+n)^\wedge(x)$ for any $x \in (A:M) \cdot (A:N)$. Since this ideal is dense in A, we conclude that $\widehat{m} + \widehat{n} = (m+n)^\wedge$ in $Q(A)$. Similarly one proves that $(am)^\wedge = a \cdot \widehat{m}$ for $a \in A$, $m \in M$. Thus $\varphi: M \to Q(A)$ is A-linear. If $m \in M$, $m \neq 0$, then by assumption $(A:m)m \neq 0$. This means that $\widehat{m} \neq 0$. Thus φ is injective. Finally, if $a \in A$ then $\widehat{a}$ coincides with a under the natural identification of A with a subring of $Q(A)$.

If $\psi: M \to Q(A)$ is a second A-module homomorphism over A, then we have, for every $m \in M$,

$$(A:m)\psi(m) = \psi((A:m)m) = (A:m)m = (A:m)\varphi(m),$$

and we conclude that $\psi(m) = \varphi(m)$, since $(A:m)$ is dense in A. $\square$

From these two lemmas we conclude

Theorem 9.4. Every Prüfer overmodule M of A embeds into $Q(A)$ over A in a unique way. $\square$

Definition 3. Let $A \subset R$ be a ring extension. An R-*overmodule* of A is an A-submodule M of R which contains A.

Theorem 4 tells us that, studying Prüfer overmodules of A, it suffices to look at $Q(A)$-overmodules of A.

Previous work in §2 now gives us the possibility to express the Prüfer property of overmodules in other ways.

Scholium 9.5. Let A be a ring and M be a $Q(A)$-overmodule of A. The following are equivalent.

(1) A is Prüfer in M.
(2) For every $x \in M$ the A-module $A + Ax$ is invertible (i.e. $Q(A)$-invertible).
(3) For every $x \in M$ the A-module $A + Ax$ is locally principal.
(4) For every $x \in M$ the A-module $A + Ax$ is a mulitplication module.

Proof. Let $x \in M$ be given. Then $[A:_{Q(A)} A + Ax] = (A:x)$ (cf. Remark 4.2). Thus the condition $(A:x) + (A:x)x = A$ means that $A + Ax$ is $Q(A)$-invertible. This explains $(1) \Leftrightarrow (2)$. The equivalence of (2), (3), (4) is stated in Proposition 2.3.   □

Theorem 9.6. A $Q(A)$-overmodule M of A is Prüfer iff the subextension $A \subset A[M]$ of $A \subset Q(A)$ is Prüfer.

Proof. If $A \subset A[M]$ is Prüfer then it follows from Theorem 1 (and Theorem 5.4, if you insist), that A is Prüfer in M. Alternatively we can work with Definition 1 and condition (8) in Theorem I.5.2.

Assume now that A is Prüfer in M. The ring $A[M]$ is the union of the rings $A_m := A[(A:m)^{-1}]$ with m running through M. By the criterion (8) in Theorem I.5.2 it suffices to verify that A is Prüfer in A_m for every $m \in M$.

Let $m \in M$ be fixed and $I := (A:m)$. The ideal I is invertible (in $Q(A)$) and $I^{-1} = A + Am$, since we have $(A:m) + (A \cdot m)m = A$. Moreover the A-submodule A of I^{-1} is distributive in I^{-1} since A is distributive in M by Theorem 1 and $I^{-1} \subset M$. We have an order preserving bijection $\mathfrak{a} \mapsto \mathfrak{a}I^{-1}$ from the partially ordered set of ideals $\mathfrak{a}$ of A to the partially ordered set of A-submodules of I^{-1}. Under this bijection I corresponds to A. We conclude that I is distributive in A.[*] We conclude from Theorem 5.7 that I is a Prüfer ideal of A, which means that A is Prüfer in $A[I^{-1}]$.   □

---

[*] Such an argument had already been used in the proof of Theorem 5.7.

Corollary 9.7. Let M be a Prüfer overmodule of A. Every finitely generated A-submodule J of M with $A \subset J$ is projective of rank one.

Proof. We may assume that M is an overmodule in $Q(A)$. Then we may regard J as an R-regular A-submodule of $R = A[M] \subset Q(A)$. Since A is Prüfer in R by Theorem 6, we know that J is R-invertible, hence $Q(A)$-invertible. The claim follows by Proposition 2.3. $\quad\square$

Corollary 9.8. An ideal I of A is Prüfer iff I is invertible and A is Prüfer in I^{-1}.

Proof. We may assume in advance that I is invertible. The claim follows from Theorem 6, applied to $M := I^{-1}$. $\quad\square$

We add still another characterization of Prüfer ideals.

Theorem 9.9. Let I be an invertible ideal of A. Then I is Prüfer iff the ring A/I^2 is arithmetical.

Proof. If I is Prüfer then A is Prüfer in $R := A[I^{-1}]$, and I is R-invertible. Also I^2 is R-invertible, and we infer from Theorem 2.8 that A/I^2 is arithmetical. {More generally A/I^n is arithmetical for every $n \in \mathbb{N}$.}

Assume now that A/I^2 is arithmetical. Let $x \in I^{-1}$ be given. We want to verify that $A + Ax$ is invertible. Then we will know that $(A: x) + (A: x)x = A$. This will prove that A is Prüfer in I^{-1} which will imply that the ideal I is Prüfer by the preceding corollary.

Let $J := I(A + Ax)$. It suffices to prove that J is invertible. We have $I \subset J \subset A$. The A-module J/I^2 is locally principal, since A/I^2 is arithmetical. Thus also J/J^2 is locally principal. It follows by Lemma 2.7 that the ideal J of A is locally principal, hence invertible (cf. Prop.2.3). $\quad\square$

If we only know that A/I is arithmetical then we cannot conclude that the ideal I is Prüfer, as the following example shows.

Example 9.10. Let $A = k[x, y]$ be the polynomial ring in two variables x, y over a field k. The ideal $I := (x)$ of A is invertible (in

$Q(A) = \mathrm{Quot}\,A)$. The ring $A/I \cong k[y]$ is arithmetical. But A/I^2 is not arithmetical, and thus I is not a Prüfer ideal of A. Indeed, taking $J := I^2 + Ay$ $K := I^2 + A(x+y)$, we have $I \cap (J+K) = I = (x)$, while $I \cap J = I \cap K = (x^2, xy)$, as is easily verified. Thus $I \cap (J+K)$ is different from $(I \cap J) + (I \cap K)$. $\qquad\square$

Definitions 4. If N is any A-module then a *distributive module extension* $N \subset M$ is an A-module M containing N as a submodule which is distributive in M, cf. [Ba₂]. A *distributive hull of N* has been defined by Barnard as a distributive module extension $N \subset D(N)$ such that for any distributive module extension $N \subset M$ there exists an A-module monomorphism $\varphi : M \to D(N)$ with $\varphi(n) = n$ for $n \in N$ [loc.cit.]. $\qquad\square$

It follows from Theorem 5.4 together with Theorems 4 and 6 above that, for any ring A, the Prüfer hull $A \subset P(A)$ is a distributive hull of the A-module A, in short $P(A) = D(A)$. Moreover we know in this case that for every distributive extension $A \subset M$ there exists a *unique homomorphism* $\varphi : M \to P(A)$ over A (and φ is injective).

Already V. Erdogdu [Er], building on the work of Barnard [Ba₁] and Davison [Dvs], has proved that the A-module A has a distributive hull with this additional property. {N.B. Not every A-module has a distributive hull, cf. [Ba₁], [Dvs].}

§10 Bezout extensions

An integral domain A is called a *Bezout domain*, if every finitely generated ideal of A is principal. Such a domain is well known to be a Prüfer domain. Indeed, every finitely generated ideal $I \neq 0$ of A is invertible in $\mathrm{Quot}\,A$. Thus the extension $A \subset \mathrm{Quot}\,A$ is Prüfer (cf. Th.2.1). In the classical literature Bezout domains appear at prominent places. For example, the ring of all algebraic integers is clearly a Bezout domain. Also the ring of holomorphic functions on a given subdomain of the complex plane is a Bezout domain, cf. [Re, p.122]. Various p-adic holomorphy rings are Bezout domains, cf. [PR, §6 and §7].

We now look for a generalization of Bezout domains in the framework of Prüfer extensions. We start with the following rather narrow minded definition of "Bezout extensions", and then will prove that two other off hand possibilities to define Bezout extensions give the same class of ring extensions, cf. (2), (3) in Theorem 2 below.

Definition 1. We call a ring extension $A \subset R$ *Bezout* if $A \subset R$ is Prüfer and every R-invertible ideal of A is principal. We then also say that A is *Bezout in R*.

Notice that, using this terminology, an integral domain A is Bezout iff the extension $A \subset \text{Quot}\,A$ is Bezout.

Proposition 10.1. If $A \subset R$ is a Bezout extension, every finitely generated R-regular A-submodule of R is principal, i.e. generated by one element.

Proof. Let I be an R-regular finitely generated A-submodule of R. Since $A \subset R$ is Prüfer, we have

$$(*) \qquad\qquad I = (A \cap I) \cdot (A + I),$$

cf. Th.1.4, (5). Moreover I and $A + I$ are R-invertible, since both these R-modules are finitely generated and R-regular. Thus $A \cap I$ is R-invertible. Since $A \subset R$ is Bezout, we have $A \cap I = As$ with some $s \in A$. For the same reason $(A + I)^{-1} = At$ with some $t \in A$. We have $Rt = R$, hence $t \in R^*$. Equation $(*)$ now gives us $I = Ast^{-1}$. $\qquad\square$

Theorem 10.2. Let $A \subset R$ be a ring extension. The following are equivalent.
(1) $A \subset R$ is Bezout.
(2) $A \subset R$ is ws, and every finitely generated R-regular ideal of A is principal.
(3) For every $x \in R$ there exists some $y \in R$ with $A + Ax = Ay$.
(4) $A \subset R$ is Prüfer, and $(A:x)$ is principal for every $x \in R \setminus A$.

Proof. (1) $\Rightarrow$ (2): $A \subset R$ is Prüfer, hence ws. If I is a finitely generated R-regular ideal of A, then I is R-invertible, hence principal.
(2) $\Rightarrow$ (1): Theorem 2.6 (or Th.2.1) tells us that $A \subset R$ is Prüfer.

(1) $\Rightarrow$ (3): This follows from Proposition 1.

(3) $\Rightarrow$ (1): We first verify condition (8) in Theorem I.5.2, and then will know that A is Prüfer in R. Let $x \in R$ be given. We have $A + Ax = Ay$ with some $y \in R$. Then $Ry = R$, hence $y \in R^*$, and $(A:x) = [A: A + Ax] = [A: Ay] = Ay^{-1}$. It follows that $(A:x)(A + Ax) = Ay^{-1} \cdot Ay = A$. Thus $(A:x) + (A:x)x = A$.

We now prove that for finitely many elements $x_1, \ldots, x_n$ of R the A-module $A + Ax_1 + \cdots + Ax_n$ is principal. We proceed by induction on n, the case $n = 1$ being done. Let $n > 1$. Then $A + Ax_1 + \cdots + Ax_{n-1} = As$ with some $s \in R^*$ by induction hypothesis. Thus $A + Ax_1 + \cdots + Ax_n = As + Ax_n = (A + Ax_n s^{-1})s$. There exists some $t \in R$ with $A + Ax_n s^{-1} = At$, and we have $A + Ax_1 + \cdots + Ax_n = Ast$.

Let finally an R-invertible ideal I of A be given. Then $I^{-1} \supset A$, and I^{-1} is finitely generated. Thus there exist elements $x_1, \ldots, x_n$ in I^{-1} such that $I^{-1} = A + Ax_1 + \cdots + Ax_n$. As we have proved this implies $I^{-1} = Ay$ with some $y \in R^*$, hence $I = Ay^{-1}$. This completes the proof that A is Bezout in R.

(1) $\Rightarrow$ (4): This is evident.

(4) $\Rightarrow$ (3): Let $x \in R \setminus A$ be given. Since $A \subset R$ is Prüfer, we have $(A:x) + (A:x)x = A$. By assumption, $(A:x) = Aa$ with some $a \in A$, and $(A + Ax)a = A$. It follows that a is a unit in R and $A + Ax = Aa^{-1}$. $\square$

Example 10.3. If v is a local Manis valuation on a ring R, i.e. v is Manis and $A := A_v$ is a local ring with maximal ideal $\mathfrak{p}_v$ (cf. I,§1), then A_v is Bezout in R.

Proof. We know by Proposition I.1.3 that the ring R is local with maximal ideal $\operatorname{supp} v$. Let $A := A_v$ and let $x \in R \setminus A$ be given. Certainly $x \notin \operatorname{supp} v$. Thus $x \in R^*$. We have $A + Ax = (Ax^{-1} + A)x$. Now $v(x) < 0$, hence $v(x^{-1}) > 0$, i.e. $x^{-1} \in \mathfrak{p}$. We conclude that $A + Ax = Ax$. According to Theorem 2 this proves that A is Bezout in R. $\square$

If A is a local ring with maximal ideal $\mathfrak{m}$, and $A \subset R$ is a ring extension, then it is evident from the definiton of Prüfer extensions in I, §5 that $A \subset R$ is Prüfer iff $(A, \mathfrak{m})$ is Manis in R. This gives us a second proof of 10.3: Every R-invertible ideal of A is principal, since A is local (cf. Prop.2.3). Thus $A \subset R$ is Bezout. We may summarize:

Scholium 10.4. Let A be a local ring with maximal ideal $\mathfrak{m}$ and $A \subset R$ a ring extension. The following are equivalent:

(1) $(A, \mathfrak{m})$ is Manis in R.
(2) $A \subset R$ is Prüfer.
(3) $A \subset R$ is Bezout.   □

Notice the close relation of our observation here to the last paragraphs of §5.

Expanding on Example 10.3 and Scholium 10.4 we ask what it means in general for a Manis valuation, or better a PM-valuation[*] $v: R \to \Gamma \cup \infty$, that A_v is Bezout in R. We assume without loss of generality that v is surjective, i.e. $\Gamma = \Gamma_v$. For any $\gamma \in \Gamma$ we denote the A_v-module $\{x \in R \mid v(x) \geq \gamma\}$ by I_γ.

Lemma 10.5. Let $v: R \twoheadrightarrow \Gamma \cup \infty$ be a surjective Manis valuation. Assume that A_v is Bezout in R. Then $v(R^*) = \Gamma$.

Proof. Let $\gamma \in \Gamma$ be given with $\gamma > 0$. We choose some $x \in R$ with $v(x) = -\gamma$. There exists some $y \in R$ with $A + Ax = Ay$. We have $Ry = R$, hence $y \in R^*$, and $v(y) = v(x) = -\gamma$, $v(y^{-1}) = \gamma$.   □

Lemma 10.6. Assume again that $v: R \twoheadrightarrow \Gamma \cup \infty$ is a surjective Manis valuation. The following are equivalent.

(1) $v(R^*) = \Gamma$.
(2) I_γ is principal for every $\gamma \in \Gamma$.
(2⁻) I_γ is principal for every negative $\gamma \in \Gamma$.

Proof. (1) $\Rightarrow$ (2): Given $\gamma \in \Gamma$ there exists some $y \in R^*$ with $v(y) = \gamma$. If $x \in I_\gamma$ then $v(xy^{-1}) \geq 0$, hence $xy^{-1} \in A$. Thus $x \in Ay$. This proves $I_\gamma = Ay$.
(2) $\Rightarrow$ (2⁻): trivial.

---

[*] We will analyze in Chapter III thoroughly what it means for a Manis valuation to be PM.

$(2^-) \Rightarrow (1)$: Let γ be a positive element of Γ. Then $I_{-\gamma} = Ay$ with some $y \in R$. Since v is surjective, we have $v(y) = -\gamma$. Now $1 \in I_\gamma$. Thus there exists some $a \in A$ with $ay = 1$, and $v(a) = \gamma$. $\qquad \square$

Proposition 10.7. Let $v \colon R \longrightarrow \Gamma \cup \infty$ be a surjective PM-valuation. The following are equivalent:

(1) $v(R^*) = \Gamma$.
(2) I_γ is principal for every $\gamma \in \Gamma$.
(2^+) I_γ is principal for every positive $\gamma \in \Gamma$.
(2^-) I_γ is principal for every negative $\gamma \in \Gamma$.
(3) A_v is Bezout in R.

Proof. Let $A := A_v$. We know that (3) implies (1) by Lemma 5, and the equivalence of (1), (2), (2^-) by Lemma 6. The implication $(2) \Rightarrow (2^+)$ is trivial. We prove that (2^+) implies (3) and then will be done.

We verify condition (4) in Theorem 2. It is here that we need to know in advance that A is Prüfer in R. Let $x \in R \setminus A$ be given. Then $v(x) = -\gamma$ with some $\gamma > 0$. For any $a \in A$ we have $ax \in A$ iff $v(a) \geq \gamma$. Thus $(A : x) = I_\gamma$. This ideal is principal by assumption. $\qquad \square$

Definition 2. We call a valuation $v \colon R \to \Gamma \cup \infty$ *Bezout-Manis*, or *BM* for short, if v is Manis and the extension $A_v \subset R$ is Bezout. $\quad \square$

Proposition 7 tells us that the BM-valuations form a very tame and agreeable subclass of the PM-valuations. They will play only a small role in the following as compared with PM-valuations. Nevertheless they seem to deserve interest on their own. Notice that every local Manis valuation is BM, as has been stated in 10.3. Some further study of BM-valuations will be made in Chapter III.

Looking for more examples of Bezout extensions we go back to the Prüfer extensions given at the end of I, §6 (I.6.14). We can readily verify that two of them are in fact Bezout extensions.

Examples 10.8. Assume either that $R = C(X)$ for some topological space X and $A = C_b(X)$, or that $R = CS(M)$ and $A = CS_b(M)$ for some semialgebraic set $M \subset k^N$ with k a real closed field and some $N \in \mathbb{N}$. Then A is Bezout in R.

Proof. We consider the case $R = C(X)$, $A = C_b(X)$. In the other case $R = CS(M)$, $A = CS_b(M)$ the proof runs exactly the same way.

We have an obvious partial ordering on R: $f \geq g$ iff $f(x) \geq g(x)$ for every $x \in X$. This makes R a lattice ordered ring ([BKW]). In particular, for each $f \in R$ we have at hands the element $|f| := f \vee (-f)$. Of course, $|f|(x) = |f(x)|$ for every $x \in X$. We call an additive subgroup L of R *absolutely convex* in R, if for every $f \in L$ and $g \in R$ with $0 \leq |g| \leq |f|$ we have $g \in L$. Notice that A is absolutely convex in R. The claim that A is Bezout in R will now be verified by the following chain of easy observations.

a) We know already from I.6.14 that A is Prüfer in R.

b) *For every $f \in R$ the ideal $(A: f)$ of A is absolutely convex in R.* Indeed, if $a \in (A: f)$ and $b \in R$ is given with $0 \leq |b| \leq |a|$, then $b \in A$ and $0 \leq |bf| \leq |af| \in A$, hence $bf \in A$, i.e. $b \in (A: f)$.

c) *Every R-invertible ideal I of A is absolutely convex in R.* Indeed, the A-module $J := I^{-1}$ is finitely generated, $J = \sum_{i=1}^{n} Af_i$ with some $f_i \in R$. Since $I \subset A$, we have $I = [A: J] = (A: J) = \bigcap_{i=1}^{n}(A: f_i)$. Each $(A: f_i)$ is absolutely convex in R, hence I too.

d) Let $f_1, \ldots, f_n$ be elements of A, and assume that and $I := \sum_{i=1}^{n} Af_i$ is regular. *Then the intersection $\bigcap_{i=1}^{n} Z(f_i)$ of the zero sets $Z(f_i) = \{x \in X \mid f_i(x) = 0\}$ is empty.* Indeed $1 = \Sigma f_i g_i$ we have an equation $1 = \sum_{i=1}^{n} f_i g_i$ with $g_i \in R$, and this implies that the $Z(f_i)$ have empty intersection. {Remark: If conversely this holds, then $g := \sum_{i=1}^{n} f_i^2$ is a unit in R and $\sum_{i=1}^{n} f_i(f_i g^{-1}) = 1$, hence I is R-regular.}

e) *Every R-invertible ideal I of A is principal.* Indeed, let $f_1, \ldots, f_n$ be a system of generators of I. Then $|f_i| \in I$ by c) and $\bigcap_{i=1}^{n} Z(f_i) = \emptyset$ by d). It follows that $g := |f_1| + \cdots + |f_n| \in R^* \cap I$. For every $i \in \{1, \ldots, n\}$ we have $\left|\frac{f_i}{g}\right| \leq 1$, hence $\frac{f_i}{g} \in A$. Thus $I = Ag$. $\qquad \square$

Returning to general theory we write down some permanence properties of the class of Bezout extensions in a similar style as done in I, §5 for Prüfer extensions.

Proposition 10.9. If $A \subset R$ is a Bezout extension and $\varphi \colon R \to D$ is any ring extension, then $\varphi(A) \subset \varphi(R)$ is again Bezout.

Proof. We verify condition (3) of Theorem 2 for the extension $\varphi(A) \subset \varphi(R)$. Let $x \in R$ be given. There exists some $y \in R$ with $A + Ax = Ay$. Applying φ we obtain
$$\varphi(A) + \varphi(A)\varphi(x) = \varphi(A)\varphi(y). \qquad \square$$

Proposition 10.10. Let $A \subset R$ be a ring extension. Assume that there exists a family $(A_\lambda \subset R_\lambda \mid \lambda \in \Lambda)$ of Bezout extensions together with ring homomorphisms $\varphi_\lambda \colon R_\lambda \to R$ such that $\varphi_\lambda(A_\lambda) \subset A$ for each $\lambda \in \Lambda$ and $R = \bigcup_{\lambda \in \Lambda} \varphi_\lambda(R_\lambda)$. Then A is Bezout in R.

Proof. We verify condition (3) of Theorem 2. Let $x \in R$ be given. We choose some $\lambda \in \Lambda$ and $x_\lambda \in R_\lambda$ with $x = \varphi_\lambda(x_\lambda)$. There exists some $y_\lambda \in R_\lambda$ such that $A_\lambda + A_\lambda x_\lambda = A_\lambda y_\lambda$. Applying φ_λ we obtain $\varphi_\lambda(A_\lambda) + \varphi_\lambda(A_\lambda)x = \varphi_\lambda(A_\lambda)y$ with $y := \varphi_\lambda(y_\lambda)$. Since $\varphi_\lambda(A_\lambda) \subset A$, this implies $A + Ax = Ay$. $\qquad \square$

Notice that the assumptions of Proposition 10 are remarkably more general than the assumptions in its counterpart Proposition I.5.13 for Prüfer extensions, and the proof is so easy.

Proposition 10.11. Let $A \subset B$ and $B \subset R$ be ring extensions. Then A is Bezout in R iff A is Bezout in B and B is Bezout in R.

Proof. a) Assume first that A is Bezout in R. We know from I, §5 that A is Prüfer in B and B is Prüfer in R. If I is a B-invertible ideal of A then I is R-invertible, hence principal. This proves that A is Bezout in B.

We now verify condition (3) of Theorem 2 for the extension $B \subset R$, and then will know that this extension is also Bezout. Let $x \in R$ be given. There exists some $y \in R$ with $A + Ax = Ay$. It follows that $B + Bx = By$.

b) Assume that both extensions $A \subset B$ and $B \subset R$ are Bezout. In order to prove that $A \subset R$ is Bezout we again appeal to condition (3) of Theorem 2. Let $x \in R$ be given. Since $B \subset R$ is Bezout, there exists some $y \in R$ with $B + Bx = By$. We have some $b \in B$ with $by = 1$. Multiplying by b we obtain $Bb + Bbx = B$. Thus $bx \in B$, and the A-submodule $Ab + Abx$ of B is B-regular. By Proposition 1 there exists some $s \in B$ with $Ab + Abx = As$. Multiplying by y be obtain $A + Ax = Asy$. Thus the extension $A \subset R$ fulfills condition (3) of Theorem 2, and we conclude that A is Bezout in R. $\square$

Proposition 10.12. Assume that R and C are overrings of a ring A in some ring extension T. Assume further that A is Bezout in R and ws in C. Then C is Bezout in the subring RC of T.

Proof. We may replace A by the overring $R \cap C$, since $R \cap C$ is Bezout in R by Proposition 11 and ws in C by Proposition I.3.7.b. Thus we assume without loss of generality that $A = R \cap C$. We know by Theorem I.5.10 that C is Prüfer in RC. Let J be an RC-invertible ideal of C. We know by Theorem 6.4 that $I := R \cap J$ is an R-invertible ideal of A and $J = IC$. Since A is Bezout in R, the ideal I is principal. Thus also J is principal. We conclude that C is Bezout in RC right from Definition 1. $\square$

We now have the means at hand to prove that, given a ring extension $A \subset R$, there exists a "Bezout hull" of A in R in complete analogy to the Prüfer hull $P(A, R)$ established in I, §5. We insert one more proposition, whose proof now is very easy.

Proposition 10.13. Let $A \subset R$ be any ring extension, and assume that B_1, B_2 are overrings of A in R such that A is Bezout both in B_1 and B_2. Then A is Bezout in $B_1 B_2$.

Proof. B_2 is Bezout in $B_1 B_2$ by Proposition 12, since A is ws in B_1 and Bezout in B_2. Furthermore A is Bezout in B_2. Now Proposition 11 gives the claim. $\square$

Theorem 10.14. Let $A \subset R$ be any ring extension. There exists a unique R-overring C of A, such that A is Bezout in C and C contains every R-overring of A in which A is Bezout.

Proof. This can be proved in exactly the same way as the analogous result for Prüfer extensions (Proof of Theorem I.5.15), now using Propositions 10 and 13.   □

Definition 3. We call the ring C described in Theorem 14 the *Bezout hull of A in R*, and write $C = \text{Bez}(A, R)$. If A is any ring, the *Bezout hull of A* is defined as the Bezout hull $\text{Bez}(A, Q(A))$ of A in its complete quotient ring $Q(A)$. We denote it more briefly by $\text{Bez}(A)$.

Remarks 10.15. i) If $A \subset B$ is any Bezout extension, there exists a unique ring homomorphism $\varphi: B \to \text{Bez}(A)$ over A, and φ is injective, as is clear by Proposition I.3.14.
ii) If R is any overring of A in $Q(A)$, then $\text{Bez}(A, R) = R \cap \text{Bez}(A)$.
iii) If $A \subset B$ and $B \subset R$ are ring extensions, the first one being Bezout, then $\text{Bez}(A, R) = \text{Bez}(B, R)$, as follows from Proposition 11 above.
iv) If x is an element of $\text{Bez}(A)$, then $A + Ax = Ay$ with y a unit of $\text{Bez}(A)$, hence $ay = 1$ for some nonzero divisor a of A. This proves that $\text{Bez}(A) \subset \text{Quot} A$. We conclude that $\text{Bez}(A)$ is the Bezout hull of A in $\text{Quot} A$.
v) Of course, also $\text{Bez}(A) = \text{Bez}(A, P(A)) = \text{Bez}(A, M(A))$.
vi) An integral domain A is a Bezout domain iff $\text{Bez}(A) = \text{Quot} A$.
vii) If A is a local ring then $\text{Bez}(A) = P(A)$, as is clear from 10.4 (or from the proof of Prop.5.10).   □

Proposition 10.16. If A is a Bezout extension then $R = S^{-1}A$ with $S := R^* \cap A$.

Proofs. This follows immediately from Proposition 4.15. A more direct proof runs as follows. Let $x \in R$ be given. Then $A + Ax = Ay$ with some $y \in R^*$. Let $s := \frac{1}{y}$. We have $As = (A + Ax)^{-1} \subset A$, hence $s \in A$. Thus $x = \frac{a}{s}$ with some $a \in A$ and $s \in S$.   □

If more generally $A \subset R$ is any ring extension, we may ask for the multiplicative set $A \cap \text{Bez}(A, R)^*$. In particular, in order to understand all the Bezout extensions of A, one should study the set $A \cap \text{Bez}(A)^*$.

Definition 4. We call an ideal I of a ring A a *Bezout ideal*, if it is invertible and A is Bezout in $A[I^{-1}]$. Then $I = As$ with some non-zero-divisor s of A. We call such an element s a *Bezout element of A*, and we denote the set of Bezout elements of A by $\beta(A)$. □

Thus $\beta(A)$ consists of all non-zero-divisors s of A such that A is Bezout in $A\left[\frac{1}{s}\right]$. Clearly $\beta(A) = A \cap \mathrm{Bez}(A)^*$. In particular $\beta(A)$ is a multiplicative subset of A, a submonoid of the monoid of all non-zero-divisors of A. This submonoid is saturated, i.e., if $a \in \beta(A)$ and $b \in A$ is a divisor of a ($a = bc$ with some $c \in A$), then $b \in \beta(A)$, as follows from Proposition 11. It contains the saturated submonoid $\mathrm{sd}(A)$ of strong divisors introduced at the end of §5. If A is local, we have $\mathrm{sd}(A) = \beta(A)$.

If every non-zero-divisor of A is a product of finitely many irreducible elements, for example if A is noetherian, then the abelian monoid $\beta(A)$ is generated by irreducible elements and the group A^*. In this case, in order to understand the Bezout extensions of A, it suffices to know the irreducible Bezout elements of A.

Example 10.17. Assume that A is a Dedekind domain. Then the irreducible Bezout elements of A are the prime elements p of A. {An element p of $A\setminus\{0\}$ is called prime, if $p|ab$ implies $p|a$ or $p|b$, i.e. if the ideal Ap is prime.} Let Ω_1 denote the set of maximal ideals of A which are principal, and let Ω_2 denote the complement $\mathrm{Max}A\setminus\Omega_1$. We have (cf. Prop.1.6) $\mathrm{Quot}A = \sum\limits_{\mathfrak{p}\in\mathrm{Max}A} A[\mathfrak{p}^{-1}]$, $\mathrm{Bez}A = \sum\limits_{\mathfrak{p}\in\Omega_1} A[\mathfrak{p}^{-1}]$, and, in the notation of §7, $\mathrm{Quot}A = \mathrm{Bez}A \times_A C$ with $C := \sum\limits_{\mathfrak{p}\in\Omega_2} A[\mathfrak{p}^{-1}]$. It follows by Proposition 12 that C is Bezout in R. □

If s is a non-zero-divisor of a ring A, then, in order to conclude that s is a Bezout element of A, then it is by no means sufficient that As is a Prüfer ideal, as shows the following

Example 10.18. Let A be a subring of a number field, which is not a principal ideal domain. Then A is a Dedekind domain, hence A is Prüfer in its quotient field F. If I is any ideal $\neq 0$ of A then, as is well known, there exists some $n \in \mathbb{N}$ such that I^n is principal, $I^n = As$. We have $A[I^{-1}] = A[s^{-1}]$. We conclude that every overring of A in F is of the form $S^{-1}A$ with S some multiplicative subset of A. {We have analyzed this situation in Prop.4.15.} But nevertheless A is not Bezout in F. □

In general we have the following characterization of Bezout elements.

Proposition 10.19. Let A be any ring and let s be a non-zero-divisor in A. Then s is a Bezout element of A iff for every $b \in A$ the ideal $As + Ab$ is principal.

Proof. Let $R := A[\frac{1}{s}] \subset \text{Quot} A$. For every $b \in A$ the ideal $As + Ab$ is regular in R, hence R-invertible. Thus, if A is Bezout in R, this ideal is principal.

Assume conversely that $As + Ab$ is principal for every $b \in A$. Given any $x \in R \setminus A$ we verify that $A + Ax$ is principal, and then will know that A is Bezout in R by Theorem 2.

We have $x = \frac{b}{s^n}$ with $n \in \mathbb{N}$ and $b \in A$, and $A + Ax = (As^n + Ab)s^{-n}$. We prove that $As^n + Ab$ is principal for every $n \in \mathbb{N}$ and $b \in A$ by induction on n, and then will be done. The claim holds for $n = 1$ by assumption. Assume the claim for some $n \geq 1$. There is some $d \in A$ with $As + Ab = Ad$. We write $s = td$, $b = ud$ with $t, u \in A$. Then $At + Au = A$, and

$$As^{n+1} + Ab = d(As^n t + Au).$$

By a very well known easy argument the equation $Au + At = A$ implies $Au + Aa = Au + Aat$ for every $a \in A$. Thus

$$As^{n+1} + Ab = d(As^n + Au),$$

and this is principal by induction hypothesis. □

Corollary 10.20. Let $\mathfrak{m}$ be a maximal ideal of a ring A. Then $\mathfrak{m}$ is a Bezout ideal iff $\mathfrak{m}$ can be generated by one element s, and s is not a zero divisor.

Proof. Clearly these conditions are necessary for $\mathfrak{m}$ to be a Bezout ideal (cf. Def.4). Proposition 19 tells us that they are also sufficient. □

Going back to I, §5 we realize that for Bezout extensions we did not yet establish analogues of the permanence properties stated in I.5.8, I.5.9, I.5.20 for Prüfer extensions. Now, the off hand analogue of

I.5.9 for Bezout extensions is clearly false. We will prove below an analogue of I.5.8 for Bezout extension under a restricted hypothesis. Concerning Proposition I.5.20, and also Proposition I.5.21, things are very easy.

Proposition 10.21. Let $(A_i \subset R_i \mid i \in I)$ be a family of ring extensions, and let $A := \prod_{i \in I} A_i$, $R := \prod_{i \in I} R_i$. The extension $A \subset R$ is Bezout iff each extension $A_i \subset R_i$, with i running through I, is Bezout.

Proof. This can be easily verified in a similar way as Prop.I.5.20, using the condition (3) in Theorem 2. $\square$

Proposition 10.22. Let $(A_i \mid i \in I)$ be a family of rings, and $A := \prod_{i \in I} A_i$. Then $\mathrm{Bez}(A) = \prod_{i \in I} \mathrm{Bez}(A_i)$. Also $\beta(A) = \prod_{i \in I} \beta(A_i)$.

Proof. One proceeds as in the proof of Prop.I.5.21. $\square$

Proposition 10.23. Assume that $A \subset R$ is a ring extension and I is an ideal of R contained in A and in the Jacobson radical of R. Then A is Bezout in R iff A/I is Bezout in R/I.

Proof. If A is Bezout in R then it is clear by Proposition 9 that A/I is Bezout in R/I. Assume now that the latter holds. Let $x \in R$ be given. Condition (3) of Theorem 2 for the extension $A/I \subset R/I$ tells us that there exists some $y \in R$ with $A + Ax = Ay + I$. We have an equation $1 = ay + b$ with $a \in A$ and $b \in I$. Since I is contained in the Jacobson radical of R, it follows that $ay \in R^*$, hence $y \in R^*$. Now
$$A + Ax = (A + Iy^{-1})y \subset (A + I)y = Ay,$$
since I is an ideal of R contained in A. Thus $A + Ax = Ay$. Condition (3) of Theorem 2 holds, hence A is Bezout in R. $\square$

By use of Proposition 21 and 23 and the corresponding propositions in I, §5 we dispose of a way to build new Bezout extensions and Prüfer extensions from old ones.

Proposition 10.24. Assume that $(A_i \mid 1 \le i \le n)$ is a finite family of subrings of a ring R. Assume further that there exists a family

$(\mathfrak{a}_i \mid 1 \leq i \leq n)$ of ideals of R such that $\mathfrak{a}_i \subset A_i$ and $\mathfrak{a}_i + \mathfrak{a}_j = R$ for $i \neq j$. Let $A := \bigcap\limits_{i=1}^{n} A_i$ and $\mathfrak{a} := \bigcap\limits_{i=1}^{n} \mathfrak{a}_i$.

a) If A_i is Prüfer in R for each $i \in \{1, \ldots, n\}$, then A is Prüfer in R.
b) If A_i is Bezout in R for each $i \in \{1, \ldots, n\}$, and $\mathfrak{a}$ is contained in the Jacobson radical $\operatorname{rad} R$, then A is Bezout in R.

Proof. 1) The natural map $\varphi \colon R/\mathfrak{a} \to \prod\limits_{i=1}^{n} R_i/\mathfrak{a}_i$ is an isomorphism of rings according to the Chinese remainder theorem. We claim that $\varphi(A/\mathfrak{a}) = \prod\limits_{i=1}^{n} A_i/\mathfrak{a}_i$. In order to prove this, we choose elements $e_1, \ldots, e_n$ of R such that $e_i \equiv 1 \mod \mathfrak{a}_i$ and $e_i \equiv 0 \mod \mathfrak{a}_j$, for $i \neq j$. Given elements $a_i \in A_i$ $(1 \leq i \leq n)$, we have $e_i a_i \equiv a_i \mod \mathfrak{a}_i$ and $e_i a_i \equiv 0 \mod \mathfrak{a}_j$ for $j \neq i$. Then $e_i a_i - a_i \in \mathfrak{a}_i \subset A_i$, hence $e_i a_i \in A_i$, and also $e_i a_i \in A_j$ for $j \neq i$. Thus $e_i a_i \in A$. It follows that

$$\varphi\Big(\sum_{i=1}^{n} e_i a_i + \mathfrak{a}\Big) = (a_i + \mathfrak{a}_i \mid 1 \leq i \leq n).$$

Thus φ maps $A/\mathfrak{a}$ onto $\prod\limits_{i=1}^{n} A_i/\mathfrak{a}_i$.

2) Assume that A_i is Prüfer in R_i for each $i \in \{1, \ldots, n\}$. Then $A_i/\mathfrak{a}$ is Prüfer in $R/\mathfrak{a}$ by Proposition I.5.7, and $\prod\limits_{i=1}^{n} A_i/\mathfrak{a}_i$ is Prüfer in $\prod\limits_{i=1}^{n} R/\mathfrak{a}_i$ by Proposition I.5.20. Applying φ^{-1} we see that $A/\mathfrak{a}$ is Prüfer in $R/\mathfrak{a}$. Now Proposition I.5.8 tells us that A is Prüfer in R.

3) If each A_i is Bezout in R and $\mathfrak{a}$ is contained in $\operatorname{rad} R$, one obtains that A is Bezout in R by fully analogeous arguments, using Propositions 9, 21, 23. $\qquad\square$

Example 10.25. Let $\mathfrak{m}_1, \ldots, \mathfrak{m}_n$ be finitely many maximal ideals of a ring R. Assume that on each field $R/\mathfrak{m}_i$ there is given a valuation $w_i \colon R/\mathfrak{m}_i \to \Gamma_i \cup \infty$. Let v_i denote the associated valuation on R with $\operatorname{supp} v_i = \mathfrak{m}_i$, i.e. the composite of w_i with the residue class map $R \to R/\mathfrak{m}_i$. The intersection $A := \bigcap\limits_{i=1}^{n} A_{v_i}$ of the valuation rings of the v_i is Prüfer in R. If R is semilocal and the $\mathfrak{m}_i$ are all the maximal ideals of R, then A is Bezout in R. In general we can say that $S^{-1}A$ is Bezout in $S^{-1}R$ with $S := A \setminus \bigcup_{i=1}^{n} \mathfrak{m}_i$, since $S^{-1}R$ is semilocal with maximal ideals $S^{-1}\mathfrak{m}_i$ $(1 \leq i \leq n)$ and each v_i extends to a valuation of $S^{-1}R$ in the obvious way (cf. I, §1). $\square$

We now strive for a theory of "Bezout modules" analogous to the theory of Prüfer modules in §9. In the following A is a fixed ring.

Definition 4. We call an *overmodule* M of A (cf. §9) *Bezout*, if for every $x \in M$ there exists some $y \in M$ with $A + Ax = Ay$. We then also say that A is *Bezout in* M.

Proposition 10.26. If an overmodule M of A is Bezout then M is Prüfer.

This follows immediately from Scholium 9.5. Here is a more direct **Second proof.** Let $x \in M$ be given. We have to verify that $(A : x) + (A : x)x = A$ (cf. §9, Def.1). There exists some $y \in M$ with $A + Ax = Ay$. Clearly $(A : x) = (A : y)$. Since $1 \in Ay$, there exists some $a \in A$ with $ay = 1$. In particular $a \in (A : y)$. If $b \in (A : y)$ is given, then $by = c \in A$. Multiplying by a we obtain $b = ca$. This proves that $(A : y) = Aa$, and we have

$$(A : x) + (A : x)x = (A : x)(A + Ax) = Aa \cdot Ay = A. \qquad \square$$

Given a Bezout overmodule M of A we now know by Theorem 9.4 that M embeds into $Q(A)$ over A in a unique way. This in studying Bezout overmodules of A we may restrict to $Q(A)$-overmodules i.e. A-submodules of $Q(A)$ which contain the ring A.

Theorem 10.27. Assume that M is a $Q(A)$-overmodule of A. Then A is Bezout in M iff A is Bezout in the ring $A[M]$ generated by M over A.

Proof. a) If $A \subset A[M]$ is a Bezout extension, then it is clear by Theorem 2 that $A[M]$ is a Bezout overmodule of A. Since M is an A-submodule of $A[M]$ containing A, it follows that M is a Bezout overmodule of A.

b) We assume that A is Bezout in M. Let $R := A[M]$. We verify that $M \subset \text{Bez}(A, R)$. This will imply that $\text{Bez}(A, R) = R$, i.e. $A \subset R$ is Bezout.

Let $x \in M$ be given. There exists some $y \in M$ with $A + Ax = Ay$. We have an element $s \in A$ with $sy = 1$. Then $s \in R^* \cap A$ and $y = s^{-1}$. Let $b \in A$ be given. Then $bs^{-1} \in Ay \subset M$. There exists some $z \in M$ with $A + A\frac{b}{s} = Az$. We have $z = \frac{c}{s}$ with $c \in A$, and we conclude that $As + Ab = Ac$. Proposition 19 now tells us that s is a Bezout element of R, i.e. A is Bezout in $A\left[\frac{1}{s}\right] = A[y]$. It follows that $A + Ax = Ay \subset A[y] \subset \text{Bez}(A, R)$, hence $x \in \text{Bez}(A, R)$. This finishes the proof that $M \subset \text{Bez}(A, R)$.   □

Corollary 10.28. Let M be a Bezout overmodule of A. Then every finitely generated submodule J of M with $A \subset J$ is free of rank one.   □

Theorem 27 gives us the means for a new description of relative Bezout hulls, as follows.

Theorem 10.29. Let $A \subset R$ be any ring extension and x an element of R. The following are equivalent.
(1) A is Bezout in $A[x]$ {i.e. $x \in \text{Bez}(A, R)$}.
(2) For every $a \in A$ the module $A + Aax$ is principal.

Proof. Of course, (1) implies (2). To prove the converse, let $x \in R$ be given obeying (2). Without loss of generality we assume that $R = A[x]$. Let $S := A \cap R^*$. We have $A + Ax = Ay$ with some $y \in R$. In particular, $1 = sy$ with some $s \in A$, hence $s \in S$. It follows that $sx \in A$ and then, that for every $z \in A[x]$ there is some $n \in \mathbb{N}$ such that $s^n z \in A$. We conclude that $R = S^{-1}A \subset \text{Quot}A$.

Now we are in business to apply Theorem 27. If $z \in A + Ax$, we have $z = b + ax$ with some $a, b \in A$, and $A + Az = A + Aax$. This module is principal. Thus the overmodule $A + Ax$ of A is Bezout. Theorem 27 tells us that the extension $A \subset A[x]$ is Bezout.   □

§11 The Prüfer extensions of a noetherian ring

The multiplicative ideal theory developed so far amply suffices to classify the Prüfer extensions of a noetherian ring.

We start with a lemma on local rings.

Lemma 11.1 [J_2, Lemma 1]. Assume that $(A, \mathfrak{m})$ is a local ring[1] and $\bigcap_{n \in \mathbb{N}} \mathfrak{m}^n = \{0\}$. Assume further that I is a distributive ideal of A and $I \neq \{0\}$. Then $I = \mathfrak{m}^n$ for some $n \in \mathbb{N}_0$. If in addition I is invertible (hence Prüfer by Th.5.7) and $I \neq A$, then A is a discrete valuation ring (hence an integral domain).

Proof. [loc.cit]. There exists some $n \in \mathbb{N}_0$ with $I \subset \mathfrak{m}^n$ but $I \not\subset \mathfrak{m}^{n+1}$. Since I is distributive, it follows that $\mathfrak{m}^{n+1} \subset I$ (cf. Prop.5.2.). Suppose that $I \neq \mathfrak{m}^n$. We choose elements $a \in \mathfrak{m}^n \setminus I$ and $b \in I \setminus \mathfrak{m}^{n+1}$. Again using Proposition 5.2 we conclude that $Ab \subset I \subset Aa$, hence $b = ac$ with some $c \in A$. Certainly $c \notin A^*$, since $Ab \neq Aa$. Thus $c \in \mathfrak{m}$. But this implies $b \in \mathfrak{m}^{n+1}$, a contradiction. We have $I = \mathfrak{m}^n$.

Assume in addition that I is also invertible and $I \neq A$. Then $\mathfrak{m}$ is invertible. We have $\mathfrak{m} = At$ with t a nonzero divisor of A (cf. Prop.2.3). If a is any element of $A \setminus \{0\}$, then we have $a \in \mathfrak{m}^r \setminus \mathfrak{m}^{r+1}$ for some $r \in \mathbb{N}_0$, and we conclude that $a = \varepsilon t^r$ with some $\varepsilon \in A^*$. Thus A is a discrete valuation ring. $\qquad\square$

Recall that, if $(A, \mathfrak{m})$ is a noetherian local ring, then $\bigcap_{n \in \mathbb{N}} \mathfrak{m}^n = \{0\}$ (Krull's intersection theorem). Thus Lemma 1 applies to noetherian local rings.

In the following *we assume that A is a noetherian ring.*

Lemma 11.2. Let $\mathfrak{p}$ be a maximal ideal of A. Then $\mathfrak{p}$ is Prüfer iff $\mathfrak{p}$ is dense in A and $A_\mathfrak{p}$ is a discrete valuation ring.

Proof. If $A_\mathfrak{p}$ is a discrete valuation ring then $\mathfrak{p}A_\mathfrak{p}$ is a free $A_\mathfrak{p}$-module of rank one. If $\mathfrak{m}$ is a maximal ideal of A different from $\mathfrak{p}$

---
[1] This means that A is a local ring with maximal ideal $\mathfrak{m}$

then $\mathfrak{p} A_{\mathfrak{m}} = A_{\mathfrak{m}}$. We conclude by Proposition 2.4 that $\mathfrak{p}$ is invertible, and by Proposition 5.2 that $\mathfrak{p}$ is distributive in A. {Condition (5) in Prop. 5.2 holds.} Thus $\mathfrak{p}$ is a Prüfer ideal (cf. Th.5.7).

Assume conversely that $\mathfrak{p}$ is a Prüfer ideal of A. Then $\mathfrak{p} A_{\mathfrak{p}}$ is a Prüfer ideal of $A_{\mathfrak{p}}$ (cf. Remark 4.20). Lemma 1 tells us that $A_{\mathfrak{p}}$ is a discrete valuation ring. $\qquad\square$

Lemma 11.3 [J_2, Th.1]. If I is a distributive ideal of A and $I \neq A$, there exist pairwise different maximal ideals $\mathfrak{m}_1, \ldots, \mathfrak{m}_r$ of A and natural numbers $k_1, \ldots, k_r$ such that

$$I = \mathfrak{m}_1^{k_1} \mathfrak{m}_2^{k_2} \ldots \mathfrak{m}_r^{k_r}.$$

The ideals $\mathfrak{m}_j^{k_j}$, $1 \leq j \leq r$, are again distributive. The set of these ideals is uniquely determined by I.

Proof. Let $I = \bigcap_{j=1}^{r} I_j$ be a minimal primary decomposition of I (which exists since A noetherian). Thus $\sqrt{I_j} = \mathfrak{m}_j$, $1 \leq j \leq r$, with pairwise different prime ideals $\mathfrak{m}_j$ of A. Let $\mathfrak{p}$ be one of the ideals $\mathfrak{m}_j$. The ideal $I_{\mathfrak{p}}$ of $A_{\mathfrak{p}}$ is distributive. Lemma 1 tells us that $I_{\mathfrak{p}} = (\mathfrak{p} A_{\mathfrak{p}})^k$ with some $k \in \mathbb{N}_0$. Since $I \subset \mathfrak{p}$, we have $k \geq 1$.

If $\mathfrak{m}$ is a maximal ideal of A containing $\mathfrak{p}$ then for the same reason $I_{\mathfrak{m}} = (\mathfrak{m} A_{\mathfrak{m}})^l$ with some $l \in \mathbb{N}$. In the ring $A_{\mathfrak{m}}$ we have $\sqrt{I_{\mathfrak{m}}} = \mathfrak{m} A_{\mathfrak{m}}$, and in the localization $A_{\mathfrak{p}}$ we have $\sqrt{I_{\mathfrak{p}}} = \mathfrak{p} A_{\mathfrak{p}}$. This forces $\mathfrak{p} = \mathfrak{m}$. Thus $\mathfrak{p} \in \operatorname{Max} A$.

If $\mathfrak{q} \in \operatorname{Max} A$ is different from $\mathfrak{p}$ then $\mathfrak{p}^k A_{\mathfrak{q}} = A_{\mathfrak{q}}$, hence trivially $\mathfrak{p}^k A_{\mathfrak{q}}$ is distributive in $A_{\mathfrak{q}}$. Also $\mathfrak{p}^k A_{\mathfrak{p}} = I_{\mathfrak{p}}$ is distributive in $A_{\mathfrak{p}}$. We conclude by Proposition 5.2 that $\mathfrak{p}^k$ is distributive in A.

We now have proved that the ideals $\mathfrak{m}_1, \ldots, \mathfrak{m}_r$ from above are maximal in A, that $(I_j)_{\mathfrak{m}_j} = \mathfrak{m}_j^{k_j} A_{\mathfrak{m}_j}$ with some $k_j \in \mathbb{N}$, and that $\mathfrak{m}_j^{k_j}$ is distributive in A ($1 \leq j \leq r$). We have $(I_j)_{\mathfrak{m}} = (\mathfrak{m}_j^{k_j})_{\mathfrak{m}}$ for every $\mathfrak{m} \in \operatorname{Max} A$, hence $I_j = \mathfrak{m}_j^{k_j}$. Also $\mathfrak{m}_i^{k_i} + \mathfrak{m}_j^{k_j} = A$ for $i \neq j$, since $\mathfrak{m}_i, \mathfrak{m}_j$ are different maximal ideals. Thus

$$I = \bigcap_{j=1}^{r} \mathfrak{m}_j^{k_j} = \prod_{j=1}^{r} \mathfrak{m}_j^{k_j}.$$

Assume finally that $\mathfrak{p}_1, \ldots, \mathfrak{p}_s$ are pairwise different maximal ideals of A and that also $I = \prod_{j=1}^{s} \mathfrak{p}_i^{l_j}$ with natural numbers $l_1, \ldots, l_s$. We have $I = \bigcap_{j=1}^{s} \mathfrak{p}_j^{l_j}$, and this is again a primary decomposition of I. By a well known uniqueness result of primary decompositions without "embedded components" (cf. [AM, Prop.4.6]) it follows that $r = s$ and the $\mathfrak{p}_j^{l_j}$ coincide with the $\mathfrak{m}_j^{k_j}$ up to numeration. Actually in the present situation this can be proved very easily by first observing that $\mathfrak{p}_1, \ldots, \mathfrak{p}_s$ are all the maximal ideals of A containing I, and then studying the localizations $A_\mathfrak{m}$ with $\mathfrak{m}$ running through these ideals. $\qquad \square$

Definition 1. Let $\Omega(A)$ denote the set of all dense maximal ideals $\mathfrak{p}$ of A such that $A_\mathfrak{p}$ is a discrete valuation ring.

By Lemma 2 we know that

$$\Omega(A) = \mathrm{Max}(A) \cap \Pi(A)$$

with $\Pi(A)$ the set of Prüfer ideals of A (cf. §4, Def.4).

From Lemma 2 and Lemma 3 we read off the following theorem.

Theorem 11.4. If I is a Prüfer ideal of A and $I \neq A$, there exist finitely many different ideals $\mathfrak{p}_1, \ldots, \mathfrak{p}_r \in \Omega(A)$ and natural numbers $k_1, \ldots, k_r \in \mathbb{N}$ such that

$$I = \mathfrak{p}_1^{k_1} \ldots \mathfrak{p}_r^{k_r}.$$

The $\mathfrak{p}_i$ and k_i are uniquely determined by I. $\qquad \square$

Corollary 11.5. Let I be a product of finitely many ideals $\mathfrak{p}^k$ with $\mathfrak{p} \in \Omega(A)$, $k \in \mathbb{N}_0$. Then every ideal J of A with $I \subset J$ is again such a product.

Proof. Since A noetherian, the ideal J is finitely generated. It follows from Corollary 5.8 that J is Prüfer. Now Theorem 4 gives the claim. $\qquad \square$

Definition 2. If M is a subset of $\Omega(A)$ then let $A(M)$ denote the subring of $Q(A)$ generated by the family of rings $A[\mathfrak{p}^{-1}]$ with $\mathfrak{p}$

running through M. Notice that, if M is finite, then $A(M) = A[I^{-1}]$ with I the product of all $\mathfrak{p} \in M$, and if M is infinite, then $A(M)$ is the union of the sets $A(M')$ with M' running through the finite subsets of M. □

Theorem 11.6. i) The ring extensions $A \subset A(M)$ with $M \subset \Omega(A)$ are precisely all Prüfer subextensions of $A \subset Q(A)$. In particular $P(A) = A(\Omega(A))$.

ii) If $M \subset \Omega(A)$ is given, then M is the set of all $A(M)$-regular prime ideals of A, in short, $M = Y(A(M)/A)$. The subsemigroups $\langle M \rangle$ of $\Pi(A)$ generated by M is the set of all $A(M)$-invertible ideals of A, in short,

$$\text{Inv}(A, A(M)) = \langle M \rangle.$$

Proof. 1) Let $M \subset \Omega(A)$ be given. The ideals in $\langle M \rangle$ are all Prüfer. Thus $A(M)$ is a Prüfer extension of A. Clearly $\langle M \rangle \subset \text{Inv}(A, A(M))$.

2) Let now $A \subset R$ be a Prüfer subextension of $A \subset Q(A)$, and $N := Y(R/A)$. If $\mathfrak{p} \in N$, then $\mathfrak{p}$ is a finitely generated prime ideal of A, since A is noetherian, and further $\mathfrak{p}R = R$. Thus $\mathfrak{p}$ is R-invertible by Theorem 1.13. This means that $\mathfrak{p}$ is invertible in $Q(A)$ and $A[\mathfrak{p}^{-1}] \subset R$. It follows that the extension $A \subset A[\mathfrak{p}^{-1}]$ is Prüfer, i.e. $\mathfrak{p}$ is Prüfer. Theorem 4 now tells us that $\mathfrak{p} \in \Omega(A)$.

This proves that $N \subset \Omega(A)$ and $A(N) \subset R$. Moreover, $N \subset Y(A(N)/A)$. But $Y(A(N)/A) \subset Y(R/A) = N$. Thus $N = Y(A(N)/A) = Y(R/A)$, and we conclude by Theorem 3.3 that $R = A(N)$. {Recall that for any ws extension $A \subset B$ the set $X(B/A)$ is the complement of $Y(B/A)$ in Spec A.}

3) Let $R = A(M)$ for some $M \subset A$. Then $M \subset N := Y(R/A)$, and $R = A(N)$, as we have proved. If $\mathfrak{p} \in N$ is given there exists finitely many elements $\mathfrak{p}_1, \ldots, \mathfrak{p}_r$ of M and some $n \in \mathbb{N}$ with $\mathfrak{p}^{-1} \subset (\mathfrak{p}_1 \ldots \mathfrak{p}_r)^{-n}$, since $\mathfrak{p}^{-1}$ is finitely generated. Thus $(\mathfrak{p}_1 \ldots \mathfrak{p}_r)^n \subset \mathfrak{p}$. It follows that $\mathfrak{p}$ is one of the ideals $\mathfrak{p}_1, \ldots, \mathfrak{p}_r$. We have $M = N$. □

Theorems 4 and 6 imply

Corollary 11.7. If $A \subset R$ is a Prüfer extension, then the group $D(A, R)$ of R-invertible A-submodules of R is a free abelian group with basis $Y(R/A)$. □

We still have to describe $D(A, R)$ as a lattice ordered group. This is easily done. We first consider the case $R = P(A)$.

Let $D(A) := D(A, P(A))$. This is an abelian group with basis $\Omega(A)$. If $I \in D(A)$ and $\mathfrak{p} \in \Omega(A)$ we have $I A_\mathfrak{p} = \mathfrak{p}^n A_\mathfrak{p}$ with some $n \in \mathbb{Z}$. We denote this number n by $v_\mathfrak{p}(I)$. We thus obtain a homomorphism

$$v_\mathfrak{p} : D(A) \longrightarrow \mathbb{Z}$$

compatible with the ordering of $D(A)$ opposite to the inclusion relation and the natural ordering of $\mathbb{Z}$, i.e.

$$I \subset J \Longrightarrow v_\mathfrak{p}(J) \leq v_\mathfrak{p}(I).$$

From the theorems 4 and 6 and their corollaries, and from the definition of the homomorphisms $v_\mathfrak{p}$, one now reads off the following facts.

Scholium 11.8.
  i) For every $I \in D(A)$ we have $v_\mathfrak{p}(I) \neq 0$ for only finitely many $\mathfrak{p} \in \Omega(A)$, and
$$I = \prod_{\mathfrak{p} \in \Omega(A)} \mathfrak{p}^{v_\mathfrak{p}(I)}.$$
 ii) For any $I, J \in D(A)$ we have $I \subset J$ iff $v_\mathfrak{p}(J) \leq v_\mathfrak{p}(I)$ for every $\mathfrak{p} \in \Omega(A)$.
iii) For any $I, J \in D(A)$ and $\mathfrak{p} \in \Omega(A)$ we have
$$v_\mathfrak{p}(I + J) = \min(v_\mathfrak{p}(I), v_\mathfrak{p}(J)),$$
$$v_\mathfrak{p}(I \cap J) = \max(v_\mathfrak{p}(I), v_\mathfrak{p}(J)),$$
and, of course,
$$v_\mathfrak{p}(IJ) = v_\mathfrak{p}(I) + v_\mathfrak{p}(J).$$
iv) Thus the lattice ordered group $D(A)$ is isomorphic to the direct sum $\mathbb{Z}^{(\Omega(A))}$ of copies of $\mathbb{Z}$ indexed by $\Omega(A)$.
 v) We also conclude that, for every $I \in D(A)$, $\mathfrak{p} \in \Omega(A)$,
$$v_\mathfrak{p}(I^{-1}) = -v_\mathfrak{p}(I).$$
vi) If $A \subset R$ is any Prüfer subextension of $A \subset P(A)$, then $D(A, R)$ is a lattice ordered subgroup of $D(A)$ and
$$D(A, R) = \{I \in D(A) \mid v_\mathfrak{p}(I) = 0 \quad \text{for every} \quad \mathfrak{p} \in \Omega(A) \backslash Y(R/A)\}.$$

$\square$

Remark 11.9. If I and J are invertible ideals of A, then we can check by the formulas in 11.8.iii that $I^{-1}J^{-1} \subset I^{-2} + J^{-2}$. This gives us again the fact, observed already in Proposition 1.6, that for any overrings B and C of A we have $BC = B + C$. It follows that, for any $M \subset \Omega(A)$,

$$A(M) = \sum_{\mathfrak{p} \in M} A[\mathfrak{p}^{-1}].$$

$\square$

Proposition 11.10. Let M and N be subsets of $\Omega(A)$. Then $A(M \cap N) = A(M) \cap A(N)$ and $A(M)A(N) = A(M) + A(N) = A(M \cup N)$.

Proof. This follows immediately from Theorem 6 (and the remark just made), since by that theorem every overring B of A in $P(A)$ is of the form $B = A(L)$ with some $L \subset \Omega(A)$, uniquely determined by B, and, for any subsets L, L' of $\Omega(A)$, $A(L) \subset A(L')$ iff $L \subset L'$. $\square$

Let $A \subset R$ be any subextension of $A \subset P(A)$, hence $R = A(X)$ with some set $X \subset \Omega(A)$. Let further $A \subset B$ a subextension of $A \subset R$, hence $B = A(M)$ with $M \subset X$. We want to determine the polar B° of B in R (cf. §7). If N is a second subset of X then Proposition 10 tells us that $A(M) \cap A(N) = A$ iff $M \cap N = \emptyset$. The largest such subset N is $X \setminus M$, and we have $A(M)A(X \setminus M) = R$, again by Proposition 10. Thus we have proved

Proposition 11.11. Let $M \subset X \subset \Omega(A)$. The polar $A(M)^\circ$ of $A(M)$ in $R := A(X)$ is $A(X \setminus M)$, and $A(X) = A(M) \times_A A(X \setminus M)$. (Recall §7, Def.3.) Thus every overring of A in R is a factor of R over A. $\square$

Corollary 11.12. The irreducible Prüfer extensions of A in $P(A)$ are precisely the rings $A[\mathfrak{p}^{-1}]$ with $\mathfrak{p}$ running through $\Omega(A)$. $\square$

Although in this chapter we usually avoid working with valuations, we now look for a relation between the set $\Omega(A)$ and the set of Manis valuations v on $P(A)$ with $A_v \supset A$. More generally we are interested in a description of the Manis valuations v on R for any Prüfer extension $A \subset R$ with $A_v \supset A$.

Definition 3. If $A \subset R$ is a Prüfer extension then $\Omega(R/A)$ denotes the set of all $\mathfrak{p} \in \Omega(A)$ with $\mathfrak{p}R = R$, i.e.

$$\Omega(R/A) = \Omega(A) \cap Y(R/A),$$

while $\Omega_R(A)$ denotes the set of all $\mathfrak{p} \in \Omega(A)$ with $\mathfrak{p}R \neq R$, i.e.

$$\Omega_R(A) = \Omega(A) \cap X(R/A). \qquad \square$$

Notice that $\Omega(A)$ is the disjoint union of these two sets (cf.§3). Notice also that for $R = P(A)$ we have $\Omega(R/A) = \Omega(A)$, $\Omega_R(A) = \emptyset$, while for $R = A$ we have $\Omega(R/A) = \emptyset$, $\Omega_R(A) = \Omega(A)$.

Theorem 11.13. Let $A \subset R$ be Prüfer.

i) The nontrivial Manis valuations v on R with $A_v \supset A$ correspond, up to equivalence, bijectively with the prime ideals $\mathfrak{p} \in \Omega(R/A)$ via $\mathfrak{p} = \mathfrak{p}_v \cap A$, $(A_v, \mathfrak{p}_v) = (A_{[\mathfrak{p}]}, \mathfrak{p}_{[\mathfrak{p}]})$. They all have the value group $\mathbb{Z}$.
ii) $\Omega(R/A) = Y(R/A)$.
iii) If $\mathfrak{p} \in \Omega(R/A)$ then $R_\mathfrak{p} = \operatorname{Quot}(A_\mathfrak{p})$.

Proof. a) We know from the definition of Prüfer extensions (I, §5) that the Manis valuations v on R with $A_v \supset A$ are given by $(A_v, \mathfrak{p}_v) = [A_{[\mathfrak{p}]}, \mathfrak{p}_{[\mathfrak{p}]}]$ with $\mathfrak{p}$ running through $\operatorname{Spec}(A) = X(R/A) \cup Y(R/A)$. If $\mathfrak{p} \in X(R/A)$, i.e. $\mathfrak{p}R \neq R$, then $A_{[\mathfrak{p}]} = R$ (cf.Th.I.3.13, there with $B = R$), thus v is trivial.
b) Let now $\mathfrak{p} \in Y(R/A)$, i.e. $R\mathfrak{p} = R$. The ideal $\mathfrak{p}$ is finitely generated since A is noetherian. Thus $\mathfrak{p}$ is R-invertible (cf.Th.1.13). We have $A[\mathfrak{p}^{-1}] \subset R$. In particular $\mathfrak{p}$ is a Prüfer ideal. Theorem 4 tells us that $\mathfrak{p} \in \Omega(A)$, hence $\mathfrak{p} \in \Omega(A) \cap Y(R/A) = \Omega(R/A)$.
c) If $\mathfrak{p} \in \Omega(R/A)$ then $A_\mathfrak{p} \subset R_\mathfrak{p}$ is a Prüfer extension (cf.e.g.Th.I.5.10), and certainly $A_\mathfrak{p} \neq R_\mathfrak{p}$. Since $A_\mathfrak{p}$ is a discrete valuation ring, it follows that $R_\mathfrak{p} = \operatorname{Quot}(A_\mathfrak{p})$. We have a discrete valuation $\tilde{v}: R_\mathfrak{p} \longrightarrow \mathbb{Z} \cup \infty$ with $(A_{\tilde{v}}, \mathfrak{p}_{\tilde{v}}) = (A_\mathfrak{p}, \mathfrak{p}A_\mathfrak{p})$. This gives us a valuation $v: R \to \mathbb{Z} \cup \infty$ with $(A_v, \mathfrak{p}_v) = (A_{[\mathfrak{p}]}, \mathfrak{p}_{[\mathfrak{p}]})$. Thus v is the Manis valuation on R with $A_v \supset A$ corresponding to $\mathfrak{p}$. Since v is nontrivial, we have $v(R) = \mathbb{Z} \cup \infty$. $\qquad \square$

Definition 4. Let $\mathfrak{p} \in \Omega(A)$. We denote the Manis valuation on $P(A)$ corresponding to $\mathfrak{p}$ by the theorem (there with $R = P(A)$) by

$v_\mathfrak{p}$, and the valuation induced by $v_\mathfrak{p}$ on the quotient field $P(A)_\mathfrak{p}$ of $A_\mathfrak{p}$ by $\tilde{v}_\mathfrak{p}$. (This is in harmony with I, §1, Def.6.)

Corollary 11.14. If $A \subset R$ is a subextension of $A \subset P(A)$, then, for any $\mathfrak{p} \in \Omega(A)$, the valuation $v_\mathfrak{p}|R$ is non trivial iff $\mathfrak{p}R = R$, i.e. $\mathfrak{p} \in \Omega(R/A)$. In this case $v_\mathfrak{p}|R$ is the Manis valuation v on R with $A_v \supset A$ corresponding to $\mathfrak{p}$, as described in Theorem 13, and $R_\mathfrak{p} = P(A)_\mathfrak{p} = P(A_\mathfrak{p})$.

Proof. $v := v_\mathfrak{p}|R$ is a Manis valuation on R with $A_v = (A_{[\mathfrak{p}]}^{P(A)}) \cap R = A_{[\mathfrak{p}]}^R$, $\mathfrak{p}_v = (\mathfrak{p}_{[\mathfrak{p}]}^{P(A)}) \cap R = \mathfrak{p}_{[\mathfrak{p}]}^R$.*) If $\mathfrak{p}R \neq R$, then $A_{[\mathfrak{p}]}^R = R$, thus v is trivial. If $\mathfrak{p}R = R$ then $A_{[\mathfrak{p}]}^R \neq R$, as we know from Theorem 13. Now v is non trivial and is the Manis valuation on R corresponding to $\mathfrak{p}$. The ring $R_\mathfrak{p}$ coincides with $P(A)_\mathfrak{p}$, since both rings are the quotient field of $A_\mathfrak{p}$ by part iii of Theorem 13. We also have $\mathrm{Quot}(A_\mathfrak{p}) = P(A_\mathfrak{p})$. $\square$

Proposition 11.15. Let R be an overring of A in $P(A)$.

i) The Manis valuations v on $P(A)$ with $A_v \supset R$ are (up to equivalence) precisely the valuations $v_\mathfrak{p}$ with $\mathfrak{p} \in \Omega_R(A)$.

ii) We have a bijection $\Omega(R) \dashrightarrow \Omega_R(A)$, which maps $\mathfrak{q} \in \Omega(R)$ to $\mathfrak{q} \cap A$.

iii) If $\mathfrak{q} \in \Omega(R)$, $\mathfrak{p} = \mathfrak{q} \cap A$, then $A_\mathfrak{p} = R_\mathfrak{q}$, $A_{[\mathfrak{p}]}^{P(A)} = R_{[\mathfrak{q}]}^{P(A)}$, and $v_\mathfrak{p} = v_\mathfrak{q}$.

Proof. If v is a Manis valuation on $P(A)$ with $A_v \supset R$ then $A_v \supset A$, hence $v = v_\mathfrak{p}$ for some $\mathfrak{p} \in \Omega(A)$. We have $\mathfrak{p} = A \cap \mathfrak{p}_v$. Also $v = v_\mathfrak{q}$ with $\mathfrak{q} := R \cap \mathfrak{p}_v$. This implies $\mathfrak{p} = \mathfrak{q} \cap A$. It follows that $\mathfrak{q} = R\mathfrak{p}$ (cf.Prop.I.4.6). In particular $\mathfrak{p} \in \Omega_R(A)$.

If we start with some $\mathfrak{p} \in \Omega_R(A)$, then $\mathfrak{q} := \mathfrak{p}R \in \mathrm{Spec}\,(R)$, and $\mathfrak{q} \cap A = \mathfrak{p}$. Let $v = v_\mathfrak{p}$. We have $A_v = A_{[\mathfrak{p}]}^{P(A)}$, and this ring coincides with $R_{[\mathfrak{q}]}^{P(A)}$ (cf.Th.I.3.13). In particular $A_v \supset R$. We have $(\mathfrak{p}_v \cap R) \cap A = \mathfrak{p}$, which implies $\mathfrak{p}_v \cap R = \mathfrak{q}$. Thus $\mathfrak{q} \in \Omega(R)$ and $v = v_\mathfrak{q}$. Since

---

*) We indicate by the superscripts $P(A)$, respectively R, in which ring extension the pair $(A_{[\mathfrak{p}]}, \mathfrak{p}_{[\mathfrak{p}]})$ has been taken, cf. the footnote in I, §1.

$A \subset R$ is ws, the natural map $A_{\mathfrak{p}} \to R_{\mathfrak{q}}$ is surjective. Now both rings are discrete valuation rings. It follows that $A_{\mathfrak{p}} = R_{\mathfrak{q}}$. □

For $I \in D(A)$, i.e. I a $P(A)$-invertible A-submodule of $P(A)$, and $\mathfrak{p} \in \Omega(A)$ we previously defined a number $v_{\mathfrak{p}}(I) \in \mathbb{Z}$ without speaking about valuations. Since $I_{\mathfrak{p}} \in D(A_{\mathfrak{p}})$, we also have such a number related to the Prüfer extension $A_{\mathfrak{p}} \subset P(A)_{\mathfrak{p}} = P(A_{\mathfrak{p}})$, which we denote by $\tilde{v}_{\mathfrak{p}}(I_{\mathfrak{p}})$. We now relate these numbers to the valuations $v_{\mathfrak{p}}$ and $\tilde{v}_{\mathfrak{p}}$.

Proposition 11.16. Let $\mathfrak{p} \in \Omega(A)$ and $I \in D(A)$. Then $v_{\mathfrak{p}}(I) = v_{\tilde{\mathfrak{p}}}(I_{\mathfrak{p}})$, and $v_{\mathfrak{p}}(I) = \min\{v_{\mathfrak{p}}(x) \mid x \in I\} = \min\{\tilde{v}_{\mathfrak{p}}(y) \mid y \in I_{\mathfrak{p}}\}$.

Proof. It is evident that $\min\{v_{\mathfrak{p}}(x) \mid x \in I\} = \min\{\tilde{v}_{\mathfrak{p}}(y) \mid y \in I_{\mathfrak{p}}\}$. Let us denote this number by n. We have $I_{\mathfrak{p}} = tA_{\mathfrak{p}}$ with some $t \in A_{\mathfrak{p}}$. Clearly $n = \tilde{v}_{\mathfrak{p}}(t)$. We also have $\mathfrak{p}A_{\mathfrak{p}} = \pi A_{\mathfrak{p}}$ with $\tilde{v}_{\mathfrak{p}}(\pi) = 1$. Now, by definition, the number $m := v_{\mathfrak{p}}(I)$ is given by $I_{\mathfrak{p}} = \mathfrak{p}^m A_{\mathfrak{p}} = (\mathfrak{p}A_{\mathfrak{p}})^m = \pi^m A_{\mathfrak{p}}$. Thus $m = n$. □

Scholium 11.17. As before let R be an overring of A in $P(A)$. By the general theory, say, in §6, we have a surjective homomorphism $D(A) \to D(R)$, $I \mapsto IR$, from the group $D(A)$ of $P(A)$-invertible A-submodules of $P(A)$ to the group $D(R)$ of $P(A)$-invertible R-submodules of $P(A) = P(R)$. The kernel of this homomorphisms is the group $D(A, R)$ of R-invertible A-submodules of R.

Let $I \in D(A)$. One easily deduces from the theory developed so far that $v_{\mathfrak{p}}(I) = v_{\mathfrak{p}R}(IR)$ for every $\mathfrak{p} \in \Omega_R(A)$, and that $I \in D(A, R)$ iff $v_{\mathfrak{p}}(I) \geq 0$ for every $\mathfrak{p} \in \Omega_R(A)$. Also $I \subset R$ iff $v_{\mathfrak{p}}(I) \geq 0$ for every $\mathfrak{p} \in \Omega_R(A)$. □

We may summarize that, up to now, we have obtained a picture completely analogous to a large part of the classical theory of Dedekind domains. {N.B. This is the special case where A is a noetherian domain and $P(A) = \mathrm{Quot}(A)$.}

We return to the theory of overrings of A in $P(A)$. Let X be a subset of $\Omega(A)$ and $R := A(X)$. We pick an element $\mathfrak{p} \in X$ and then have a unique subset M of X such that $A_{[\mathfrak{p}]}^R = A(M)$. We want to determine this set M.

Proposition 11.18. For every $\mathfrak{p} \in X$ we have $A^R_{[\mathfrak{p}]} = A(X \setminus \{\mathfrak{p}\})$. The rings $A^R_{[\mathfrak{p}]}$ with $\mathfrak{p}$ running through X are precisely all maximal proper subrings of R containing A. For every $\mathfrak{p} \in X$

$$R = A[\mathfrak{p}^{-1}] \times_A A^R_{[\mathfrak{p}]}.$$

Proof. Let $\mathfrak{p} \in X$ be given and $v := v_{\mathfrak{p}}|R$. Then $A_v = A^R_{[\mathfrak{p}]}$. For a prime ideal $\mathfrak{q} \in X$ we have $A[\mathfrak{q}^{-1}] \subset A^R_{[\mathfrak{p}]}$ iff $v(A[\mathfrak{q}^{-1}]) \geq 0$ iff $v(\mathfrak{q}^{-1}) \geq 0$. But anyway $v(\mathfrak{q}^{-1}) = -v(\mathfrak{q}) \leq 0$. Thus $A[\mathfrak{q}^{-1}] \subset A^R_{[\mathfrak{p}]}$ iff $v(\mathfrak{q}) = 0$. We have $\mathfrak{p}_v \cap A = \mathfrak{p}$. Thus $v(\mathfrak{q}) = 0$ iff $\mathfrak{q} \not\subset \mathfrak{p}$ iff $\mathfrak{q} \neq \mathfrak{p}$. From the theory above (cf. Remark 9) it is now evident that $A^R_{[\mathfrak{p}]} = A(X \setminus \{\mathfrak{p}\})$. The other statements in the proposition now are obvious by Theorem 6 and Proposition 11. □

Up to now we did not make use of the contents of §10 on Bezout extensions. There remains the task to describe the Bezout hull $\mathrm{Bez}(A)$ of a given noetherian ring A in the present framework. This is indeed possible. The following proposition in essence is a straightforward generalization of the observations in 10.17 about Bezout extensions of Dedekind domains.

Let Ω_1 denote the set of *principal* ideals $\mathfrak{p} \in \Omega(A)$, and let Ω_2 denote the complement $\Omega(A) \setminus \Omega_1$. {N.B. We know by Corollary 10.20 that the $\mathfrak{p} \in \Omega_1$ are just all maximal ideals of A which are principal and dense in A.}

Proposition 11.19. The Bezout ideals I of A with $I \neq A$ are the products $\mathfrak{p}_1^{k_1} \ldots \mathfrak{p}_r^{k_r}$ of powers of finitely many elements $\mathfrak{p}_1 \ldots \mathfrak{p}_r$ of Ω_1. The Bezout hull of A is

$$\mathrm{Bez}(A) = A(\Omega_1) = \sum_{\mathfrak{p} \in \Omega_1} A(\mathfrak{p}^{-1}).$$

Further $P(A) = \mathrm{Bez}(A) \times_A C$ with $C := A(\Omega_2) = \sum_{\mathfrak{p} \in \Omega_2} A(\mathfrak{p}^{-1})$, and also $P(A) = \mathrm{Bez}(C)$. □

The proof is by now easy and can safely be left to the reader.

§12 Invertible hulls for modules over noetherian rings

As before we assume that A is a noetherian ring. We retain the notations $\Omega(A)$, $\Omega(R/A)$, $\Omega_R(A)$ etc. from §11.

We now study A-submodules of $P(A)$ which no longer have to be $P(A)$-invertible. But we always assume that these modules are finitely generated. Notice that this is not a restriction of generality for a module contained in A (= ideal of A), since A is noetherian. We will exhibit a class of such modules, called "amenable modules", such that every module in this class has an "invertible hull" $\mathfrak{a}^*$. This is an $P(A)$-invertible A-module containing $\mathfrak{a}$ and closely related to $\mathfrak{a}$, see below. For technical reasons we will have to work more generally in an arbitrary Prüfer extension $A \subset R$ instead of the maximal one $A \subset P(A)$.

Definition 1. If $\mathfrak{a}$ is a finitely generated A-submodule of $P(A)$ and $\mathfrak{p} \in \Omega(A)$ we define

$$v_\mathfrak{p}(\mathfrak{a}) := \min\{v_\mathfrak{p}(x) \mid x \in \mathfrak{a}\} \in \mathbb{Z} \cup \infty. \qquad \square$$

Notice that this minimum exists: If $\mathfrak{a} = Ax_1 + \cdots + Ax_r$, then

$$v_\mathfrak{p}(\mathfrak{a}) = \min\{v_\mathfrak{p}(x_i) \mid 1 \le i \le r\}.$$

Notice also, that, if $\mathfrak{a}$ is $P(A)$-invertible, then this number $v_\mathfrak{p}(\mathfrak{a})$ is the same as the number $v_\mathfrak{p}(\mathfrak{a})$ defined in §11, due to Proposition 11.16. In particular then $v_\mathfrak{p}(\mathfrak{a}) \in \mathbb{Z}$.

Remark 12.1. If $\mathfrak{p} \in \Omega(A)$ then, for every finitely generated $A_\mathfrak{p}$-submodule $\mathfrak{b}$ of $P(A_\mathfrak{p}) = \mathrm{Quot}(A_\mathfrak{p})$, we have a number $\tilde{v}_\mathfrak{p}(\mathfrak{b}) \in \mathbb{Z} \cup \infty$, applying Definition 1 to $\mathfrak{p}A_\mathfrak{p} \in \Omega(A_\mathfrak{p})$. Clearly $\tilde{v}_\mathfrak{p}(\mathfrak{b}) = n < \infty$ iff $\mathfrak{b} = \mathfrak{p}^n A_\mathfrak{p}$, and $\tilde{v}_\mathfrak{p}(\mathfrak{b}) = \infty$ iff $\mathfrak{b} = 0$. It is also evident that, for $\mathfrak{a}$ as above, we have $v_\mathfrak{p}(\mathfrak{a}) = \tilde{v}_\mathfrak{p}(\mathfrak{a})$.

Proposition 12.2. Let $\mathfrak{a}$ be any finitely generated A-submodule of $P(A)$.

  i) There exist only finitely many $\mathfrak{p} \in \Omega(A)$ with $v_\mathfrak{p}(\mathfrak{a}) < 0$.

ii) $\mathfrak{a} \subset A$ iff $v_\mathfrak{p}(\mathfrak{a}) \geq 0$ for every $\mathfrak{p} \in \Omega(A)$.

iii) If R is an overring of A in $P(A)$ then $\mathfrak{a} \subset R$ iff $v_\mathfrak{p}(\mathfrak{a}) \geq 0$ for every $\mathfrak{p} \in \Omega_R(A)$.

Proof. i): $A + \mathfrak{a}$ is a finitely generated $P(A)$-regular A-submodule of $P(A)$, hence a $P(A)$-invertible module. Thus $v_\mathfrak{p}(A + \mathfrak{a}) \neq 0$ for only finitely many $\mathfrak{p} \in \Omega(A)$. Since $v_\mathfrak{p}(A + \mathfrak{a}) = \min(v_\mathfrak{p}(A), v_\mathfrak{p}(\mathfrak{a})) = \min(0, v_\mathfrak{p}(\mathfrak{a}))$, this gives the first claim.

ii): We know from I, §5 (Remark I.5.5) that A is the intersection of the rings $A_{[\mathfrak{p}]}^{P(A)}$ with $\mathfrak{p}$ running through $\mathrm{Max}(A)$. If $\mathfrak{p} \notin \Omega(A)$ we have $A_{[\mathfrak{p}]}^{P(A)} = P(A)$, while, if $\mathfrak{p} \in \Omega(A)$, we have $A_{[\mathfrak{p}]}^{P(A)} = A_{v_\mathfrak{p}}$. This gives the second claim.

iii) We have a bijection $\Omega_R(A) \xrightarrow{\sim} \Omega(R)$ which sends $\mathfrak{p} \in \Omega_R(A)$ to $\mathfrak{p}R$. Moreover the valuations $v_\mathfrak{p}$ and $v_{\mathfrak{p}R}$ on $P(A)$ are the same (cf.Prop.11.15). Finally $v_\mathfrak{p}(\mathfrak{a}) = v_{\mathfrak{p}R}(\mathfrak{a}R)$. Thus the third claim iii) follows from the second one. $\qquad\square$

Given a Prüfer extension $A \subset R$ we now strive for a description of the class of all finitely generated A-modules $\mathfrak{a} \subset R$ such that $v_\mathfrak{p}(\mathfrak{a}) < \infty$ for every $\mathfrak{p} \in \Omega(R/A)$ and $v_\mathfrak{p}(\mathfrak{a}) \neq 0$ for only finitely many $\mathfrak{p} \in \Omega(R/A)$. Notice that the $P(A)$-invertible A-modules belong to this class.

Definition 2. For every $\mathfrak{p} \in \Omega(A)$ we define a new prime ideal $\tilde{\mathfrak{p}}$ by

$$\tilde{\mathfrak{p}} := \bigcap_{n \in \mathbb{N}} \mathfrak{p}^n = A \cap \mathrm{supp}\, v_\mathfrak{p}. \qquad\square$$

Remarks 12.3. i) Of course, $\tilde{\mathfrak{p}} \subsetneq \mathfrak{p}$.

ii) The valuation $v_\mathfrak{p}|A \colon A \to \mathbb{Z} \cup \infty$ is the composite of the localisation map $A \to A_\mathfrak{p}$ with the valuation $\tilde{v}_\mathfrak{p}|A_\mathfrak{p} \colon A_\mathfrak{p} \to \mathbb{Z} \cup \infty$. Since $A_\mathfrak{p}$ is a discrete valuation ring, it follows that $\tilde{\mathfrak{p}} = ker(A \to A_\mathfrak{p})$. In particular $\tilde{\mathfrak{p}} A_\mathfrak{p} = 0$. If A is an integral domain, this forces $\tilde{\mathfrak{p}} = 0$.

iii) The support of the valuation $v_\mathfrak{p}$ is the ideal $\tilde{\mathfrak{p}}_{[\mathfrak{p}]}$ of $A_{[\mathfrak{p}]}$, and – for the same reason as before – $\tilde{\mathfrak{p}}_{[\mathfrak{p}]}$ is the kernel of the natural map $A_{[\mathfrak{p}]} \to A_\mathfrak{p}$.

iv) If $\mathfrak{q}$ is a prime ideal of A with $\mathfrak{q} \subsetneq \mathfrak{p}$, then $\mathfrak{q} = \tilde{\mathfrak{p}}$. This follows from the fact that $\dim A_\mathfrak{p} = 1$. Indeed, we have $\mathfrak{q}A_\mathfrak{p} \subsetneq \mathfrak{p}A_\mathfrak{p}$, hence $\mathfrak{q}A_\mathfrak{p} = 0 = \tilde{\mathfrak{p}}A_\mathfrak{p}$. Taking preimages in A we obtain $\mathfrak{q} = \tilde{\mathfrak{p}}$. $\qquad\square$

We fix a Prüfer extension $A \subset R$.

Definition 3. i) We call an A-submodule $\mathfrak{a}$ of R *amenable in R,* (or *R-amenable*) if $\mathfrak{a}$ is finitely generated and $\mathfrak{a} \not\subset \tilde{\mathfrak{p}}$ for every $\mathfrak{p} \in \Omega(R/A)$, i.e. $v_\mathfrak{p}(\mathfrak{a}) < \infty$ for every $\mathfrak{p} \in \Omega(R/A)$.
ii) In the case $R = P(A)$ we call an R-amenable module more briefly *"amenable"*. $\qquad\square$

Example 12.4. If A is an integral domain, then $\tilde{\mathfrak{p}} = 0$ for every $\mathfrak{p} \in \Omega(A)$. Thus, if $\mathfrak{a} \subset R$ is finitely generated and $\mathfrak{a} \neq 0$, then $\mathfrak{a}$ is amenable in R. $\qquad\square$

Proposition 12.5. If $\mathfrak{a}$ is amenable in R, then $v_\mathfrak{p}(\mathfrak{a}) \neq 0$ only for finitely many prime ideals $\mathfrak{p} \in \Omega(R/A)$.

Proof. We know from Proposition 2 that $v_\mathfrak{p}(\mathfrak{a}) < 0$ for only finitely many $\mathfrak{p} \in \Omega(A)$. In order to prove that $v_\mathfrak{p}(\mathfrak{a}) > 0$ for only finitely many $\mathfrak{p} \in \Omega(R/A)$, we may replace $\mathfrak{a}$ by $\mathfrak{a} \cap A$, and now assume that $\mathfrak{a} \subset A$.

If $\mathfrak{p} \in \Omega(R/A)$ and $v_\mathfrak{p}(\mathfrak{a}) > 0$ then $\mathfrak{a} \subset \mathfrak{p}$. But $\mathfrak{a} \not\subset \tilde{\mathfrak{p}}$ since $\mathfrak{a}$ is amenable. Thus $\mathfrak{p}$ is minimal among the prime ideals containing $\mathfrak{a}$. Since $A/\mathfrak{a}$ is noetherian, there exist only finitely many such $\mathfrak{p}$. $\qquad\square$

Definition 4. For every R-amenable A-submodule $\mathfrak{a}$ of R we define an invertible A-submodule $I_R(\mathfrak{a})$ by

$$I_R(\mathfrak{a}) := \prod_{\mathfrak{p} \in \Omega(R/A)} \mathfrak{p}^{v_\mathfrak{p}(\mathfrak{a})}.$$

Notice that this makes sense by the preceding proposition. We call $I_R(\mathfrak{a})$ the *R-invertible hull* of $\mathfrak{a}$. This terminology will be justified by Theorem 8 below. $\qquad\square$

Convention. If there is no doubt, which Prüfer extension $A \subset R$ is under consideration, then we usually write more briefly $\mathfrak{a}^*$ instead of $I_R(\mathfrak{a})$. This we also do now.

Remarks 12.6. i) If $\mathfrak{a}$ is R-invertible, i.e. $\mathfrak{a} \in D(A, R)$, then Scholium 11.8 (cf. there i) and vi)) tells us that $\mathfrak{a}^* = \mathfrak{a}$. Conversely, if $\mathfrak{a} = \mathfrak{a}^*$ then $\mathfrak{a} \in D(A, R)$.

ii) If $\mathfrak{a}$ and $\mathfrak{b}$ are finitely generated A-submodules of R, then clearly the statements ii), iii) in Scholium 11.8 remain true for $\mathfrak{a}, \mathfrak{b}$ instead of I, J there. Thus, if $\mathfrak{a}$ and $\mathfrak{b}$ are R-amenable, we have the following facts:

a) $\mathfrak{a} \subset \mathfrak{b} \Rightarrow \mathfrak{a}^* \subset \mathfrak{b}^*$
b) $(\mathfrak{a} + \mathfrak{b})^* = \mathfrak{a}^* + \mathfrak{b}^*$
c) $(\mathfrak{a} \cap \mathfrak{b})^* = \mathfrak{a}^* \cap \mathfrak{b}^*$
d) $(\mathfrak{a}\mathfrak{b})^* = \mathfrak{a}^*\mathfrak{b}^*$. $\qquad\qquad\square$

Lemma 12.7. If $I \in D(A, R)$, then $I = \{x \in R \mid v_{\mathfrak{p}}(x) \geq v_{\mathfrak{p}}(I)$ for every $\mathfrak{p} \in \Omega(R/A)\}$.

Proof. Certainly I is contained in the right hand set. Let now $x \in R$ be given with $v_{\mathfrak{p}}(x) \geq v_{\mathfrak{p}}(I)$ for every $\mathfrak{p} \in \Omega(A)$. Then $v_{\mathfrak{p}}(xI^{-1}) = v_{\mathfrak{p}}(x) - v_{\mathfrak{p}}(I) \geq 0$ for every $\mathfrak{p} \in \Omega(R/A)$ (cf. Scholium 11.8.v). Thus $xI^{-1} \subset A$ (cf. Prop.2.ii), which implies $x \in I$. $\quad\square$

Theorem 12.8. Let $\mathfrak{a}$ be an R-amenable A-submodule of R.

i) The R-invertible hull $\mathfrak{a}^*$ is the smallest module $I \in D(A, R)$ containing $\mathfrak{a}$.
ii) $(\mathfrak{a}^*)^{-1} = [A:_R \mathfrak{a}]$. Also $\mathfrak{a}^* = [A:_R [A:_R \mathfrak{a}]]$.

Proof. i): By the preceding lemma we have

$$\mathfrak{a}^* = \{x \in R \mid v_{\mathfrak{p}}(x) \geq v_{\mathfrak{p}}(\mathfrak{a}) \quad \text{for every} \quad \mathfrak{p} \in \Omega(R/A)\}.$$

Thus certainly $\mathfrak{a} \subset \mathfrak{a}^*$. If $I \in D(A, R)$ and $\mathfrak{a} \subset I$ then $\mathfrak{a}^* \subset I^* = I$.
ii): Let $x \in R$ be given. Then $x \in [A:_R \mathfrak{a}]$ iff $\mathfrak{a}x \subset A$. This holds iff $v_{\mathfrak{p}}(x\mathfrak{a}) \geq 0$ for every $\mathfrak{p} \in \Omega(R/A)$ (cf. Lemma 7 with $I = A$). Now $v_{\mathfrak{p}}(x\mathfrak{a}) = v_{\mathfrak{p}}(x) + v_{\mathfrak{p}}(\mathfrak{a})$. Thus $x \in [A:_R \mathfrak{a}]$ iff $v_{\mathfrak{p}}(x) \geq -v_{\mathfrak{p}}(\mathfrak{a})$ for every $\mathfrak{p} \in \Omega(R/A)$. Let $J := (\mathfrak{a}^*)^{-1}$. We have $v_{\mathfrak{p}}(J) = -v_{\mathfrak{p}}(\mathfrak{a})$ for every $\mathfrak{p} \in \Omega(R/A)$. Invoking again Lemma 7, we see that $x \in [A:_R \mathfrak{a}]$ iff $x \in J$. This proves $J = [A:_R \mathfrak{a}]$. Applying this to the ideal J instead of $\mathfrak{a}$, we obtain $[A:_R J] = (J^*)^{-1} = J^{-1} = \mathfrak{a}^*$. $\quad\square$

Here is a converse of Theorem 8.

Proposition 12.9. Let $\mathfrak{a}$ be a finitely generated A-submodule of R. Assume there exists a minimal $I \in D(A, R)$ containing $\mathfrak{a}$. Then $\mathfrak{a}$ is amenable in R and $I = \mathfrak{a}^*$.

Proof. Let $\mathfrak{b} := \mathfrak{a}I^{-1}$. This is an ideal of A such that there does not exist an R-invertible ideal $J \neq A$ of A with $\mathfrak{b} \subset J$. It follows that $\mathfrak{b} \not\subset \mathfrak{p}$, hence $v_\mathfrak{p}(\mathfrak{b}) = 0$, for every $\mathfrak{p} \in \Omega(R/A)$. We see from Definitions 3 and 4 that $\mathfrak{b}$ is R-amenable and $\mathfrak{b}^* = A$. It follows that $\mathfrak{a}$ is R-amenable and $\mathfrak{a}^* = (I\mathfrak{b})^* = I^*\mathfrak{b}^* = I$. □

Proposition 12.10. Let $A \subset B$ be a subextension of $A \subset R$, and let C denote the polar B° of B with respect to R. Thus $R = B \times_A C$ (cf. Prop.11.11). Let $\mathfrak{a}$ be an R-amenable A-submodule of B. Then $I_R(\mathfrak{a}) \subset B$. Both the A-modules $\mathfrak{a}$ and $I_R(\mathfrak{a})$ are B-amenable and $I_B(\mathfrak{a}) = I_B I_R(\mathfrak{a})$. $I_B(\mathfrak{a})$ is the B-component (cf. §8, Def.1) of $I_R(\mathfrak{a})$.

Proof. 1) The ring B is the union of modules I^{-1} with I running through the set $\mathrm{Inv}(A, B)$ of B-invertible ideals of A. If I and J are such ideals then $IJ \in \mathrm{Inv}(A, B)$ and $(IJ)^{-1}$ contains both I^{-1} and J^{-1}. Since $\mathfrak{a}$ is finitely generated, it follows that $\mathfrak{a} \subset I^{-1}$ for some $I \in \mathrm{Inv}(A, B)$. This implies $I_R(\mathfrak{a}) \subset I^{-1}$, as we have proved above, hence $I_R(\mathfrak{a}) \subset B$.

2) Since $\Omega(B/A) \subset \Omega(R/A)$ we have $\mathfrak{a} \not\subset \tilde{\mathfrak{p}}$ for every $\mathfrak{p} \in \Omega(B/A)$. Thus $\mathfrak{a}$ is B-amenable. For the same reason, $I_R(\mathfrak{a})$ is B-amenable. Now $v_\mathfrak{p}(\mathfrak{a}) = v_\mathfrak{p}(I_R(\mathfrak{a}))$ for every $\mathfrak{p} \in \Omega(B/A)$. Thus $I_B(\mathfrak{a}) = I_B I_R(\mathfrak{a})$.

3) We have $I_R(\mathfrak{a}) = \prod_{\mathfrak{p} \in \Omega(R/A)} \mathfrak{p}^{v_\mathfrak{p}(\mathfrak{a})}$ and $I_B(\mathfrak{a}) = \prod_{\mathfrak{p} \in \Omega(B/A)} \mathfrak{p}^{v_\mathfrak{p}(\mathfrak{a})}$. As is clear from §11 (Prop.11.11, Th.11.13.ii), $\Omega(R/A)$ is the disjoint union of $\Omega(B/A)$ and $\Omega(C/A)$. Let $\mathfrak{c} := \prod_{\mathfrak{p} \in \Omega(C/A)} \mathfrak{p}^{v_\mathfrak{p}(\mathfrak{a})}$. Then $I_R(\mathfrak{a}) = I_B(\mathfrak{a}) \cdot \mathfrak{c}$. If $\mathfrak{p} \in \Omega(C/A)$ then $\mathfrak{p}C = C$. Thus $\mathfrak{c}C = C$. This gives us $I_R(\mathfrak{a})C = I_B(\mathfrak{a})C$. Also $I_B(\mathfrak{a})B = B$. Recalling just Definition 1 of §8, we see that $I_B(\mathfrak{a})$ is the B-component of $I_R(\mathfrak{a})$. □

We now add some items to the transfer theory from §6 in the present noetherian situation. Let $A \subset C$ be a ws extension. Both C and $P(A)$ have unique embeddings over A into $Q(A)$. We regard them as overrings of A in $Q(A)$. The subring $T := CP(A) = C \otimes_A P(A)$ of $Q(A)$ is Prüfer over C, hence $T \subset P(C)$.

The ring C is again noetherian. Indeed, if J is an ideal of C, then $J \cap A$ is finitely generated and $J = C(J \cap A)$ by Proposition I.4.6. Thus also J is finitely generated.

We first want to analyze the relations between the sets of prime ideals $\Omega(C)$ and $\Omega(A)$ and the Manis valuations corresponding to these prime ideals. Notice that, in the special case that $C \subset P(A)$, this has already been done in §11, cf. in particular Proposition 11.15.

Definition 5. Let $\Omega_C(A)$ denote the set of all $\mathfrak{p} \in \Omega(A)$ with $\mathfrak{p}C \neq C$, i.e. $\Omega_C(A) = \Omega(A) \cap X(C/A)$.

Theorem 12.11. (Recall that $T := C \cdot P(A)$.)

i) Let $\mathfrak{q} \in \Omega(T/C)$. Then $\mathfrak{p} := \mathfrak{q} \cap A \in \Omega_C(A)$. The natural map $A_{\mathfrak{p}} \to C_{\mathfrak{q}}$ is an isomorphism, in short $A_{\mathfrak{p}} = C_{\mathfrak{q}}$. Also $P(A)_{\mathfrak{p}} = T_{\mathfrak{q}} = P(C)_{\mathfrak{q}} = \mathrm{Quot}(A_{\mathfrak{p}})$. The valuation $v_{\mathfrak{p}} : P(A) \to \mathbb{Z} \cup \infty$ (cf. §11, Def.4) is the restriction of $v_{\mathfrak{q}} : P(C) \to \mathbb{Z} \cup \infty$ to the subring $P(A)$ of $P(C)$, and $\tilde{v}_{\mathfrak{p}} = \tilde{v}_{\mathfrak{q}}$.

ii) The map $\Omega(T/C) \to \Omega_C(A)$, $\mathfrak{q} \mapsto \mathfrak{q} \cap A$, is a bijection with inverse map $\mathfrak{p} \mapsto \mathfrak{p}C$.

iii) If $P(A) \cap C = A$ then $\Omega_C(A) = \Omega(A)$.

iv) If $\mathfrak{a}$ is any finitely generated A-submodule of $P(A)$ then $v_{\mathfrak{p}}(\mathfrak{a}) = v_{\mathfrak{p}C}(\mathfrak{a}C)$ for every $\mathfrak{p} \in \Omega_C(A)$.

Proof. a) We start with a prime ideal $\mathfrak{q} \in \Omega(T/C)$. Let $w := v_{\mathfrak{q}}|T$ be the corresponding Manis valuation on T. We have $A_w = C_{[\mathfrak{q}]}^T$, $\mathfrak{p}_w = \mathfrak{q}_{[\mathfrak{q}]}^T$. The localization $C_{\mathfrak{q}}$ is a discrete valuation ring. Moreover $T_{\mathfrak{q}} = P(C)_{\mathfrak{q}} = \mathrm{Quot}(C_{\mathfrak{q}})$, as has been stated in Corollary 11.14.

Let $\mathfrak{p} := \mathfrak{q} \cap A$ and $v := v_{\mathfrak{q}}|P(A)$. We have $\mathfrak{p}_v \cap A = (\mathfrak{p}_w \cap C) \cap A = \mathfrak{q} \cap A = \mathfrak{p}$. Thus v is non trivial. Since w has the value group $\mathbb{Z}$, also v has the value group $\mathbb{Z}$, and $v(P(A)) = \mathbb{Z} \cup \{\infty\}$. We conclude that $\mathfrak{p} \in \Omega(A)$ and that v is the valuation $v_{\mathfrak{p}}$ on $P(A)$ corresponding to $\mathfrak{p}$. Further $\mathfrak{p}C = \mathfrak{q}$, in particular $\mathfrak{p} \in \Omega_C(A)$.

b) Conversely let $\mathfrak{p} \in \Omega_C(A)$ be given. Then $\mathfrak{q} := \mathfrak{p}C$ is a prime ideal of C. The natural map $\varphi_{\mathfrak{p}} : A_{\mathfrak{p}} \to C_{\mathfrak{q}}$ is surjective, since $A \subset C$ is ws. $A_{\mathfrak{p}}$ is a discrete valuation ring and $P(A)_{\mathfrak{p}}$ is its quotient field. The epimorphism $\varphi_{\mathfrak{p}} : A_{\mathfrak{p}} \twoheadrightarrow C_{\mathfrak{q}}$ extends to a homomorphism $\psi_{\mathfrak{p}} : P(A)_{\mathfrak{p}} \to T_{\mathfrak{q}}$. Now $\psi_{\mathfrak{p}}$ is injective, since $P(A)_{\mathfrak{p}}$ is a field and $T_{\mathfrak{q}} \neq \{0\}$. Thus also $\varphi_{\mathfrak{p}}$ is injective, hence an isomorphism. In short, $A_{\mathfrak{p}} = C_{\mathfrak{q}}$. It now is also clear that $\psi_{\mathfrak{p}}$ is surjective. (Recall that $T = P(A) \cdot C$.) Thus $\psi_{\mathfrak{p}}$ is an isomorphism. In short, $P(A)_{\mathfrak{p}} = T_{\mathfrak{q}}$.

We have $\mathfrak{q}T = T$, since $P(A)\mathfrak{p} = P(A)$. Thus $\mathfrak{q}$ is dense in C. We claim that $\mathfrak{q}$ is a maximal ideal of C. Indeed, let $\mathfrak{m}$ be a maximal ideal of C containing $\mathfrak{q}$. Then $\mathfrak{m} \cap A \supset \mathfrak{p}$. Since $\mathfrak{p}$ is maximal, we have $\mathfrak{m} \cap A = \mathfrak{p}$, and this forces $\mathfrak{m} = C(\mathfrak{m} \cap A) = \mathfrak{q}$.

We conclude (just recalling Definitions 1 and 3 in §11) that $\mathfrak{q} \in \Omega(T/C)$. From Proposition 11.15 we know that $T_{\mathfrak{q}} = P(C)_{\mathfrak{q}}$. Thus $P(A)_{\mathfrak{p}} = P(C)_{\mathfrak{q}}$. We conclude that $\tilde{v}_{\mathfrak{p}} = \tilde{v}_{\mathfrak{q}}$. For $\mathfrak{a}$ a finitely generated A-submodule of $P(A)$ we have $v_{\mathfrak{p}}(\mathfrak{a}) = \tilde{v}_{\mathfrak{p}}(\mathfrak{a}A_{\mathfrak{p}}) = \tilde{v}_{\mathfrak{q}}(\mathfrak{a}C_{\mathfrak{q}}) = v_{\mathfrak{q}}(\mathfrak{a}C)$.

c) Assume finally that $P(A) \cap C = A$. By the transfer theory in §6 we know that we have a bijection $\mathrm{Inv}(A, P(A)) \to \mathrm{Inv}(C, T)$, $I \mapsto IC$. Thus for $\mathfrak{p} \in \Omega(A)$ certainly $\mathfrak{p}C \neq C$. This means that now $\Omega_C(A) = \Omega(A)$. $\qquad\square$

Remark 12.12. We learned in §6 that we have a surjective group homomorphism from $D(A) = D(A, P(A))$ to $D(C, T)$, sending $I \in D(A)$ to IC. It is now clear from Theorem 11 that for any $\mathfrak{p} \in \Omega_C(A)$ and associated $\mathfrak{q} := \mathfrak{p}C \in \Omega(T/C)$ we have $v_{\mathfrak{p}}(I) = v_{\mathfrak{q}}(IC)$. $\qquad\square$

Proposition 12.13. Let $\mathfrak{p} \in \Omega_A(C)$. Then $(\mathfrak{p}C)^{\sim} = \tilde{\mathfrak{p}}C$.

Proof. Let $\mathfrak{q} := \mathfrak{p}C$. We have $v_{\mathfrak{p}} = v_{\mathfrak{q}}|P(A)$, hence

$$\tilde{\mathfrak{p}} := A \cap \mathrm{supp}\,(v_{\mathfrak{p}}) = A \cap (C \cap \mathrm{supp}\, v_{\mathfrak{q}}) = A \cap \tilde{\mathfrak{q}}.$$

This implies $\tilde{\mathfrak{q}} = \tilde{\mathfrak{p}}C$ (cf.Prop.I.4.6). $\qquad\square$

Let $A \subset R$ be a fixed subextension of $A \subset P(A)$. Then $C \subset RC$ is a subextension of $C \subset T$.

Proposition 12.14. Assume that $\mathfrak{a}$ is an R-amenable A-submodule of R. Then $\mathfrak{a}C$ is an RC-amenable C-submodule of RC and $I_{RC}(\mathfrak{a}C) = I_R(\mathfrak{a})C$.

Proof. Due to Theorem 11.ii we have a bijection $\Omega_C(A) \xrightarrow{\sim} \Omega(T/C)$, $\mathfrak{p} \mapsto \mathfrak{p}C$. The set $\Omega_C(R/A) := \Omega_C(A) \cap \Omega(R/A)$ is mapped onto the subset $\Omega(RC/C)$ of $\Omega(T/A)$, since for $\mathfrak{p} \in \Omega_C(A)$ we have $\mathfrak{p}R = R$ iff $\mathfrak{p}RC = RC$, as can be easily deduced from Theorem 6.5. The A-module $\mathfrak{a}$ is R-amenable iff $v_{\mathfrak{p}}(\mathfrak{a}) < \infty$ for every $\mathfrak{p} \in \Omega(R/A)$. Now $v_{\mathfrak{p}}(\mathfrak{a}) = v_{\mathfrak{p}C}(\mathfrak{a}C)$ for every $\mathfrak{p} \in \Omega_C(A)$, as has been stated in Theorem

11.iv.   Thus $v_{\mathfrak{q}}(\mathfrak{a}C) < \infty$ for every $\mathfrak{q} \in \Omega(RC/C)$, i.e. $\mathfrak{a}C$ is RC-amenable. {N.B. This could also be deduced from Proposition 13.}
If $\mathfrak{p} \in \Omega(R/A) \setminus \Omega_C(A)$ then $\mathfrak{p}C = C$. We thus obtain

$$
I_{RC}(\mathfrak{a}C) = \prod_{\mathfrak{p} \in \Omega_C(R/A)} (\mathfrak{p}C)^{v_{\mathfrak{p}C}(\mathfrak{a}C)} = \prod_{\mathfrak{p} \in \Omega_C(R/A)} (\mathfrak{p}C)^{v_{\mathfrak{p}}(\mathfrak{a})} =
$$

$$
= \left(\prod_{\mathfrak{p} \in \Omega_C(R/A)} \mathfrak{p}^{v_{\mathfrak{p}}(\mathfrak{a})} \right) \cdot C = \left(\prod_{\mathfrak{p} \in \Omega(R/A)} \mathfrak{p}^{v_{\mathfrak{p}}(\mathfrak{a})} \right) \cdot C = I_R(\mathfrak{a})C. \qquad \square
$$

We have reached a point where we can give an argument why it seems to be necessary to study invertible hulls $I_R(\mathfrak{a})$ for R an overring of A in $P(A)$ instead of just $R = P(A)$: Let $A \subset C$ be a ws extension, as before. If $\mathfrak{a}$ is an amenable ($= P(A)$-amenable) A-submodule of $P(A)$ then $\mathfrak{a}C$ is a T-amenable C-submodule of $T := P(A)C$, and $I_T(\mathfrak{a}C) = I_{P(A)}(\mathfrak{a}) \cdot C$ by the proposition just proved. But there seems to be no reason in general, why $\mathfrak{a}C$ should be amenable in $P(C)$. Thus we cannot even define $I_{P(C)}(\mathfrak{a}C)$.

How about a theory of invertible hulls if A is no longer noetherian? The results obtained here, in particular the formulas in Theorem 8.ii, strongly suggest that a good such theory is possible. We defer this to part II of the book, when more tools are available.

Chapter III:
PM-valuations and valuations of weaker type

Summary:

In this chapter we study valuations. Our primary goal is to understand PM-valuations (PM = "Prüfer-Manis", cf.I §6, Def.1). From an algebraic viewpoint these are the really good valuations on rings, while Manis valuations still allow some pathologies. A non trivial PM-valuation $v: R \to \Gamma \cup \infty$ is characterized, up to equivalence, by the ring extension $A_v \subset R$. {More generally this holds for Manis valuations, cf. I §2.} We had named these Prüfer extensions $A_v \subset R$ "PM-extensions" (I, §6). Given a Prüfer extension $A \subset R$ we in §1 study the family of all R-overrings B of A which are PM in R. But then we stop to look more closely at the families of PM-valuations arising here, deferring this central topic to part II of the book. Instead we try to elucidate the structure of a single PM-valuation in §2 and §3.

Many arguments and results in §1 – §3 can already be found in Griffin's seminal paper [G₂] (often presented there in a very brief way). We usually have refrained from documenting this in detail.

In §4 we introduce the class of "tight" valuations. These are a special sort of Manis valuations which are still more general than PM-valuations. Various arguments about PM-valuations seem to become more transparent if we separate what is true more generally for tight valuations, and where we really need the stronger PM-property.

Given a pair $(A, \mathfrak{p})$ consisting of a ring A and a prime ideal $\mathfrak{p}$ of A, we prove in §5 the existence of a unique maximal ring extension $A \subset C$ such that there lives a (unique) PM-valuation v on C with $A = A_v$ and $\mathfrak{p} = \mathfrak{p}_v$. We call C the *PM-hull* of $(A, \mathfrak{p})$ and write $C = \mathrm{PM}(A, \mathfrak{p})$. Analogously we construct a *TV-hull* $\mathrm{TV}(A, \mathfrak{p})$ of $(A, \mathfrak{p})$, where v is a tight valuation instead of a PM-valuation, and, under a necessary restriction on $(A, \mathfrak{p})$, a *Manis valuation hull* $\mathrm{MV}(A, \mathfrak{p}, R)$ within a given ring extension $A \subset R$. In sections §6 – §8 we explore and describe these valuation hulls in various ways.

In §6 – §8 we also discuss the possibility to improve a given valuation
$v: R \to \Gamma \cup \infty$ successively to a Manis, a tight, and a PM-valuation
by special restriction (cf. Def.11 in I, §1) to rings between A_v and
R. Then, in §9, we study what happens if we compose such a valuation with a ring homomorphism. Finally, in the last section §10,
we resume the topic of transfer principles from II, §6, now studying
valuations.

In this chapter up to §9 only easy parts of Chapter II will be used:
II, §1, the beginnings of II, §3, and some elementary facts about
tight extensions from II, §4. In the last section III, §10 we have to
rely on II, §6 of course. (To be honest, there is one further excep-
tion, Theorem 3.5, which relies on the difficult Theorem II.2.8. But
Theorem 3.5 will not be used later in Chapter III.)

§1 The PM-overrings in a Prüfer extension

In this section $A \subset R$ is a fixed ring extension, which usually will be
weakly surjective. Recall that an R-*overring* of A means a subring B
of R with $A \subset B$. Recall also that we use the abbreviation "ws" for
"weakly surjective" (I, §3) and the abbreviation "PM" for "Prüfer-
Manis" (I, §6).

We start with two easy lemmas, the first one being evident.

Lemma 1.0. Let v be a valuation on R and $A := A_v$, $\mathfrak{p} := \mathfrak{p}_v$. Then
$A = A_{[\mathfrak{p}]}$ and $\mathfrak{p} = \mathfrak{p}_{[\mathfrak{p}]}$. □

Lemma 1.1. Assume that A is ws in R. Let $\mathfrak{p}$ be a prime ideal of
A. Then $\mathfrak{p}R \neq R$ iff $A_{[\mathfrak{p}]} = R$. In this case $\mathfrak{p}R = \mathfrak{p}_{[\mathfrak{p}]}$.

Proof. $\mathfrak{p}_{[\mathfrak{p}]}$ is a prime ideal of $A_{[\mathfrak{p}]}$ and $\mathfrak{p}_{[\mathfrak{p}]} \cap A = \mathfrak{p}$. If $A_{[\mathfrak{p}]} = R$ then
we conclude from I, §4 (Prop.I.4.6) that $\mathfrak{p}_{[\mathfrak{p}]} = \mathfrak{p}R$ and in particular
$\mathfrak{p}R \neq R$. Conversely, if $\mathfrak{p}R \neq R$ then we learn from Theorem I.3.13,
there with $B = R$, that $R = A_{[\mathfrak{p}]}$. □

Theorem 1.2. Assume that A is Prüfer in R. Then the map $\mathfrak{p} \mapsto$
$A_{[\mathfrak{p}]}$ from the set $Y(R/A)$ of R-regular prime ideals of A into the set of

R-overrings of A is a *bijection* from $Y(R/A)$ to the set of proper PM-subrings B of R containing A. {Here "proper" means that $B \neq R$.} If B is such a subring of R then $B = A_{[\mathfrak{p}]}$ with $\mathfrak{p} := \mathfrak{p}_B \cap A$.

Proof. For every prime ideal $\mathfrak{p}$ of A the ring $A_{[\mathfrak{p}]}$ is PM in R (as stated already in I, §6), and if $\mathfrak{p} \in Y(R/A)$ then $A_{[\mathfrak{p}]} \neq R$ by Lemma 1. Let $\mathfrak{p} \in Y(R/A)$ and $\mathfrak{q} \in Y(R/A)$ be given with $A_{[\mathfrak{p}]} = A_{[\mathfrak{q}]} =: B$. Then the pairs $(B, \mathfrak{p}_{[\mathfrak{p}]})$ and $(B, \mathfrak{q}_{[\mathfrak{q}]})$ are Manis in R. We conclude from I, §2 that $\mathfrak{p}_{[\mathfrak{p}]} = \mathfrak{p}_B = \mathfrak{q}_{[\mathfrak{q}]}$.[*] Intersecting back with A we obtain $\mathfrak{p} = A \cap \mathfrak{p}_B = \mathfrak{q}$. This proves the injectivity of our map.

Let B be a proper subring of R which contains A and is PM in R. Let $\mathfrak{q} := \mathfrak{p}_B$ and $\mathfrak{p} := \mathfrak{q} \cap A$. The pair $(B, \mathfrak{q})$ is Manis in R. By Theorem I.3.13 (2) we have $A_{[\mathfrak{p}]} = B_{[\mathfrak{q}]}$, and by Lemma 0 we have $B_{[\mathfrak{q}]} = B$. Thus $B = A_{[\mathfrak{p}]}$. Again by Lemma 1, $\mathfrak{p} \in Y(R/A)$.   □

Theorem 1.3. Assume that $(A, \mathfrak{p})$ is a Manis pair in R and A is Prüfer in R. Then every R-regular prime ideal $\mathfrak{r}$ of A is contained in $\mathfrak{p}$, and $\mathfrak{r} = \mathfrak{r}_{[\mathfrak{r}]} = \mathfrak{r}_{[\mathfrak{p}]}$.

Proof. The pair $(A_{[\mathfrak{r}]}, \mathfrak{r}_{[\mathfrak{r}]})$ is Manis in R. Let v and w be Manis valuations on R with $A_v = A$, $\mathfrak{p}_v = \mathfrak{p}$, $A_w = A_{[\mathfrak{r}]}$, $\mathfrak{p}_w = \mathfrak{r}_{[\mathfrak{r}]}$. We want to prove that $\mathfrak{p}_w \subset \mathfrak{p}$. Then we can conclude from $\mathfrak{r}_{[\mathfrak{r}]} \subset \mathfrak{p} \subset A$ and $\mathfrak{r} \subset \mathfrak{r}_{[\mathfrak{p}]} \subset \mathfrak{r}_{[\mathfrak{r}]}$ that $\mathfrak{r} = A \cap \mathfrak{r}_{[\mathfrak{r}]} = \mathfrak{r}_{[\mathfrak{r}]} \subset \mathfrak{p}$ and $\mathfrak{r} = \mathfrak{r}_{[\mathfrak{p}]}$, and we will be done.

Since $\mathfrak{r}$ is R-regular, $A_w \neq R$. It follows that the support $\mathfrak{q}_w$ of w is different from $\mathfrak{p}_w$, thus $\mathfrak{p}_w \not\subset \mathfrak{q}_w$. Suppose that $\mathfrak{p}_w \not\subset \mathfrak{p}$. Then there exists some $a \in \mathfrak{p}_w$ with $a \notin \mathfrak{q}_w$, $a \notin \mathfrak{p}$. {For this conclusion we only need that $\mathfrak{p}_w, \mathfrak{q}_w, \mathfrak{p}$ are additive subgroups of R.} We have $w(a) > 0$, $w(a) \neq \infty$.

Since w is Manis we can pick some $a' \in R$ with $aa' \in A_{[\mathfrak{r}]} \setminus \mathfrak{r}_{[\mathfrak{r}]}$. We choose $d \in A \setminus \mathfrak{r}$ with $daa' \in A$. Since $a \notin \mathfrak{p}$ we have $v(a) \leq 0$. But $v(daa') \geq 0$. Thus $v(da') \geq 0$, i.e. $da' \in A$, and $a' \in A_{[\mathfrak{r}]}$, i.e. $w(a') \geq 0$. This contradicts the fact that $w(a') = -w(a) < 0$. We conclude that $\mathfrak{p}_w \subset \mathfrak{p}$.   □

---

[*] Recall the notation $\mathfrak{p}_B$ from Definition 2 in I, §2.

Corollary 1.4. If A is PM in R, $A \neq R$, then $\mathfrak{p}_A$ is the unique R-regular maximal ideal of A and $A_{[\mathfrak{p}_A]} = A$.

Proof. Let $\mathfrak{p} := \mathfrak{p}_A$. By Lemma 0 we have $A_{[\mathfrak{p}]} = A \neq R$. By Lemma 1 we conclude that $\mathfrak{p}$ is R-regular. Now Theorem 3 tells us that $\mathfrak{p}$ is the unique R-regular maximal ideal of A. $\square$

Theorem 3 also gives us a supplement to Theorem 2. Namely, it turns out that the bijection considered there is an anti-isomorphism of partially ordered sets, both orderings being given by the inclusion relation.

Proposition 1.5. Assume that A is Prüfer in R. Let $\mathfrak{p}$ and $\mathfrak{q}$ be prime ideals of A. Assume that $\mathfrak{q}$ is R-regular. Then $A_{[\mathfrak{p}]} \subset A_{[\mathfrak{q}]}$ iff $\mathfrak{q} \subset \mathfrak{p}$.

Proof of the nontrivial direction. Assume that $A_{[\mathfrak{p}]} \subset A_{[\mathfrak{q}]}$. Let $\mathfrak{q}' := A_{[\mathfrak{p}]} \cap \mathfrak{q}_{[\mathfrak{q}]}$. This is a prime ideal of $A_{[\mathfrak{p}]}$ with $\mathfrak{q}' \cap A = \mathfrak{q}$, hence $\mathfrak{q}' = \mathfrak{q} A_{[\mathfrak{p}]}$ by Theorem I.4.8. {Alternatively we may invoke Prop.I.4.6.} From $\mathfrak{q}R = R$ we conclude that $\mathfrak{q}'R = R$. Now Theorem 3 tells us that $\mathfrak{q}' \subset \mathfrak{p}_{[\mathfrak{p}]}$. Intersecting with A we obtain $\mathfrak{q} \subset \mathfrak{p}$. $\square$

In connection with this proof we state

Remark 1.6. If A is Prüfer in R and $\mathfrak{p}, \mathfrak{q}$ are prime ideals of A with $\mathfrak{q} \subset \mathfrak{p}$, then $\mathfrak{q} A_{[\mathfrak{p}]} = \mathfrak{q}_{[\mathfrak{p}]} = \mathfrak{q}_{[\mathfrak{q}]} \cap A_{[\mathfrak{p}]}$.

Proof. $\mathfrak{q} \subset (\mathfrak{q}A_{[\mathfrak{p}]}) \cap A \subset \mathfrak{q}_{[\mathfrak{p}]} \cap A \subset \mathfrak{q}_{[\mathfrak{q}]} \cap A_{[\mathfrak{p}]} \cap A = \mathfrak{q}$. Thus $(\mathfrak{q}A_{[\mathfrak{p}]}) \cap A = \mathfrak{q}_{[\mathfrak{p}]} \cap A = \mathfrak{q}_{[\mathfrak{q}]} \cap A_{[\mathfrak{p}]} \cap A$. The claim follows from Proposition I.4.6, since A is ws in $A_{[\mathfrak{p}]}$. $\square$

Definition 1. For any ring extension $A \subset R$ we denote by $\Omega(R/A)$ the set of maximal ideals of A which are R-regular, i.e. $\Omega(R/A) :=$ Max $A \cap Y(R/A)$. $\square$

Notice that $\Omega(R/A)$ is the set of maximal elements of $Y(R/A)$ under the inclusion relation.

If A is ws in R then $Y(R/A)$ is empty iff $A = R$ (cf. Lemma II.3.2). We start out for a characterization of the Prüfer extensions $A \subset R$ such that $\Omega(R/A)$ consists of one element.

Proposition 1.7. Assume that A is ws in R and $A \neq R$. Then, with $\Omega := \Omega(R/A)$,

$$A = \bigcap_{\mathfrak{p} \in \Omega} A_{[\mathfrak{p}]}.$$

Proof. A is the intersection of the rings $A_{[\mathfrak{p}]}$ with $\mathfrak{p}$ running through $\mathrm{Max}\, A$ (cf. I.5.5). But if $\mathfrak{p} \notin \Omega$ then $A_{[\mathfrak{p}]} = R$ (cf. Lemma 1). $\square$

Theorem 1.8. Assume that A is Prüfer in R and $A \neq R$. The ring A is Manis in R iff $\Omega(R/A)$ consists of one element $\mathfrak{p}$, and then $\mathfrak{p} = \mathfrak{p}_A$.

Proof. If A is Manis in R then Corollary 4 above tells us that $\Omega(R/A) = \{\mathfrak{p}_A\}$. Conversely, if $\Omega(R/A) = \{\mathfrak{p}\}$, then Proposition 7 tells us that $A = A_{[\mathfrak{p}]}$. It follows that $\mathfrak{p} = \mathfrak{p}_{[\mathfrak{p}]}$. The pair $(A, \mathfrak{p}) = (A_{[\mathfrak{p}]}, \mathfrak{p}_{[\mathfrak{p}]})$ is Manis in R by the definition of Prüfer extensions. Thus $\mathfrak{p} = \mathfrak{p}_A$. $\square$

In passing we state an amplification of Proposition 7 to A-modules, which will be useful later.

Lemma 1.9. Assume that $A \subset R$ is ws and that I is an R-regular A-submodule of R. Then, for every prime ideal $\mathfrak{p}$ of A which is *not* R-regular, we have $I_{[\mathfrak{p}]} = I A_{[\mathfrak{p}]} = R$.

Proof. Let $\mathfrak{p} \in \mathrm{Spec}\, A$, $\mathfrak{p}R \neq R$. By Lemma 1 (or by I, §3) we have $A_{[\mathfrak{p}]} = R$. Thus $I A_{[\mathfrak{p}]} = IR = R$. Since $I A_{[\mathfrak{p}]} \subset I_{[\mathfrak{p}]}$, also $I_{[\mathfrak{p}]} = R$. $\square$

Proposition 1.10. Assume that A is ws in R and $A \neq R$. For every R-regular A-submodule I of R we have, with $\Omega := \Omega(R/A)$,

$$I = \bigcap_{\mathfrak{p} \in \Omega} I A_{[\mathfrak{p}]} = \bigcap_{\mathfrak{p} \in \Omega} I_{[\mathfrak{p}]}.$$

Proof. It is trivial to state that $I \subset \bigcap_{\Omega} I A_{[\mathfrak{p}]} \subset \bigcap_{\Omega} I_{[\mathfrak{p}]}$. By the preceding lemma and by I.5.5 we have $\bigcap_{\Omega} I_{[\mathfrak{p}]} = \bigcap_{\mathfrak{p} \in \mathrm{Max}\, A} I_{[\mathfrak{p}]} = I$. This gives the result. $\square$

Remark 1.11. Assume again that A is ws in R and $A \neq R$. Then, for *any* A-submodule I of R,

$$I = \bigcap_{\mathfrak{p} \in \Omega} IA_{[\mathfrak{p}]} = IR \cap \bigcap_{\mathfrak{p} \in \Omega} I_{[\mathfrak{p}]}.$$

Proof. For every $\mathfrak{p} \in \operatorname{Max} A$ we have $I \subset IA_{[\mathfrak{p}]} \subset I_{[\mathfrak{p}]}$, and I is the intersection of the sets $I_{[\mathfrak{p}]}$ (cf. I.5.5). Since $A_{[\mathfrak{p}]} = R$ for $\mathfrak{p} \notin \Omega$, it follows that

$$I = \bigcap_{\mathfrak{p} \in \operatorname{Max} A} IA_{[\mathfrak{p}]} = \bigcap_{\mathfrak{p} \notin \Omega} IR \cap \bigcap_{\mathfrak{p} \in \Omega} IA_{[\mathfrak{p}]} = \bigcap_{\mathfrak{p} \notin \Omega} IR \cap \bigcap_{\mathfrak{p} \in \Omega} I_{[\mathfrak{p}]}.$$

Since Ω is not empty we may omit IR in the third term. $\qquad\square$

We finally write down a consequence of Theorem 8 and our theory of distributive ideals in II, §5.

Proposition 1.12. Let I be a maximal ideal of a ring A which is also invertible. Then A is PM in $A[I^{-1}]$.

Proof. Let $\mathfrak{m}$ be any maximal ideal of A. If $\mathfrak{m} \neq I$ then $I_\mathfrak{m} = A_\mathfrak{m}$. If $\mathfrak{m} = I$ then $I_\mathfrak{m}$ is the unique maximal ideal of $A_\mathfrak{m}$. Proposition II.5.2 now tells us that the ideal I of A is distributive, since clearly condition (5) there is fulfilled. We conclude by Theorem II.5.7 that I is a Prüfer ideal, i.e. A is Prüfer in $R := A[I^{-1}]$.

R is the union of the sets I^{-n}, $n \in \mathbb{N}$. Thus, if $\mathfrak{m}$ is an R-regular maximal ideal of A, we have some $n \in \mathbb{N}$ such that $1 \in \mathfrak{m}I^{-n}$, hence $I^n \subset \mathfrak{m}$. It follows that $I \subset \mathfrak{m}$, hence $I = \mathfrak{m}$. Thus $\mathfrak{m}$ is the unique R-regular maximal ideal of A, and we conclude by Theorem 8 that A is PM in R. $\qquad\square$

§2 The regular modules in a PM-extension

Notations. In the following v is a valuation on a ring R with value group Γ, and $A := A_v$, $\mathfrak{p} := \mathfrak{p}_v$. For every $\gamma \in \Gamma$ we denote the set of all $x \in R$ with $v(x) \geq \gamma$ by I_γ.[*]

---

[*] We will write more precisely $I_{\gamma,v}$ instead of I_γ, if necessary.

Notice that I_γ is a v-convex A-submodule of R. If $x \in R \setminus A$, hence $v(x) = -\gamma$ with $\gamma > 0$, we have

$$(A\!:\!x) = \{a \in A \mid v(a) + v(x) \geq 0\} =$$
$$= \{a \in A \mid v(a) \geq \gamma\} = I_\gamma.$$

Thus the ideals $(A\!:\!x)$ with $x \in R \setminus A$ are precisely the A-modules I_γ with $\gamma > 0$. Recalling Theorem I.3.13 we obtain the following

Proposition 2.1. A is ws in R iff $I_\gamma R = R$ for every $\gamma \in \Gamma$ with $\gamma > 0$. Then every v-convex A-submodule I of R different from $\operatorname{supp} v$ is R-regular.

Indeed, the last statement is evident, since every such module I contains I_γ for some $\gamma > 0$. $\qquad\qquad\qquad\qquad\qquad\qquad\qquad\qquad$ □

We strive for an understanding of all R-regular A-submodules of R in the case that v is PM.

If v is Manis then we know from Chapter I (Prop. I.1.15) that an A-submodule I of R is v-convex iff I contains $\operatorname{supp} v$ and $I_{[\mathfrak{p}]} = I$.

Theorem 2.2. If the valuation v is PM, then the R-regular A-submodules of R are precisely all v-convex submodules I of R with $I \neq \operatorname{supp} v$.

Proof. We know from Proposition 1 that the v-convex submodules $I \neq \operatorname{supp} v$ are R-regular. Let now an R-regular A-module $I \subset R$ be given. Then I is the intersection of the modules $I_{[\mathfrak{q}]}$ with $\mathfrak{q}$ running through $\Omega(R/A)$ (Prop. 1.10). But in the present case $\Omega(R/A) = \{\mathfrak{p}\}$, as we have seen in §1 (Cor. 1.4). Thus $I = I_{[\mathfrak{p}]}$, and we conclude that I is v-convex. $\qquad\qquad\qquad\qquad\qquad\qquad\qquad\qquad$ □

Definition. For M an additive subgroup of R we denote by M^v the v-convex hull of M, i.e. the smallest subset of R which is v-convex and contains M.

Clearly M^v is the set of all $z \in R$ with $v(z) \geq v(x)$ for some $x \in M$. Observe that M^v is an A-submodule of R. For any $x \in R$ with $v(x) = \gamma \neq \infty$ we have $(Ax)^v = I_\gamma$. From Theorem 2 we obtain immediately

Corollary 2.3. Assume that v is PM. Let I be an additive subgroup of R with $IR = R$. Then $I^v = AI$.

Proof. Of course, $AI \subset I^v$. But AI is an R-regular A-submodule of R. Thus AI is v-convex. This forces $AI = I^v$. □

Notice also that Proposition I.1.15, cited above, implies the following

Remark 2.4. Assume that the valuation v is Manis. If I is any A-submodule of R, then $I^v = (I + \operatorname{supp} v)_{[\mathfrak{p}]}$. □

We now expound the rather striking information provided by Theorem 2 for the special classes of prime ideals and invertible modules in a PM-extension.

Theorem 2.5. Assume that v is PM. Let $\mathfrak{r}$ be a prime ideal of A. The following are equivalent.

(1) $\mathfrak{r}$ is R-regular.
(2) $\mathfrak{r}$ is v-convex and $\mathfrak{r} \neq \operatorname{supp} v$.
(3) $\operatorname{supp} v \subsetneqq \mathfrak{r} \subset \mathfrak{p}$.
(4) $\mathfrak{r} \subset \mathfrak{p}$, but $\mathfrak{r} \not\subset \operatorname{supp} v$.

Proof. The equivalence of (1) and (2) is covered by Theorem 2. The implications $(2) \Rightarrow (3) \Rightarrow (4)$ are evident. We now prove $(4) \Rightarrow (2)$ as follows, cf. [G₂, p.415].

Let $x \in \mathfrak{r} \setminus \operatorname{supp} v$ and $y \in A$ be given with $v(y) \geq v(x)$. We verify that $y \in \mathfrak{r}$. The we will know that $\mathfrak{r}$ is v-convex. There exists some s in $R \setminus \operatorname{supp} v$ with $v(sx) = 0$, i.e. $sx \in A \setminus \mathfrak{p}$. This implies $v(sy) \geq 0$, i.e. $sy \in A$. It follows that $(sx)y = (sy)x \in \mathfrak{r}$. Since $sx \in A \setminus \mathfrak{p} \subset A \setminus \mathfrak{r}$, we conclude that $y \in \mathfrak{r}$. □

Theorem 2.6. If v is PM, the R-invertible A-submodules of R are precisely the sets I_γ, $\gamma \in \Gamma$. We have $I_\gamma^{-1} = I_{-\gamma}$.

Proof. For any $\gamma \in \Gamma$ we have $I_\gamma I_{-\gamma} \subset A$. From Proposition 1 we know that I_γ and $I_{-\gamma}$ are R-regular. Thus also $I_\gamma I_{-\gamma}$ is R-regular, hence v-convex by Theorem 2. Since $I_\gamma I_{-\gamma}$ contains an element x with $v(x) = 0$ this forces $I_\gamma I_{-\gamma} = A$.

Conversely, if I is an R-invertible A-module then I is certainly R-regular, hence v-convex. I is also finitely generated. Thus $I = (Ax)^v$ for some $x \in I$. We have $I = I_\gamma$ with $\gamma = v(x)$. $\square$

Remark. Part of this theorem can be generalized greatly, cf. Proposition 4.6 below.

Corollary 2.7. Assume again that v is PM. Let γ, δ be elements of Γ with $\gamma \leq \delta$, and let x be an element of R with $v(x) = \gamma$. Then $I_\delta + Ax = I_\gamma$.

Proof. I_δ is R-invertible. Since A is Prüfer in R, also $I_\delta + Ax$ is R-invertible (cf. Th. II.1.13). Theorem 6 tells us that $I_\delta + Ax = I_\gamma$. {We can also argue as follows: $I_\delta + Ax$ is v-convex in R by Prop.1 and Th.2. This implies $I_\delta + Ax = I_\gamma$.} $\square$

We now start out to prove a statement about the A-modules I_γ which is much sharper than Corollary 7. We need an easy lemma, valid for an arbitrary valuation v.

Lemma 2.8. Let $x \in A$ and $v(x) \neq \infty$, and let y be an element of R with $v(y) = -v(x)$. Then $(Ax)^v = [A: A + Ay] = (A: y).$[*]

Proof. $[A: A + Ay] = \{z \in R \mid z(A + Ay) \subset A\} = \{z \in A \mid zy \in A\} = \{z \in A \mid v(z) + v(y) \geq 0\} = \{z \in A \mid v(z) \geq v(x)\} = (A: y).$ $\square$

Theorem 2.9. Assume again that v is PM. Let γ and δ be elements of Γ with $\gamma \leq \delta$. Let x be an element of R with $v(x) = \gamma$. Then there exists some $u \in I_\delta$ with $Au + Ax = I_\gamma$.

Proof. a) We first consider the case $\gamma > 0$. We choose $y \in R$ with $v(y) = -\gamma$. By Corollary 7 we have $I_\delta + Axy = A$. Thus there exists $u \in I_\delta$ and $a \in A$ with $u + axy = 1$. Since $v(u) > 0$ we have $v(axy) = 0$. Thus $z := ay$ has the value $v(z) = -\gamma$. By Lemma 8, $I_\gamma = (Ax)^v = [A: A + Az] = (A + Az)^{-1}$. Now the relation $u \cdot 1 + x \cdot z = 1$ gives us the claim $Au + Ax = I_\gamma$ (cf. Remark II.1.10.a.)

---

[*] Recall the definition of $[A:I]=[A:_R I]$ from II, §1.

b) We deal with the remaining case $\gamma \leq 0$. If v is trivial, there is nothing to be proved. Thus we may assume without loss of generality that $\delta > 0$. Corollary 7 tells us that $I_\delta + Ax = I_\gamma$. We choose $c \in A$ with $v(c) = \delta - \gamma$. As already proved above there exists some $u \in I_\delta$ with $Au + Acx = I_\delta$. This implies $I_\gamma = I_\delta + Ax = Au + Ax$.   □

Comment. A guiding line for our study here is the idea that PM-valuations are "nearly as good" as valuations on fields, and in this respect are considerably better than just Manis valuations. Theorem 9 gives an illustration that this idea is not completely silly. Recall that, if R is a field, then for two elements x, y of R we have either $Ax \subset Ay$ or $Ay \subset Ax$, and $Ax \subset Ay$ iff $v(x) \leq v(y)$. Also every finitely generated A-submodule of R is generated by just one element. The simplicity of these properties seems to be a major reason why valuations on fields are loved and used much in algebraic and analytic geometry, and even more in real algebraic and real analytic geometry.

Now, if R is a ring, things are not that simple. Of course, we may pass from v to the valuation $\hat{v}$ on the residue class field $\mathrm{Quot}\,(R/\mathrm{supp}\,v)$, but then subtle phenomena in the relation between $\hat{v}$ and R tend to be hidden. Now, if v is PM, we have some comfort in working with v instead of $\hat{v}$. Due to Theorem 9 every finitely generated R-regular A-submodule I of R is generated by two elements x, u, where in addition x can be choosen to be any element of R with $v(x) = \mathrm{Min}_{z \in I} v(z)$, and then u can be choosen such that $v(u) \geq \delta$ for any $\delta \in \Gamma$. Also, if x, y are elements of R with $v(x) \leq v(y)$, then $Ay \subset Ax + I_\delta$ with $\delta \in \Gamma$ being as big as we want. Thus Ay is "nearly" contained in Ax. Of course, the word "nearly" here is to be understood with a big grain of salt. The valuation v cannot distinguish elements in $\mathrm{supp}\,v$.   □

§3 More ways to characterize PM-extensions, and a look at BM-extensions

In §1 (Theorem 1.8) we obtained an important new characterization for a ring extension to be PM (already knowing that the extension is Prüfer). We now add more characterization theorems for the PM property as a consequence of previous results. After that we study a special class of PM extensions and use this for still another characterization of PM-extensions. Finally we work on Bezout-Manis extensions, to be defined below.

Theorem 3.1. Let $A \subset R$ be a ring extension with A integrally closed in R. Then A is PM in R iff the set of R-overrings of A is a chain, i.e. is totally ordered by inclusion.

Proof. i) Assume that A is PM in R. Let B and C be R-overrings of A. Then B and C are R-regular A-modules. Both B and C are v-convex in R by Theorem 2.2. Thus $B \subset C$ or $C \subset B$.
ii) Assume now that any two R-overrings of A are comparable. Given two elements x, y of R we have $A[x] \subset A[y]$ or $A[y] \subset A[x]$. This implies that $xy \in A[x] + A[y]$. Now Theorem II.1.7 tells us that $A \subset R$ is Prüfer. If $\mathfrak{p}$ and $\mathfrak{q}$ are R-regular prime ideals of A then $A_{[\mathfrak{p}]} \subset A_{[\mathfrak{q}]}$ or $A_{[\mathfrak{q}]} \subset A_{[\mathfrak{p}]}$. This implies that $\mathfrak{q} \subset \mathfrak{p}$ or $\mathfrak{p} \subset \mathfrak{q}$ (Prop. 1.5). In particular A has only one maximal R-regular ideal. Theorem 1.8 tells us that A is PM in R. □

Corollary 3.2. If A is PM in R then every R-overring B of A is again PM in R.

Proof. The R-overrings of A form a chain. Thus also the R-overrings of B form a chain. □

Theorem 3.3. Let $A \subset R$ be a ring extension. The extension $A \subset R$ is PM iff for every R-overring B of A (with $B \neq R$) the set $R \setminus B$ is multiplicatively closed.

Proof. If $A \subset R$ is PM then, as just proved, every R-overring B of A is PM in R and thus $R \setminus B$ is multiplicatively closed. Conversely, if $B \neq R$ and $R \setminus B$ is multiplicatively closed, then B is integrally closed in R, as has already been stated in Theorem I.2.1. If this holds for every R-overring B of A then $A \subset R$ is Prüfer by Theorem I.5.2. □

Theorem 3.4. Let v be a non trivial Manis valuation on a ring R with value group Γ. Let $A := A_v$, $\mathfrak{p} := \mathfrak{p}_v$, and $I_\gamma := \{x \in R \mid v(x) \geq \gamma\}$ where $\gamma \in \Gamma$. The following are equivalent.

(1) A is Prüfer in R (hence PM).

(2) For every $x \in R$ the A-module $A + Ax$ is v-convex in R.

(3) A is ws in R and $\mathfrak{p}$ is the unique R-regular maximal ideal of A.

(3') A is ws in R and $\mathfrak{p}$ is the unique R-regular maximal ideal of A containing $\operatorname{supp} v$.

(4) The R-regular ideals of A are precisely the v-convex ideals different from $\operatorname{supp} v$.

(5) If $\gamma \in \Gamma$, $\gamma > 0$, and $x \in A$, $v(x) = 0$, then $I_\gamma + Ax = A$.

(6) For every R-overring B of A we have $B_{[\mathfrak{p}]} = B$.

(7) If $\mathfrak{m}$ is a maximal ideal of A different from $\mathfrak{p}$, and $\mathfrak{r}$ is a prime ideal of A contained in $\mathfrak{m} \cap \mathfrak{p}$, then $\mathfrak{r} \subset \operatorname{supp} v$.

Proof. Implication $(1) \Rightarrow (2)$ is clear from §2 (Th.2.2), $(1) \Rightarrow (3)$ is clear from §1 (Cor. 1.4), and $(1) \Rightarrow (4)$ is again clear from §2 (Th. 2.2). The implication $(4) \Rightarrow (5)$ is evident (cf. Cor.2.7), and $(3) \Rightarrow (3')$ is trivial.

$(2) \Rightarrow (1)$: The A-modules $A + Ax$, with x running through R, form a chain, since they are v-convex. From this one deduces easily that the R-overrings of A form a chain. {If B, C are A-overrings with $B \not\subset C$ and $x \in B \setminus C$ then $C \subset A + Ax$.} Theorem 1 tells us that A is PM in R.

$(3') \Rightarrow (1)$: Let $x \in R \setminus A$ be given and $I := (A:x) + x(A:x)$. Clearly $\operatorname{supp} v \subset I$. We verify that $I = A$ and then will be done by Theorem I.5.2. Since A is ws in R, the ideal $(A:x)$ is R-regular. A fortiori I is R-regular. Since v is Manis there exists some $y \in \mathfrak{p}$ with $xy \in A \setminus \mathfrak{p}$. We have $y \in (A:x)$. Thus $x(A:x) \not\subset \mathfrak{p}$, a fortiori $I \not\subset \mathfrak{p}$. Since by assumption $\mathfrak{p}$ is the only R-regular maximal ideal of A, we conclude that $I = A$.

$(5) \Rightarrow (1)$: Let $x \in R \setminus A$ be given. We again verify that $(A:x) + x(A:x) = A$, and then will be done. We have $(A:x) = I_\gamma$ with $\gamma = -v(x) > 0$ (cf. the beginning of §2). We choose $z \in (A:x)$ with $v(z) = \gamma$. Then $v(xz) = 0$. By assumption (5) it follows that $I_\gamma + Axz = A$, hence $I_\gamma + xI_\gamma = A$.

$(3) \Rightarrow (6)$: Let B be an R-overring of A. For every $x \in B$ the A-module $A + Ax$ is v-convex in R. It follows that B is v-convex in R. According to Proposition I.1.15 (or 2.4) this means $B = B_{[\mathfrak{p}]}$.

$(6) \Rightarrow (1)$: Let B be an R-overring of A. We have $B = B_{[\mathfrak{p}]}$, and this means that B is v-convex. Since all R-overrings of A are v-convex, they form a chain. Theorem 1 tells us that A is PM in R.

$(1) \Rightarrow (7)$: Let $\mathfrak{m}$ be a maximal ideal of A different from $\mathfrak{p}$, and $\mathfrak{r}$ a prime ideal of A contained in $\mathfrak{m} \cap \mathfrak{p}$. The ideal $\mathfrak{m}$ is not R-regular (cf.(3)), hence $\mathfrak{r}$ is not R-regular. But $\mathfrak{r} \subset \mathfrak{p}$. Theorem 2.5 tells us that $\mathfrak{r} \subset \operatorname{supp} v$.

$(7) \Rightarrow (1)$ (cf. [G$_2$, p.418 f]): The pair $(A_{[\mathfrak{p}]}, \mathfrak{p}_{[\mathfrak{p}]}) = (A, \mathfrak{p})$ is Manis in R. Let $\mathfrak{m}$ be a maximal ideal of A different from $\mathfrak{p}$. We prove that $A_{[\mathfrak{m}]} = R$, which implies that $(A_{[\mathfrak{m}]}, \mathfrak{m}_{[\mathfrak{m}]})$ is again a (trivial) Manis pair in R. Then we will know that $A \subset R$ is Prüfer.

Let $x \in R$ be given. We have to verify that $x \in A_{[\mathfrak{m}]}$. This is trivial if $x \in A$. Assume now that $x \in R \setminus A$, i.e. $v(x) < 0$. The set

$$H := \{\gamma \in \Gamma_v \mid n|\gamma| \le -v(x) \quad \text{for every} \quad n \in \mathbb{N}\}$$

is a convex subgroup of Γ_v, and $H \ne \Gamma_v$. Let $\mathfrak{r}$ denote the corresponding v-convex prime ideal of A, i.e.

$$\mathfrak{r} = \{a \in A \mid v(a) \notin H\} = \{a \in A \mid \exists n \in \mathbb{N} \quad \text{with} \quad v(a^n) > -v(x)\}.$$

We have $\mathfrak{r} \ne \operatorname{supp} v$ and $\mathfrak{r} \subset \mathfrak{p}$. By assumption (7) this implies $\mathfrak{r} \not\subset \mathfrak{m}$. We choose some $s \in \mathfrak{r} \setminus \mathfrak{m}$ and then some $n \in \mathbb{N}$ with $v(s^n) > -v(x)$. We have $s^n x \in A$, hence $x \in A_{[\mathfrak{m}]}$.   □

Remark. The implication $(1) \Rightarrow (7)$ can also be proved in the following different way. Let $\mathfrak{m}$ be a maximal ideal of R different from $\mathfrak{p}$ and $\mathfrak{r}$ a prime ideal contained in $\mathfrak{m} \cap \mathfrak{p}$. Since $\mathfrak{m}$ is not R-regular, we have $A_{[\mathfrak{m}]} = R$. Suppose that $\mathfrak{r} \not\subset \operatorname{supp} v$. We choose some $x \in \mathfrak{r} \setminus \operatorname{supp} v$ and then some $y \in R$ with $xy \in A \setminus \mathfrak{p}$. We finally choose some $s \in A \setminus \mathfrak{m}$ with $sy \in A$, which is possible since $A_{[\mathfrak{m}]} = R$. Now $xsy = x(sy) \in \mathfrak{r}$. On the other hand, $xsy = (xy)s \in A \setminus \mathfrak{r}$, since $xy \in A \setminus \mathfrak{p} \subset A \setminus \mathfrak{r}$ and $s \in A \setminus \mathfrak{m} \subset A \setminus \mathfrak{r}$. This contradiction proves that $\mathfrak{r} \subset \operatorname{supp} v$.   □

In Theorem 4 the criteria (5) and (7) deserve special interest, since they do not involve the extension R of A but only the restriction $v|A: A \to \Gamma_+ \cup \infty$ of the valuation v. In the next section we will see how the extension R of A and the valuation v can be recovered from $v|A$. The criteria (3), (3'), (4), (7) are essentially contained in Griffin's paper [G$_2$], cf. there Propositions 12 and 14.

Using a major result from Chapter II we obtain still another criterion for a ring extension to be PM.

Theorem 3.5. A ws ring extension $A \subset R$ is PM iff the set of R-regular ideals of A is a chain.

Proof. If $A \subset R$ is PM with associated PM-valuation v, the R-regular ideals of A are v-convex, as just stated (Theorem 4), and thus form a chain.

Assume now that $A \subset R$ is ws and the set $\mathcal{F}$ of R-regular ideals of A is a chain. Then $\mathcal{F}$ is a distributive lattice, and Theorem II.2.8 tells us that $A \subset R$ is Prüfer. A has a unique R-regular maximal ideal. It follows that $A \subset R$ is PM, since condition (3) in Theorem 4 holds. {Alternatively we may argue by Prop. 1.7 that $A = A_{[\mathfrak{p}]}$, with $\mathfrak{p}$ the maximal R-regular ideal of A, and thus A is Manis in R.} $\quad\square$

In I, §1 we have studied the class of *valuations with maximal support*, i.e. valuations $v: R \to \Gamma \cup \infty$ such that $\operatorname{supp} v = \mathfrak{q}$ is a maximal ideal of R. Every such valuation is Manis, since the associated valuation $\bar{v}: R/\mathfrak{q} \to \Gamma \cup \infty$ is a valuation on a field. In Proposition I.1.11 we have characterized valuations with maximal support within the broader class of Manis valuations in various ways. We now can add an interesting facet to this.

Proposition 3.6. Let $v: R \to \Gamma \cup \infty$ be a valuation, $A := A_v$, $\mathfrak{p} := \mathfrak{p}_v$, $\mathfrak{q} := \operatorname{supp} v$. Then $\mathfrak{q}$ is a maximal ideal of R iff v is Manis and $\mathfrak{q} + Ax = A$ for every $x \in A \setminus \mathfrak{p}$. Every such valuation is PM.

Proof. a) Assume first that v has maximal support $\mathfrak{q}$. Then v is Manis, $A/\mathfrak{q}$ is a Krull valuation domain, and $\mathfrak{p}/\mathfrak{q}$ is the maximal ideal of the local ring $A/\mathfrak{q}$. It follows that $Ax + \mathfrak{q} = A$ for every $x \in A \setminus \mathfrak{p}$. This implies condition (5) in Theorem 4 above. We conclude that v is PM.
b) Assume now that v is Manis and $Ax + \mathfrak{q} = A$ for every $x \in A \setminus \mathfrak{p}$. It follows that every ideal $I \neq A$ of A with $\mathfrak{q} \subset I$ is contained in $\mathfrak{p}$. Thus $\mathfrak{p}$ is the unique maximal ideal of A containing $\mathfrak{q}$. Proposition I.1.11 tells us that $\mathfrak{q} \in \operatorname{Max} R$. $\quad\square$

Remark 3.7. That a valuation v with maximal support $\mathfrak{q}$ is PM can be also seen as follows, applying only results from Chapter I: We use the same notations as before. $\bar{v}: R/\mathfrak{q} \to \Gamma \cup \infty$ is a Krull valuation with valuation ring $A/\mathfrak{q}$. Certainly $\bar{v}$ is Manis, hence v is

Manis. Also $A/\mathfrak{q}$ is a Prüfer domain, hence the extension $A/\mathfrak{q} \subset R/\mathfrak{q}$ is Prüfer. This implies that $A \subset R$ is Prüfer by Proposition I.5.8. $\square$

The valuations with maximal support are the "easy" PM-valuations. On a given ring R one gets them all by "pulling back" the valuations on the residue class fields of the maximal ideals to the ring R. Condition (5) in Theorem 4 above, which characterizes PM valuations in general, is a much more subtle matter than the condition $Ax + \mathfrak{q} = A$ for $x \in A \setminus \mathfrak{p}$.

It is not difficult now to give examples of PM-valuations which do *not* have maximal support. With a little more theory this will be even easier. We postpone an example to §10 (Ex.10.6). The contents of §4 – §9 are not necessary to understand that example.

In Proposition I.1.11 we gave various characterizations of valuations with maximal support. We want to add still another criterion. For this we need a beautiful lemma due to McAdam [McA] (in the case of domains).

Lemma 3.8. Let $A \subset R$ be a ring extension and x an element of R. Assume that A is integrally closed in $A[x]$. Assume further that the ideal $I := (A : x)$ of A is finitely generated and dense in A {i.e. $Iy \neq 0$ for every $y \neq 0$ in A}. Assume finally that $Ix \subset \sqrt{I}$. Then $x \in A$.

Proof. Since the ideal Ix of A is again finitely generated, there exists some $n \in \mathbb{N}$ with $(Ix)^n \subset I$. We proceed by induction on n. If $n = 1$, then $Ix \subset I$, and we conclude by a well known key fact from commutative algebra that x is integral over A, hence $x \in A$ (cf. [Bo, V §1, Th.1]). Assume now that $n > 1$. From $I(I^{n-1}x^n) \subset I$ we infer that every element in $I^{n-1}x^n$ is integral over A, hence $(Ix)^{n-1}x \subset A$. Thus $(Ix)^{n-1} \subset (A : x) = I$, and it follows by the induction hypothesis that $x \in A$. $\square$

Proposition 3.9. Let $A \subset R$ be a ring extension with $A \neq R$. Let $\mathfrak{p}$ be a prime ideal of A and $\mathfrak{q}$ the conductor of A in R. The following are equivalent.

   i) $(A, \mathfrak{p})$ is PM in R and $\mathfrak{q}$ is a maximal ideal of R.

ii) A is ws and integrally closed in R, and $A_{[\mathfrak{p}]} = A$. For every $x \in R$ the ideal $(A{:}\,x)$ is finitely generated. The set of prime ideals of A containing $\mathfrak{q}$ is totally ordered by inclusion.

Proof. The implication (i) $\Rightarrow$ (ii) is by now obvious {cf. Prop.I.1.11 for the last statement in (ii)}.

ii) $\Rightarrow$ i): All the conditions in (ii) remain true if we replace $A, \mathfrak{p}, R$ by $A/\mathfrak{q}$, $\mathfrak{p}/\mathfrak{q}$, $R/\mathfrak{q}$, and if we know (i) for $A/\mathfrak{q}$, $\mathfrak{p}/\mathfrak{q}$ and $R/\mathfrak{q}$ then (i) follows for $A, \mathfrak{p}, R$. {Notice that $A/\mathfrak{q}$ has in $R/\mathfrak{q}$ the conductor $\{0\}$, and that $(A{:}\,x)/\mathfrak{q} = (A/\mathfrak{q} : \overline{x})$ for any $x \in R$ and $\overline{x}$ its image in $R/\mathfrak{q}$.} Thus we may assume that $\mathfrak{q} = \{0\}$. Now the set Spec A of prime ideals is totally ordered by inclusion. In particular, A is a local domain. We denote its maximal ideal by $\mathfrak{m}$.

We first prove for a given $x \in R \setminus A$ that $x \in R^*$ and $x^{-1} \in A$. The ideal $I{:}= (A{:}\,x)$ is finitely generated and $IR = R$. Thus I is dense in A. The ideals $\sqrt{I}$ and $\sqrt{Ix}$ of A are prime, hence comparable. Suppose that $\sqrt{Ix} \subset \sqrt{I}$. Then $Ix \subset \sqrt{I}$, and Lemma 8 applies. It follows that $x \in A$, a contradiction. Thus $\sqrt{I} \subset \sqrt{Ix}$. We have $IR = R$, hence $\sqrt{Ix}R = R$. A fortiori $\sqrt{IxR} = R$, hence $IxR = R$. This implies $xR = R$, i.e. $x \in R^*$.

One easily verifies that $(A{:}\,x^{-1}) = Ix$. Indeed, if $a \in A$ is given, then $ax^{-1} \in A$ iff $a \in Ax$ iff $a \in (Ax) \cap A$, i.e. $a \in (A{:}\,x)x$. The ideal $J{:}= (A{:}\,x^{-1})$ is finitely generated and dense in A by hypothesis (ii), and $Jx^{-1} = I \subset \sqrt{Ix} = \sqrt{J}$. Applying Lemma 8 anew we see that $x^{-1} \in A$.

Now Theorem I.2.5 tells us that $(A, \mathfrak{m})$ is Manis in R and $\mathfrak{m} \setminus \{0\}$ is the set of inverses x^{-1} of all $x \in R \setminus A$. But if $x \in R \setminus A$ then $(A{:}\,x) = Ax^{-1} \subset \mathfrak{p}$, since $A_{[\mathfrak{p}]} = A$. Thus $\mathfrak{m} = \mathfrak{p}$. From Proposition I.1.11 we conclude that $\mathfrak{q}$ is a maximal ideal of R. As said above this implies that A is Prüfer in R. $\qquad\square$

From this Proposition one easily deduces a general criterion for a valuation to be PM.

Theorem 3.10. Let v be a special valuation on a ring R and $A{:}= A_v$, $\mathfrak{p}{:}= \mathfrak{p}_v$, $\mathfrak{q}{:}= \mathrm{supp}\,v$. Then v is PM iff the following holds:
a) A is ws in R. b) For every $x \in R$ the ideal $(A{:}\,x)$ is finitely generated. c) $\mathfrak{p}$ is the unique R-regular maximal ideal of A containing

$\mathfrak{q}$. d) The set of prime ideals $\mathfrak{r}$ of A with $\mathfrak{q} \subset \mathfrak{r} \subset \mathfrak{p}$ is totally ordered by inclusion.

Proof. We may assume that v is not trivial, i.e. $A \neq R$. If v is PM then certainly the conditions a) – d) are fulfilled.

Let us now assume that a) – d) are true. Since v is special, the conductor of A in R is $\mathfrak{q}$ (Prop.I.2.2). By Theorem 4 above it suffices to verify that v is Manis. {Use condition (3′) there.} This means the same as that the localisation $\tilde{v}$ of v, i.e. the valuation induced by v on $R_{\mathfrak{p}}$, is Manis. We have $A_{\tilde{v}} = A_{\mathfrak{p}}$, $\mathfrak{p}_{\tilde{v}} = \mathfrak{p}_{\mathfrak{p}}$, $\mathrm{supp}\,\tilde{v} = \mathfrak{q}_{\mathfrak{p}}$, and we see immediately that $\tilde{v}$ fulfills again the conditions a) – d). Thus we may assume that A is local with maximal ideal $\mathfrak{p}$. It is then clear that all the conditions (ii) in Proposition 9 are fulfilled. We obtain that $\mathfrak{q}$ is a maximal ideal of R {and that $(A, \mathfrak{p})$ is PM in R}. It is now evident that v is Manis (cf. Proposition 6 above or Prop.I.1.7).
$\square$

One may well ask whether all conditions a) – d) in the theorem are necessary to conclude that v is PM. In particular, is it possible to omit condition d), i.e. is d) a consequence of a) – c)? We do not know the answer.

In the proof of Theorem 10 we have arrived at the following result, which should be noted separately.

Corollary 3.11. Let v be a special valuation on a ring R and $A := A_v$, $\mathfrak{p} := \mathfrak{p}_v$, $\mathfrak{q} := \mathrm{supp}\,v$. Assume that for every $x \in R$ the ideal $(A : x)$ is R-regular and finitely generated. Assume further that the prime ideals $\mathfrak{r}$ with $\mathfrak{q} \subset \mathfrak{r} \subset \mathfrak{p}$ form a chain. Then v is Manis. $\square$

Already in I, §2 we observed that a nontrivial Manis valuation v on a ring R is uniquely determined by the subring A_v of R, up to equivalence. We now write down a stronger uniqueness statement for PM-valuations.

Theorem 3.12. Let v be a non-trivial PM-valuation on a ring R. If w is a special valuation on R with $A_w = A_v$, then w is equivalent to v.

Proof. Let $A := A_v = A_w$. We have $\mathfrak{p}_v = \mathfrak{p}_A := \{x \in R \mid \exists y \in R \setminus A \text{ with } xy \in A\}$, and certainly $\mathfrak{p}_w \supset \mathfrak{p}_A$. Since v is PM, we know by

Corollary 1.4 that $\mathfrak{p}_A$ is a maximal ideal of A. This forces $\mathfrak{p}_w = \mathfrak{p}_A$. Since both v and w are special, we also have $\operatorname{supp} v = \operatorname{supp} w = \mathfrak{q}_A$ (in the notation of I, §2, cf. there Proposition I.2.2).

We now verify that w is Manis. Then we will know by I, §2 that $w \sim v$. Let $x \in R$ be given with $w(x) \neq \infty$, hence also $v(x) \neq \infty$. Since v is Manis, there exists some $y \in R$ with $v(y) = -v(x)$. We have $xy \in A_v \setminus \mathfrak{p}_v = A_w \setminus \mathfrak{p}_w$. Thus $w(xy) = 0$, hence $w(y) = -w(x)$. This proves that $w(R) \setminus \{\infty\}$ is a group.   □

In the following we focus on the special class of PM-extensions corresponding to the BM-valuations (= Bezout-Manis valuations) introduced in II, §10.

Definition 1. We call a ring extension $A \subset R$ *Bezout-Manis* (or *BM* for short), if there exists a BM-valuation v on R with $A_v = R$. In other terms, $A \subset R$ is BM iff $A \subset R$ is Bezout and PM.   □

It seems desirable to extract characterizations of BM-extensions from the various characterizations of PM-extensions gained up to now. Many such characterizations can be readily written down. For example, if we add to the conditions a) – d) in Theorem 10 the further condition that $(A:x)$ is principal for every $x \in R$, then we have obtained a characterization of BM-extensions due to Theorem II.10.2 and Theorem 10 above.

We now give a characterization of BM-extensions which adds new facets to the contents of II, §10. We start with the notion of "Marot extensions".

Definition 2. We call a ring extension $A \subset R$ *Marot*, if for every $x \in R$ (or $x \in R \setminus A$) the A-module $A + Ax$ is generated by a set of units of R. We then also say that the ring A is *Marot in R*.   □

If $A \subset R$ is a Bezout extension and $x \in R$, we have $A + Ax = Ay$ with some $y \in R^*$. Thus Bezout extensions certainly are Marot. Notice also that, if A is Marot in R, then every R-overring of A is Marot in R.

The rings A with A Marot in $\operatorname{Quot} A$ are the "Marot rings", as defined for example in Huckaba's book [Huc], cf. there §7. Marot rings have been introduced by Jean Marot in the late sixties under the label "rings with property (P)", cf. [Ma$_1$], [Ma$_2$]. They nowadays play a prominent role in multiplicative ideal theory, as is testified by Huckaba's book [Huc]. (We have mentioned Marot rings briefly in Chapter I, §6, Example 9.) Maybe the more general Marot extensions are useful as well. We refrain here from a systematic study of these extensions, employing them only as a mean for a better understanding of BM-extensions, in particular their relation to PM-extensions.

Theorem 3.13 (cf. [Huc, p.35f] for $R = \operatorname{Quot} A$). Let $A \subset R$ be a Marot extension. The following are equivalent.

(1) A is BM in R.
(2) A is PM in R.
(3) There exists a valuation v on R with $A_v = A$.
(4) $R \setminus A$ is closed under multiplication.
(5) $s^{-1} \in A$ for every unit s of R with $s \notin A$.

Proof. The implications $(1) \Rightarrow (2) \Rightarrow (3) \Rightarrow (4) \Rightarrow (5)$ are trivial. We prove $(5) \Rightarrow (1)$ in several steps.
a) If $s, t \in R^*$ then $As \subset At$ or $At \subset As$, since either $\frac{s}{t} \in A$ or $\frac{t}{s} \in A$.
b) If $x \in R \setminus A$, the A-module $A + Ax$ is principal. Indeed, since $A \subset R$ is Marot, we have units $s_1, \ldots, s_n$ of R with $A + Ax = As_1 + As_2 + \cdots + As_n$. By a) we know that any two of the modules As_i are comparable. Thus $A + Ax = As_i$ for some $i \in \{1, \ldots, n\}$. This proves that A is Bezout in R.
c) Let M_1 and M_2 be two R-overmodules of A. We verify that $M_1 \subset M_2$ or $M_2 \subset M_1$. Assume that $M_1 \not\subset M_2$. There exist a finitely generated A-overmodule $M_1' \subset M_1$ with $M_1' \not\subset M_2$. Since A is Bezout in R, we have $M_1' = As$ with some $s \in R^*$. We can write $M_2 = \bigcup_{i \in I} L_i$, where each L_i is a finitely generated R-overmodule of A, hence $L_i = At_i$ with $t_i \in R^*$. Since $M_1' \not\subset M_2$ we have $M_1' \not\subset L_i$ for each $i \in I$. By step a) we conclude that $L_i \subset M_1'$ for each i, hence $M_2 \subset M_1'$. A fortiori $M_2 \subset M_1$.
d) In particular any two R-overrings of A are comparable. Now Theorem 1 tells us that A is PM in R. Since A is also Bezout in R, we conclude that A is BM in R. □

Corollary 3.14. Every Marot extension is convenient.

Proof. If $A \subset R$ is Marot and B is an R-overring of A, then $B \subset R$ is Marot. If in addition $R \setminus B$ is closed under multiplication, then we know by Theorem 13 that $B \subset R$ is PM.   □

We mentioned this fact already in I, §6, Example 9 in the special case $R = \mathrm{Quot}\, A$.

We will use Theorem 12 for giving examples of BM-valuations. This needs some preparations. We first exhibit a class of ring extensions which are easily seen to be Marot.

Definition 3. We call a ring extension $A \subset R$ *additively regular*, if for every $x \in R$ there exists some $a \in A$ such that $x + a$ is a unit in R.

This definition generalizes the notion of an additively regular *ring* occuring in the literature. A ring A is called additively regular if, in our terminology, the extension $A \subset \mathrm{Quot}\, A$ is additively regular, cf. [Huc, p.32]. We mentioned additively regular rings in I, §6, Example 8.

Remark 3.15 (cf. [Huc, Th.7.2]). Every additively regular extension $A \subset R$ is Marot. Indeed, if $x \in R$ is given, and a is an element of A with $x + a \in R^*$, then $A + Ax$ is generated as an A-module by the units 1 and $x + a$.   □

Given any ring R we introduce a certain overring $R(x)$ of the ring $R[x]$ of polynomials in one variable x in its total rings of quotients $\mathrm{Quot}\, R[x]$, cf. e.g. [Gi, §33]. The definition runs as follows. The *contents* $c(f)$ of a polynomial $f = a_0 + a_1 x + \cdots + a_n x^n$ is the ideal generated by its coefficients, $c(f) = \sum_{i=0}^{n} R a_i$. The polynomial f is called *unimodular* if $c(f) = R$. Given two polynomials $f, g \in R[x]$, we have the general formula

$$c(f)^{m+1} c(g) = c(f)^m c(fg)$$

with $m := \deg g$, cf. [Gi, Th.28.1]. By this formula it is evident, that the set S of unimodular polynomials in $R[x]$ is closed under multiplication, and also, that every $f \in S$ is a nonzero divisor. $R(x)$ is defined as the subring $S^{-1} R[x]$ of $\mathrm{Quot}\, R[x]$.

Proposition 3.16. Let R_0 denote the prime subring $\mathbb{Z} \cdot 1_R$ of R. The extension $R_0[x] \subset R(x)$ is additively regular, hence Marot.

Proof. Let $\xi \in R(x)$ be given. We write $\xi = \frac{f}{g}$ with $f, g \in R[x]$, $c(g) = R$. Let $n := \deg f$. Then

$$\frac{f}{g} + x^{n+1} = \frac{f + x^{n+1}g}{g}$$

is a unit in $R(x)$, since $c(f + x^{n+1}g) = c(f) + c(g) = R$.   □

Given a valuation $v: R \to \Gamma \cup \infty$ on R, we extend v to a valuation $v': R[x] \to \Gamma \cup \infty$ on $R[x]$ by the formula

$$v'(\sum_{i=0}^{n} a_i x^i) = \min(v(a_0), \ldots, v(a_n))$$

cf. [Bo, Chap.VI, Lemma 10.1)]. {The proof there, that v' is a valuation, works over any ring R instead of a field.} If f is a unimodular polynomial then certainly $c(f)$ is not contained in $\operatorname{supp} v$, hence $v'(f) \neq \infty$. Thus v' extends uniquely to a valuation $v^*: R(x) \to \Gamma \cup \infty$ by the formula

$$v^*(\frac{f}{g}) = v'(f) - v'(g)$$

with $f, g \in R[x]$, $c(g) = R$ (cf. I, §1).

Proposition 3.17. If the valuation v is special, then v^* is Bézout-Manis.

Proof. It is evident that v' is special, and then, that v^* is special. We have $v^*(x) = 0$. Thus $R_0[x]$ is certainly contained in A_{v^*}. It now follows from Lemma 15, that A_{v^*} is Marot in $R(x)$, and then by Theorem 13, that A_{v^*} is BM in $R(x)$. Since v^* is special, we conclude by Theorem 12 that v^* is Manis, hence BM.   □

Thus every special valuation on a ring R can be extended to a BM-valuation on $R(x)$ in a natural way.

In all our arguments for proving Proposition 16, starting with the proof of Proposition 15, we can replace the ring $R(x)$ by the subring $R\langle x\rangle$ consisting of the fractions $\frac{f}{g}$ with $f, g \in R[x]$ and g monic (i.e. having highest coefficient 1). We thus see that $R_0[x]$ is Marot in $R\langle x\rangle$, and then obtain as above:

Corollary 3.18. For every special valuation v on a ring R the restriction $v^*|R\langle x\rangle$ of the valuation v^* from above is BM.   □

§4 Tight valuations

Definition 1. We call a valuation v on a ring R *tight*, if for every $x \in R$ with $v(x) \neq \infty$ the A_v-module $\{y \in R \mid v(y) \geq v(x)\}$ is R-invertible.

Examples 4.1 a) Every PM-valuation is tight, cf. Theorem 2.6.
b) Every valuation $v: R \to \Gamma \cup \infty$ with $v(R^*) = \Gamma$ is tight. Indeed, let $x \in R$ be given with $v(x) = \gamma \neq \infty$. Choose $y \in R^*$ with $v(y) = \gamma$. Then $\{z \in R \mid v(z) \geq \gamma\} = A_v y$. This A_v-module is R-invertible, $(A_v y)^{-1} = A_v y^{-1}$.   □
c) Let A be any factorial domain, e.g. a polynomial ring $k[x_1, \ldots, x_n]$ over a field k. Let p be a prime element of A, and let R denote the localization $A[\frac{1}{p}]$, which is a subring of the quotient field $K := \operatorname{Quot} A$. Let finally $v: A[\frac{1}{p}] \to \mathbb{Z} \cup \infty$ be the restriction of the p-adic valuation on K to $A[\frac{1}{p}]$. Thus $v(p^n f) = n$, if $n \in \mathbb{Z}$ and $f \in A$ is not divisible by p. We have $v(R^*) = \mathbb{Z}$. Thus v is tight. Clearly $A_v = A$, $\mathfrak{p}_v = pA$. If pA is not a maximal ideal of A, then certainly v is not PM (cf.Cor.1.4). If A is the polynomial ring $k[x_1, \ldots, x_n]$ from above and $n \geq 2$, this holds for every prime element p.
d) Let k be an integral domain and L a finitely generated projective k-module of rank one. For any $n \in \mathbb{N}$ we denote the n-fold tensor product $L \otimes_k L \otimes_k \cdots \otimes_k L$ by L^n. Further we set $L^\circ := k$, $L^{-1} := \operatorname{Hom}_k(L, k)$, $L^{-n} := (L^{-1})^n$ for $n \in \mathbb{N}$. We build the $\mathbb{Z}$-graded k-algebra $R := \bigoplus_{n \in \mathbb{Z}} L^n$ with the obvious multiplication coming from the natural isomorphisms $L^n \otimes_k L^m \xrightarrow{\sim} L^{n+m}$. The multiplication is commutative, as can be seen by localizing at the prime ideals

of k, and R is a domain. Let $A := \bigoplus_{n \geq 0} L^n$, $I := \bigoplus_{n > 0} L^n = LA$.
Then A is a subring of R, and I is an R-invertible ideal of A with
$I^{-1} = L^{-1}A = \bigoplus_{n \geq -1} L^n$. For every $x \in R$ we put $v(x) := \sup\{n \in \mathbb{Z} \mid x \in I^n\} \in \mathbb{Z} \cup \infty$. It is easily checked that v is a valuation on R
with $A_v = A$, $\mathfrak{p}_v = I$, $\operatorname{supp} v = \{0\}$. {Notice that, if $x \in I^n \setminus I^{n+1}$,
$y \in I^m \setminus I^{m+1}$, then $xy \in I^{n+m} \setminus I^{n+m+1}$.} We have $I_{n,v} = I^n$ for
every $n \in \mathbb{Z}$. Thus v is tight. If k is not a field, then I is not a
maximal ideal of A, hence v is certainly not PM.   □

The last two examples seem to indicate, that already a valuation
theory over rings fitting the needs of classical algebraic geometry
should not restrict to PM-valuations but sometimes admit tight valuations.

We will prove soon that every tight valuation is Manis. Thus tight
valuations have their place between Manis and PM-valuations. We
here regard tight valuations as an auxiliary notion to understand
better the nature of PM-valuations, starting from Manis valuations.
Our first major goal is to make explicit that a tight valuation v is
completely determined by its restriction to A_v, cf. Scholium 4.7 and
Theorem 4.8 below. After that we will study coarsenings and special
restrictions (cf. I, §1) of a tight valuation v. In the special case that
v is PM this will give us a valuation theoretic description of the rings
B with $A_v \subset B \subset R$.

We return to the notations at the beginning of §2: v is a valuation on
a ring R with value group Γ, $A := A_v$, $\mathfrak{p} := \mathfrak{p}_v$, $I_\gamma := \{x \in R \mid v(x) \geq \gamma\}$ for $\gamma \in \Gamma$. We denote the support of v by $\mathfrak{q}$.

Proposition 4.2. Let $\gamma \in \Gamma$ be given. Assume that there exists
some $x \in R$ with $v(x) = \gamma$. Assume further that I_γ is R-invertible.
Then $I_\gamma^{-1} = I_{-\gamma}$, and there exists some $y \in R$ with $v(y) = -\gamma$. Also
$I_\gamma I_\alpha = I_{\gamma+\alpha}$ for every $\alpha \in \Gamma$, and $I_\gamma \mathfrak{q} = \mathfrak{q}$.

Proof. a) It is obvious that $I_\gamma I_\alpha \subset I_{\gamma+\alpha}$ for every $\alpha \in \Gamma$. In
particular $I_\gamma I_{-\gamma} \subset A$. Thus $I_{-\gamma} \subset I_\gamma^{-1}$. On the other hand, $x I_\gamma^{-1} \subset A$. Since $v(x) = \gamma$, we conclude that $v(z) \geq -\gamma$ for every $z \in I_\gamma^{-1}$,
i.e. $I_\gamma^{-1} \subset I_{-\gamma}$. This proves $I_\gamma^{-1} = I_{-\gamma}$.
b) The A-module I_γ^{-1} is finitely generated, $I_\gamma^{-1} = Ay_1 + Ay_2 + \cdots + Ay_r$ with $v(y_1) \leq v(y_2) \leq \cdots \leq v(y_r)$. As just proved, $v(y_1) \geq -\gamma$.

Suppose that $v(y_1) > -\gamma$. Then $v(z) > -\gamma$ for every $z \in I_\gamma^{-1}$, hence $v(a) > \gamma + (-\gamma) = 0$ for every $a \in I_\gamma I_\gamma^{-1}$. But $1 \in I_\gamma I_\gamma^{-1}$. This contradiction proves that $v(y_1) = -\gamma$.

c) For every $\alpha \in \Gamma$ we have $I_\gamma I_\alpha \subset I_{\gamma+\alpha}$. Also $I_{-\gamma} I_{\gamma+\alpha} \subset I_\alpha$. Multiplying by I_γ we obtain $I_{\gamma+\alpha} \subset I_\gamma I_\alpha$. Thus $I_{\gamma+\alpha} = I_\gamma I_\alpha$.

d) We have $I_\alpha \mathfrak{q} \subset \mathfrak{q}$ for every $\alpha \in \Gamma$. In particular $I_\gamma \mathfrak{q} \subset \mathfrak{q}$ and $I_{-\gamma} \mathfrak{q} \subset \mathfrak{q}$. Multiplying the second inclusion relation by I_γ we obtain $\mathfrak{q} \subset I_\gamma \mathfrak{q}$, hence $I_\gamma \mathfrak{q} = \mathfrak{q}$. $\qquad\square$

Corollary 4.3.

i) If v is tight then v is Manis.

ii) Conversely, if $v(R) \supset \Gamma_+$ and I_γ is R-invertible for every $\gamma \in \Gamma_+$, then v is tight.

iii) Also, if $v(R) \supset \Gamma_-$ and I_γ is R-invertible for every $\gamma \in \Gamma_-$, then v is tight.

Proof. If v is tight, it is evident from the second statement in Proposition 2 that $v(R) \setminus \{\infty\}$ is a group. Assertions (ii) and (iii) follow from the first and second statement in Proposition 2. $\qquad\square$

Corollary 4.4. If v is tight then R is the union of the subrings $A[I_\gamma^{-1}]$ with γ running through $\Gamma \cap v(A)$, and A is tight[*] in R.

Proof. R is the union of the sets $I_\gamma^{-1} = I_{-\gamma}$ with γ running through $\Gamma \cap v(A)$. Thus R is also the union of the subrings $A[I_\gamma^{-1}]$ with $\gamma \in \Gamma \cap v(A)$. A is tight in each extension $A[I_\gamma^{-1}]$, hence in R. $\qquad\square$

We give an example of a Manis valuation which is not tight.

Example 4.5. Let k be a field and $R := k[x, y]$ with two indeterminates x, y. We take on the subring $k[x]$ of R the valuation $u : k[x] \to \mathbb{Z}$ corresponding to the prime element x of $k[x]$, i.e. $u(x^n f(x)) = n$ if $x \nmid f(x)$. Then we define a valuation $v : R \to \mathbb{Z} \cup \infty$ by the formula

$$v\left(\sum_{i=0}^{n} f_i(x) y^i\right) = \min_{0 \le i \le n} \{u(f_i(x)) - i\}$$

---

[*] Recall Definition 1 in II, §2.

and, of course, $v(0) = \infty$, cf. [Bo, VI §10, Lemma 1]. Since $v(x) = 1$ and $v(y) = -1$ we have $v(R\backslash\{0\}) = \mathbb{Z}$. Thus v is Manis. Let $A: = A_v$, $\mathfrak{p}: = \mathfrak{p}_v$. One easily verifies that $\mathfrak{p} = Ax$. We have $R\mathfrak{p} = Rx \neq R$. Thus A is not ws in R (cf. Prop.2.1). A fortiori, v is not tight. □

We add an observation which will be useful later on.

Proposition 4.6. Assume that v is any valuation. Let I be an R-invertible A-submodule of R. Then there exists some $\gamma \in \Gamma \cap v(R)$ with $I = I_\gamma$.

Proof. We have $IR = R$. Thus I is not contained in the conductor $\mathfrak{q}$ of A in R. Write $I = Ax_1 + \cdots + Ax_r$ with $v(x_1) \leq v(x_2) \leq \cdots \leq v(x_r)$. We have $\gamma: = v(x_1) \neq \infty$. If $y \in R$ is given, then $yI \subset A$ iff $v(y) \geq -\gamma$. Thus $I^{-1} = [A:I] = I_{-\gamma}$. Write $I^{-1} = Ay_1 + \cdots + Ay_s$ with $-\gamma \leq v(y_1) \leq v(y_2) \leq \cdots \leq v(y_s)$. We have $-\gamma = v(y_1)$ since otherwise II^{-1} would be contained in $\mathfrak{p}$, a contradiction. Now Proposition 2 tells us that $I = I_\gamma$. □

Definition 2. If v is any valuation then $\mathfrak{H}(v)$ denotes the set of v-convex ideals $(Ax)^v$ of $A: = A_v$ with x running through $A \backslash \mathfrak{q}$. Thus $\mathfrak{H}(v) = \{I_\gamma \mid \gamma \in \Gamma_+ \cap v(R)\}$.

Scholium 4.7. If v is Manis the system $\mathfrak{H}(v)$ is completely determined by the triple $(A, \mathfrak{p}, \mathfrak{q})$. Indeed, for every $x \in A$ we have $(Ax)^v = (Ax + \mathfrak{q})_{[\mathfrak{p}]} = \{y \in A \mid sy \in Ax + \mathfrak{q} \text{ for some } s \in A \backslash \mathfrak{p}\}$ (Remark 2.4). If v is tight, then Proposition 6 tells us that $\mathfrak{H}(v)$ coincides with the set $\mathrm{Inv}\,(A, R)$ of all R-invertible ideals of A. Moreover the extension R of A is canonically isomorphic over A to the subring $A_{[\mathfrak{H}(v)]}$ of the tight hull $T(A)$ (cf. notations in II, §4, Def.3), as is now evident from II, §4.

We conclude that if v_1 and v_2 are tight valuations on extensions $R_1 \subset T(A)$, $R_2 \subset T(A)$ of a given ring A, and if $A = A_{v_1} = A_{v_2}$, $\mathfrak{p}_{v_1} = \mathfrak{p}_{v_2}$, $\mathrm{supp}\,v_1 = \mathrm{supp}\,v_2$, then $R_1 = R_2$. Moreover, the valuations v_1 and v_2 are equivalent, since they both give the same Manis pair $(A, \mathfrak{p}_{v_1})$ in R_1. Thus a tight valuation v is determined by the triple $(A_v, \mathfrak{p}_v, \mathrm{supp}\,v)$ in a strong sense. □

Theorem 4.8. Let $u: A \to \Gamma \cup \infty$ be a valuation on a ring A with $u(A) = \Gamma_+ \cup \infty$. Assume that for every $\gamma \in \Gamma_+$ the ideal $J_\gamma: = \{x \in A \mid u(x) \geq \gamma\}$ of A is invertible.

i) Then $J_\alpha J_\beta = J_{\alpha+\beta}$ for all $\alpha, \beta \in \Gamma_+$. Thus $\mathfrak{H}(u)$ is a multiplicative subset of $\operatorname{Inv} A$.

ii) Let R denote the tight subextension $A_{[\mathfrak{H}(u)]}$ of the tight hull $T(A)$ (cf. II, §4). The valuation u extends in a unique way to a valuation $v \colon R \to \Gamma \cup \infty$. This valuation is tight, and $v(R) = \Gamma \cup \infty$, $A_v = A$.

Assertion i) in this theorem is evident from Proposition 2. In order to prove assertion ii) we need an easy lemma (which will be also useful later on).

Lemma 4.9. Let $u \colon A \to \Gamma \cup \infty$ be a valuation on a ring A and $\mathfrak{q} := \operatorname{supp} u$. Let $A \subset B$ be any ring extension. Then u extends in a unique way to a valuation $w \colon A_{[\mathfrak{q}]}^B \to \Gamma \cup \infty$ on the saturation $A_{[\mathfrak{q}]}^B$ of A in B with respect to the multiplicative set $A \setminus \mathfrak{q}$ (I, §1, Def.10).

Proof. u gives us a valution $\tilde{u} \colon A_\mathfrak{q} \to \Gamma \cup \infty$ in a natural way (I, §1, Def. 6). Composing $\tilde{u}$ with the natural homomorphism $A_{[\mathfrak{q}]}^B \to A_\mathfrak{q}$ we obtain a valuation w extending u. If $x \in A_{[\mathfrak{q}]}^B$ there exists some $s \in A \setminus \mathfrak{q}$ with $sx \in A$. Thus $u(sx) = u(s) + w(x)$. We conclude that w is the only valuation on $A_{[\mathfrak{q}]}^B$ extending u.   □

Proof of Theorem 4.8.ii. a) We work in the ring extension $T(A)$ of A. Let $\mathfrak{q}$ denote the support of u and $A_{[\mathfrak{q}]} := A_{[\mathfrak{q}]}^{T(A)}$. For every $\gamma \in \Gamma_+$ we have $J_\gamma^{-1} \subset A_{[\mathfrak{q}]}$. Thus $R = A_{[\mathfrak{H}(u)]} \subset A_{[\mathfrak{q}]}$. By Lemma 9 above u extends to a valuation $w \colon A_{[\mathfrak{q}]} \to \Gamma$. The restriction v of w to R is a valuation on R extending u. We see as in the proof of Lemma 7, that v is the only such valuation.

b) We verify that $A_v = A$. Let $x \in A_v$ be given. There exists some $\gamma \in \Gamma_+$ with $x \in J_\gamma^{-1}$. If $y \in J_\gamma$ then $xy \in A$. Since $v(x) \geq 0$, we have $u(xy) = v(x) + u(y) \geq \gamma$. We conclude that $xJ_\gamma \subset J_\gamma$. Multiplying by J_γ^{-1} we obtain $xA \subset A$, hence $x \in A$.

c) For $\gamma \in \Gamma$ let $I_\gamma := \{x \in R \mid v(x) \geq \gamma\}$. Since $A_v = A$ we have $I_\gamma = J_\gamma$ if $\gamma \in \Gamma_+$. Thus I_γ is R-invertible for these γ. Corollary 3 tells us that v is tight.   □

We now study the processes of coarsening and special restriction (cf. I, §1) for a tight valuation.

Notations 4.10. Let $v \colon R \to G \cup \infty$ be any valuation and H a convex subgroup of G. We denote the subring of R consisting of all

$x \in R$ with $v(x) \geq h$ for some $h \in H$ by $A_{v,H}$ and the prime ideal of $A_{v,H}$ consisting of all $x \in R$ with $v(x) > h$ for every $h \in H$ by $\mathfrak{p}_{v,H}$. We further denote the special restriction of v to $A_{v,H}$ (cf. I, §1, Def.11) by v_H. Thus $v_H(x) = v(x)$ if $x \in A_{v,H} \setminus \mathfrak{p}_{v,H}$, i.e. $v(x) \in H$, and $v_H(x) = \infty$ if $x \in \mathfrak{p}_{v,H}$. □

Remarks 4.11. These rings $A_{v,H}$ are precisely the v-convex subrings of R. Notice that $A_{v,H}$ is the valuation ring A_w of the coarsening $w = v/H$ of v and $\mathfrak{p}_{v,H}$ is the center $\mathfrak{p}_w$ of w. If the special restriction v_H is Manis then $A_{v,H} = A^R_{[\mathfrak{p}_{v,H}]}$, as is checked easily. v_H has the support $\mathfrak{p}_{v,H}$ and v/H has the same support as v. If v is Manis with value group G then the v-convex subrings of R correspond *bijectively* with the convex subgroups H of G, and v_H is Manis with value group H for every such H. □

Proposition 4.12. Assume that $v: R \twoheadrightarrow \Gamma \cup \infty$ is a tight surjective valuation and H is a convex subgroup of Γ. Let $B := A_{v,H}$. Then $w := v/H$ is a tight valuation on R with $A_w = B$, $\mathfrak{p}_w = \mathfrak{p}_{v,H}$, and $u := v_H = v|_B$ is a tight valuation on B with $A_u = A_v$, $\mathfrak{p}_u = \mathfrak{p}_v$.

Proof. We know already that $(A_w, \mathfrak{p}_w) = (A_{v,H}, \mathfrak{p}_{v,H})$ and $(A_u, \mathfrak{p}_u) = (A_v, \mathfrak{p}_v)$. It is clear from the definitions and Proposition 2 that

$$B = \bigcup_{h \in H_+} I_{-h,v} = \bigcup_{h \in H_+} I_{h,v}^{-1},$$

and, for any $\gamma \in \Gamma$, that

$$I_{\gamma+H,w} = \bigcup_{h \in H} I_{\gamma+h,v} = \bigcup_{h \in H_+} I_{\gamma-h,v} = I_\gamma B.$$

The B-module $I_\gamma B$ is R-invertible, since $(I_\gamma B)(I_{-\gamma} B) = B$. Thus w is tight. We have $u(B) = H \cup \infty$ and $A_u = A_v$, $I_{h,u} = I_{h,v}$ for $h \in H$. The A_v-modules $I_{h,v}$ with $h \in H$ are B-invertible. Thus u is tight. □

Theorem 4.13. Assume again that $v: R \twoheadrightarrow \Gamma \cup \infty$ is a tight surjective valuation. Let B be an overring of A_v in R. The following are equivalent.

  i) The extension $A_v \subset B$ is tight.

ii) B is v-convex in R.

iii) There exists a (unique) convex subgroup H of Γ with $B = A_{v,H}$.

Proof. The equivalence ii) $\Leftrightarrow$ iii) has been stated above (cf. Remark 11).

iii) $\Rightarrow$ i): We have just seen (Prop.12) that there exists a tight valuation u on B with $A_u = A_v$. This implies that extension $A_v \subset B$ is tight.

i) $\Rightarrow$ iii): We look at the set Φ consisting of all $\gamma \in \Gamma_+$ with $I_{-\gamma} := I_{-\gamma,v} \subset B$. It is evident that $0 \in \Phi$ and Φ is convex in Γ_+. If γ and δ are elements of Φ then $I_{-\gamma-\delta} = I_{-\gamma}I_{-\delta} \subset \Phi$. Thus Φ is closed in Γ under addition. We conclude that $\Phi = H_+$ with H a convex subgroup of Γ, and $A_{v,H} \subset B$. If an element $x \in B \setminus A_v$ is given there exists a B-invertible ideal I of A_v with $x \in I^{-1}$, since $A_v \subset B$ is tight (cf. II, §2, Def.1). A fortiori I is R-invertible, and Proposition 6 above tells us that $I = I_\gamma$ for some $\gamma \in \Gamma_+$. We have $x \in I_{-\gamma} \subset B$. Thus $\gamma \in H_+$ and $x \in A_{v,H}$. We conclude that $A_{v,H} = B$. $\square$

Corollary 4.14. Assume that v is a PM-valuation on R with value group Γ. The R-overrings B of A_v correspond uniquely with the convex subgroups H of Γ and with the coarsenings w of v (up to equivalence) via $B = A_{v,H} = A_w$.

Proof. This is immediate from Theorem 13 since A_v is tight in every R-overring B. $\square$

From the corollary it is evident that the overrings in a PM-extension form a chain, a fact already proved in §3 in a different way (Th.3.1).

Scholium 4.15. Let $A \subset R$ be a non trivial PM extension and v a Manis (hence PM) valuation on R with $A_v = A$. Let $\mathfrak{p} := \mathfrak{p}_v$, $\mathfrak{q} := \operatorname{supp} v$. Recall from I, §2 that $\mathfrak{p} = \mathfrak{p}_A^R$, $\mathfrak{q} = \mathfrak{q}_A^R$ {cf. Def. 2 in I, §2; we write more precisely $\mathfrak{p}_A^R$, $\mathfrak{q}_A^R$ instead of $\mathfrak{p}_A, \mathfrak{q}_A$}. Let B be an R-overring of A and w the coarsening of v with $B = A_w$. We have $B = A_{[\mathfrak{r}]}$ with $\mathfrak{r} := \mathfrak{p}_w$, and these ideals $\mathfrak{r}$ are precisely the v-convex prime ideals of A. The special restriction $u := v|_B$ is PM and $A_u = A$, $\mathfrak{p}_u = \mathfrak{p}$, $\operatorname{supp} u = \mathfrak{r}$. Thus u and w are the PM-valuations corresponding to the PM-extensions $A \subset B$ and $B \subset R$. {We know a priori from Theorem 3.1 that the extensions $A \subset B$ and $B \subset R$ are PM}. We have $\mathfrak{p}_A^B = \mathfrak{p}$, $\mathfrak{q}_A^B = \mathfrak{r}$, $\mathfrak{p}_B^R = \mathfrak{r}$, $\mathfrak{q}_B^R = \mathfrak{q}$. Recall also that the prime ideals $\mathfrak{r}$ occuring here are precisely all prime ideals $\mathfrak{r}$ of A with $\mathfrak{q} \subset \mathfrak{r} \subset \mathfrak{p}$. (Prop. I.1.10). $\qquad\qquad\qquad\qquad\qquad\square$

§5 Existence of various valuation hulls

In the following $(A, \mathfrak{p})$ is a pair consisting of a ring A and a prime ideal $\mathfrak{p}$ of A, and $A \subset R$ is a ring extension of A.

Definition 1. We call $(A, \mathfrak{p})$ *tightly valuating* (abbreviated: tv) in R if $(A, \mathfrak{p})$ is Manis in R and the associated Manis valuation of R is tight. We call $(A, \mathfrak{p})$ *Prüfer-Manis* (abbreviated: PM) *in R* if $(A, \mathfrak{p})$ is Manis in R and A is Prüfer in R, i.e. the associated Manis valuation is PM. We then also say that $(A, \mathfrak{p})$ is a *TV-pair* resp. *PM-pair in R*.

We start out to prove that there exists a unique maximal R-overring C of A such that $(A, \mathfrak{p})$ is PM in C. Later we will prove the same fact for "tv" instead of "PM".

Looking for an extension $A \subset R'$ such that $(A, \mathfrak{p})$ is PM in R' we may always assume in advance that $\mathfrak{p}$ is a maximal ideal of A, since otherwise the PM property forces $A = R'$ (cf. Cor.1.4).

Proposition 5.1. Let $A \subset B_1$ and $A \subset B_2$ be subextensions of $A \subset R$ such that $(A, \mathfrak{p})$ is PM both in B_1 and B_2. Then $B_1 \subset B_2$ or $B_2 \subset B_1$.

Proof. We may assume that $A \neq B_1$ and $A \neq B_2$. Then $\mathfrak{p}$ is a maximal ideal of A. Let $B := B_1 B_2$. We prove that $(A, \mathfrak{p})$ is PM in B. Then the claim of the proposition will follow since the B-overrings of A form a chain (Th.3.1). We know from I, §5 that A is Prüfer in B (Cor.I.5.11). Since A is ws in B and $A \neq B$ there exist prime ideals of A which are B-regular (cf. Lemma II.3.2). Let $\mathfrak{q}$ be such a prime ideal. We verify that $\mathfrak{q} \subset \mathfrak{p}$. It then will follow that $\mathfrak{p}$ is the unique B-regular maximal ideal of A, and we are done by Theorem 1.8.

Suppose that $\mathfrak{q} B_1 \neq B_1$ and $\mathfrak{q} B_2 \neq B_2$. Since A is ws both in B_1 and B_2 it follows that $B_1 \subset A_{[\mathfrak{q}]}$ and $B_2 \subset A_{[\mathfrak{q}]}$ (cf.Th.I.3.13), hence

$B \subset A_{[\mathfrak{q}]}$. This contradicts the assumption $\mathfrak{q}B = B$. We conclude that $\mathfrak{q}B_1 = B_1$ or $\mathfrak{q}B_2 = B_2$. Now Theorem 1.3 tells us that $\mathfrak{q} \subset \mathfrak{p}$. □

Lemma 5.2. Let $\mathfrak{B}$ be a chain of R-overrings of A and C the union of the rings $B \in \mathfrak{B}$. If $(A, \mathfrak{p})$ is Manis in every $B \in \mathfrak{B}$ then $(A, \mathfrak{p})$ is Manis in C.

This follows immediately from the construction in I.1.20. Another proof of the lemma runs as follows.

Proof. Let $x \in C \setminus A$ be given. We choose some $B \in \mathfrak{B}$ with $x \in B$. Since $(A, \mathfrak{p})$ is Manis in B there exists some $x' \in \mathfrak{p}$ with $xx' \in A \setminus \mathfrak{p}$. This proves that $(A, \mathfrak{p})$ is Manis in C (Th.I.2.4). □

Theorem 5.3. There exists an R-overring C of A such that the following holds: If B is any R-overring of A then $(A, \mathfrak{p})$ is PM in B iff $B \subset C$.

Proof. Let $\mathfrak{B}$ denote the set of all R-overrings B of A such that $(A, \mathfrak{p})$ is PM in B. We have $A \in \mathfrak{B}$, thus $\mathfrak{B}$ is not empty. Let C denote the union of the sets $B \in \mathfrak{B}$. Proposition 1 tells us that $\mathfrak{B}$ is a chain. Thus C is a subring of R containing A. Lemma 2 tells us that $(A, \mathfrak{p})$ is Manis in C. Every $B \in \mathfrak{B}$ is contained in the Prüfer hull $P(A, R)$ of A in R (cf. I, §5). Thus $C \subset P(A, R)$, and we conclude that A is Prüfer in C and $(A, \mathfrak{p})$ is PM in C. If B is a C-overring of A then $(A, \mathfrak{p})$ is Manis in B and A is Prüfer in B, hence $B \in \mathfrak{B}$. □

Definition 2. We call this ring C the *PM-hull of* $(A, \mathfrak{p})$ *in* R and denote it by $PM(A, \mathfrak{p}, R)$. For the PM-hull of $(A, \mathfrak{p})$ in the complete ring of quotients $Q(A)$ we write more briefly $PM(A, \mathfrak{p})$. We call $PM(A, \mathfrak{p})$ the *PM-hull of the pair* $(A, \mathfrak{p})$.

Notice that $PM(A, \mathfrak{p})$ is also the PM-hull of $(A, \mathfrak{p})$ in the Prüfer hull $P(A)$ (and in $T(A)$ or $M(A)$ as well). Of course, if $\mathfrak{p}$ is not maximal in A, then $PM(A, \mathfrak{p}) = A$.

From Theorem 3 and the theory of weakly surjective hulls in I, §3 (Prop.I.3.14) the following is obvious.

Scholium 5.4. Let $A \subset B$ be any ring extension such that $(A, \mathfrak{p})$ is PM in B. Then there exists a unique homomorphism $\varphi: B \to PM(A, \mathfrak{p})$ over A, and φ is injective. □

We now give a description "from above" of the PM-hull $\text{PM}(A, \mathfrak{p}, R)$ in the case that $A \subset R$ is Prüfer.

Theorem 5.5. Assume that A is Prüfer in R and $\mathfrak{p}$ is a maximal ideal of A. Then $\text{PM}(A, \mathfrak{p}, R)$ is the intersection of the subrings $A_{[\mathfrak{m}]}$ of R with $\mathfrak{m}$ running through all (R-regular) maximal ideals of A different from $\mathfrak{p}$. {This means $\text{PM}(A, \mathfrak{p}, R) = R$ if there are no such ideals $\mathfrak{m}$.}

Proof. If $\mathfrak{p}$ is the only maximal ideal of A then A is local, hence $(A, \mathfrak{p})$ is PM in R (cf. e.g. Th.1.8), and $\text{PM}(A, \mathfrak{p}, R) = R$. Assume now that A has maximal ideals $\mathfrak{m} \neq \mathfrak{p}$, and let C denote the intersection of the rings $A_{[\mathfrak{m}]}$ with $\mathfrak{m}$ running through these ideals. We will prove directly, i.e. without using Theorem 3, that $(A, \mathfrak{p})$ is PM in C and that C contains every R-overring B of A with $(A, \mathfrak{p})$ PM in B.

If $\mathfrak{m} \in \text{Max} A$, $\mathfrak{m} \neq \mathfrak{p}$, then $C \subset A_{[\mathfrak{m}]}$, hence $\mathfrak{m} C \neq C$. If also $\mathfrak{p} C \neq C$, then A has no C-regular maximal ideals, and we conclude that $C = A$ (cf. §1 or Lemma II.3.2). In this case certainly $(A, \mathfrak{p})$ is PM in C. Otherwise $\mathfrak{p}$ is the only C-regular maximal ideal of A. Then we know by Theorem 1.8 that $(A, \mathfrak{p})$ is PM in C.

Let now B be an R-overring of A, such that $(A, \mathfrak{p})$ is PM in B and $A \neq B$. Then $\mathfrak{p}$ is the only B-regular maximal ideal of A. Thus $B \subset A_{[\mathfrak{m}]}$ for every maximal ideal $\mathfrak{m} \neq \mathfrak{p}$ of A (cf. Th.I.3.13), and we conclude that $B \subset C$. $\qquad \square$

Remark 5.6. If $A \subset R$ is any ring extension then, applying the proof of Theorem 5 to the extension $A \subset P(A, R)$, we obtain a new proof of Theorem 3. $\qquad \square$

Still another proof of Theorem 3 will be given in part II of the book.

Lemma 5.7. Assume that B_1 and B_2 are overrings of A in R and that there exist Manis valuations v_1 and v_2 on B_1, B_2 respectively with $A_{v_1} = A_{v_2} = A$, $\mathfrak{p}_{v_1} = \mathfrak{p}_{v_2} = \mathfrak{p}$.
i) Then $B_{1[\mathfrak{p}]} \subset B_{2[\mathfrak{p}]}$ or $B_{2[\mathfrak{p}]} \subset B_{1[\mathfrak{p}]}$.
ii) Assume that $B_{1[\mathfrak{p}]} \subset B_{2[\mathfrak{p}]}$. Then $\mathfrak{H}(v_1) \subset \mathfrak{H}(v_2)$ (cf. Def. 2 in §4) and $\text{supp } v_1 \supset \text{supp } v_2$.

Proof. We assume without loss of generality that the valuations $v_i (i = 1, 2)$ are surjective, $v_i(B_i) = \Gamma_i \cup \infty$. Let $\tilde{v}_i : (B_i)_{\mathfrak{p}} \twoheadrightarrow \Gamma_i \cup \infty$ denote the localization of v_i and $v_i' : (B_i)_{[\mathfrak{p}]} \twoheadrightarrow \Gamma_i \cup \infty$ the "restriction" of $\tilde{v}_i$ to $(B_i)_{[\mathfrak{p}]}$, which is the unique extension of v_i to $(B_i)_{[\mathfrak{p}]}$. Of course, all these valuations are again Manis, and $\tilde{v}_1$, $\tilde{v}_2$ even are PM. {N.B. Every local Manis valuation is PM, as follows right from the definition of Prüfer extensions I, §5, Def.1.} Proposition 1 tells us that $B_{1\mathfrak{p}} \subset B_{2\mathfrak{p}}$ or $B_{2\mathfrak{p}} \subset B_{1\mathfrak{p}}$, and this implies that $B_{1[\mathfrak{p}]} \subset B_{2[\mathfrak{p}]}$ or $B_{2[\mathfrak{p}]} \subset B_{1[\mathfrak{p}]}$. We assume that $B_{1[\mathfrak{p}]} \subset B_{2[\mathfrak{p}]}$. We use the new notations $C := B_{1[\mathfrak{p}]}$, $D := B_{2[\mathfrak{p}]}$, $w := v_2'$. The valuation v_1' is equivalent to the special restriction $u := w|_C$, since also u is Manis (cf. Prop.I.1.17) and $A_u = A_w (= A_{[\mathfrak{p}]})$, $\mathfrak{p}_u = \mathfrak{p}_w (= \mathfrak{p}_{[\mathfrak{p}]})$. Replacing v_1 by an equivalent valuation we assume that $v_1' = u$. Now Γ_1 is a convex subgroup of Γ_2. Let $I \in \mathfrak{H}(v_1)$ be given. We have an element $\alpha \in (\Gamma_1)_+$ with $I = I_{\alpha, v_1}$. Then $(I_{\alpha, v_1})_{[\mathfrak{p}]} = I_{\alpha, u} = I_{\alpha, w} = (I_{\alpha, v_2})_{[\mathfrak{p}]}$. But both ideals I_{α, v_1} and I_{α, v_2} of A are saturated in A with respect to the multiplicative set $A \setminus \mathfrak{p}$. Thus $I_{\alpha, v_1} = I_{\alpha, v_2}$. This proves that $\mathfrak{H}(v_1) \subset \mathfrak{H}(v_2)$. Since $\operatorname{supp} v_i$ is the intersection of all ideals in $\mathfrak{H}(v_i)$, it follows that $\operatorname{supp} v_1 \supset \operatorname{supp} v_2$. $\qquad\square$

Proposition 5.8. Let B_1 and B_2 be R-overrings of A. Assume that $(A, \mathfrak{p})$ is tv both in B_1 and B_2. Then $B_1 \subset B_2$ or $B_2 \subset B_1$.

Proof. Let v_i denote the tight valuation on B_i corresponding to the pair $(A, \mathfrak{p})$ $(i = 1, 2)$. By the preceding lemma we may assume that $\mathfrak{H}(v_1) \subset \mathfrak{H}(v_2)$. Since v_i is tight we have $B_i := A_{[\mathfrak{H}(v_i)]}^R$ (cf. Scholium 4.7). Thus $B_1 \subset B_2$. $\qquad\square$

Lemma 5.9. Assume that $\mathfrak{B}$ is a non empty set of R-overrings of A and that $(A, \mathfrak{p})$ is tv in every $B \in \mathfrak{B}$. let C denote the union of the sets $B \in \mathfrak{B}$. Then C is an R-overring of A and $(A, \mathfrak{p})$ is tv in C.

Proof. Proposition 8 tells us that $\mathfrak{B}$ is totally ordered by inclusion. Thus C is a subring of R containing A. For notational convenience we choose an indexing $(B_i \mid i \in I)$ of the set $\mathfrak{B}$ with I a totally ordered index set and $B_i \subset B_j$ whenever $i < j$. For every $i \in I$ we choose a tight valuation $v_i : B_i \twoheadrightarrow \Gamma_i \cup \infty$ with $(A_{v_i}, \mathfrak{p}_{v_i}) = (A, \mathfrak{p})$. Using the construction in I.1.20 we obtain a surjective Manis valuation $v : C \twoheadrightarrow \Gamma \cup \infty$ such that $v|_{B_i} \sim v_i$ for every $i \in I$. We now replace

v_i by $v|_{B_i}$ for every $i \in I$. Then we have $v|_{B_i} = v_i$, and Γ_i is convex subgroup of Γ for every $i \in I$. Moreover, if $i < j$ then $\Gamma_i \subset \Gamma_j$, and Γ is the union of the subgroups Γ_i.

Let $\gamma \in \Gamma$ be given with $\gamma \leq 0$. We choose some $i \in I$ with $\gamma \in \Gamma_i$, and then have $I_{\gamma,v} = I_{\gamma,v_i}$. Since v_i is tight, the ideal $I_{\gamma,v}$ is invertible. Now Corollary 4.3 (or already Prop.4.2) tells us that v is tight. Thus $(A, \mathfrak{p})$ is tv in C. $\square$

Choosing for $\mathfrak{B}$ the set of all R-overrings B of A such that $(A, \mathfrak{p})$ is tv in B we obtain

Theorem 5.10. There exists an R-overring C of A such that the following holds: $(A, \mathfrak{p})$ is tv in C, and C contains every other R-overring B of A such that $(A, \mathfrak{p})$ is tv in B. $\square$

Definition 3. We call this ring C the *tight valuation hull* (abbreviated: *TV-hull*) of $(A, \mathfrak{p})$ in R, and we denote it by $\mathrm{TV}(A, \mathfrak{p}, R)$. In the case $R = Q(A)$ we write more shortly $\mathrm{TV}(A, \mathfrak{p})$ for this extension of A and call it the *tight valuation hull of the pair* $(A, \mathfrak{p})$. $\square$

Of course, $\mathrm{TV}(A, \mathfrak{p}) = \mathrm{TV}(A, \mathfrak{p}, T(A)) = \mathrm{TV}(A, \mathfrak{p}, M(A))$. If $A \subset B$ is any ring extension such that $(A, \mathfrak{p})$ is tv in B then there exists a unique homomorphism $\varphi \colon B \to \mathrm{TV}(A, \mathfrak{p})$ over A, and φ is injective.

We add observations on the transitivity of the properties PM and tv.

Proposition 5.11. Let B and C be subrings of R with $A \subsetneqq B \subsetneqq C$. Let $\mathfrak{q}$ denote the conductor of A in B and $\mathfrak{r}$ the conductor of B in C. Assume that $(A, \mathfrak{p})$ is Manis in B and $(B, \mathfrak{q})$ is Manis in C.

i) Then $(A, \mathfrak{p})$ is Manis in C and $\mathfrak{r}$ is the conductor of A in C.
ii) If $(A, \mathfrak{p})$ is PM in B and $(B, \mathfrak{q})$ is PM in C, then $(A, \mathfrak{p})$ is PM in C.
iii) If $(A, \mathfrak{p})$ is PM in B and $(B, \mathfrak{q})$ is tv in C, then $(A, \mathfrak{p})$ is tv in C.

Proof. A moment of reflection on Proposition I.2.8 reveals that assertion (i) follows from that proposition. But we prefer to give another proof of (i), actually a variation of some of the arguments in the proof of I.2.8, which perhaps can be grasped more easily.

Let $x \in C \setminus A$ be given. We look for some $y \in \mathfrak{p}$ with $xy \in A \setminus \mathfrak{p}$. If $x \in B$ then such an element y exists since $(A, \mathfrak{p})$ is Manis in B. Assume now that $x \in C \setminus B$. Since $(B, \mathfrak{q})$ is Manis in R there exists some $u \in \mathfrak{q}$ with $xu \in B \setminus \mathfrak{q}$. Since $\mathfrak{q}$ is the conductor of A in B and $(A, \mathfrak{p})$ is Manis in B there exists some $v \in B \setminus \mathfrak{q}$ with $xuv \in A \setminus \mathfrak{p}$. We have $uv \in \mathfrak{q} \subset \mathfrak{p}$, and we are done. Thus $(A, \mathfrak{p})$ is Manis in R. Let $\mathfrak{r}'$ denote the conductor of A in C. We have $\mathfrak{r}' \subset \mathfrak{r}$. On the other hand $\mathfrak{r} \subset \mathfrak{q} \subset A$. Thus $\mathfrak{r} \subset \mathfrak{r}'$. We conclude that $\mathfrak{r}' = \mathfrak{r}$.

Assertion (i) being proved, assertion (ii) is now evident, since the extension $A \subset C$ is Prüfer if both $A \subset B$ and $B \subset C$ are Prüfer (cf. Th.I.5.6).

It remains to prove assertion (iii). Let $v \colon C \twoheadrightarrow \Gamma \cup \infty$ be a surjective Manis valuation, unique up to equivalence, with $A_v = A$, $\mathfrak{p}_v = \mathfrak{p}$, $\operatorname{supp} v = \mathfrak{r}$. Let further w be a Manis valuation on R with $A_w = B$, $\mathfrak{p}_w = \mathfrak{q}$. Theorem I.2.6 tells us that w is coarser than v. Replacing w by an equivalent valuation we have $w = v/H$ with H a convex subgroup of Γ. On the other hand, the special restriction $u := v|_B = v_H$ is a Manis valuation on B with $A_u = A$, $\mathfrak{p}_u = \mathfrak{p}$, $\operatorname{supp} u = \mathfrak{q}$. We have $\operatorname{supp} w = \operatorname{supp} v = \mathfrak{r}$.

We now assume that u is PM and w is tight. Let $\gamma \in \Gamma$ be given. We have to verify that $I_\gamma := I_{\gamma, v}$ is C-invertible, and then will know that v is tight.

For every $h \in H$ we have $I_{h,v} = I_{h,u}$. Thus $I_h := I_{h,v}$ is B-invertible, a fortiori C-invertible. We conclude by Proposition 4.2 that $I_\gamma I_h = I_{\gamma + h}$, $I_{-\gamma} I_h = I_{-\gamma + h}$, for every $\gamma \in \Gamma$. {Of course, $I_{-\gamma} := I_{-\gamma, v}$}. It now follows that $I_{\gamma + H, w} = I_\gamma B$ and $I_{-\gamma + H, w} = I_{-\gamma} B$ (cf. the proof of Prop.4.12). Since w is tight, we obtain $I_\gamma I_{-\gamma} B = B$. We choose finitely generated A-submodules J_1 of I_γ and J_2 of $I_{-\gamma}$ such that $J_1 J_2 B = B$ and moreover J_1 contains an element x with $v(x) = \gamma$. Then $J_1 J_2$ is a finitely generated B-regular ideal of A. Since A is Prüfer in B, this ideal is B-invertible (Th.II.1.8), hence C-invertible. Thus J_1 is C-invertible. It follows by Proposition 4.6 that $J_1 = I_\alpha$ with some element $\alpha \geq \gamma$ of Γ. But J_1 contains the element x with $v(x) = \gamma$. Thus $J_1 = I_\gamma$, and we conclude that I_γ is C-invertible. $\square$

Proposition 5.12. Assume that B is an R-overring of A different from A. Let $\mathfrak{q}$ denote the conductor of A in B.

i) If $(A, \mathfrak{p})$ is tv in B then $(B, \mathfrak{q})$ is tv in $\mathrm{TV}(A, \mathfrak{p}, R)$, and $\mathrm{TV}(A, \mathfrak{p}, R) \subset \mathrm{TV}(B, \mathfrak{q}, R)$.

ii) If $(A, \mathfrak{p})$ is PM in B then $\mathrm{TV}(A, \mathfrak{p}, R) = \mathrm{TV}(B, \mathfrak{q}, R)$ and $\mathrm{PM}(A, \mathfrak{p}, R) = \mathrm{PM}(B, \mathfrak{q}, R)$.

Proof. We prove the assertions about the TV-hulls, leaving the (easier) proof of the assertion on the PM-hulls to the reader. Let $C := \mathrm{TV}(A, \mathfrak{p}, R)$ and $D := \mathrm{TV}(B, \mathfrak{q}, R)$. Assume that $(A, \mathfrak{p})$ is tv in B. Then $B \subset C$. Since the extension $A \subset B$ is tight, we conclude from Theorem 4.13 and Proposition 4.12 that $(B, \mathfrak{q})$ is tv in C. Thus $C \subset D$. If $(A, \mathfrak{p})$ is PM in B then we know by Proposition 11 that $(A, \mathfrak{p})$ is tv in D, hence $D \subset C$, and we conclude that $C = D$. $\square$

How about the existence of a "Manis valuation hull" in analogy to the hulls $\mathrm{PM}(A, \mathfrak{p}, R)$ and $\mathrm{TV}(A, \mathfrak{p}, R)$? We are only able to deduce a weaker result.

Definition 4. We call the pair $(A, \mathfrak{p})$ *saturated in R* if $A^R_{[\mathfrak{p}]} = A$, hence also $\mathfrak{p}^R_{[\mathfrak{p}]} = \mathfrak{p}$.

Notice that, if there exists a valuation v on R with $A_v = A$, $\mathfrak{p}_v = \mathfrak{p}$, then $(A, \mathfrak{p})$ is saturated in R.

Lemma 5.13. Let B be an R-overring of A, such that $(A, \mathfrak{p})$ is saturated in B. Then $(A, \mathfrak{p})$ is Manis in B iff $(A_{\mathfrak{p}}, \mathfrak{p}_{\mathfrak{p}})$ is Manis in $B_{\mathfrak{p}}$.

Proof. If $v: B \twoheadrightarrow \Gamma \cup \infty$ is a Manis valuation on B with $A_v = A$, $\mathfrak{p}_v = \mathfrak{p}$ then, of course, the localization $\tilde{v}: B_{\mathfrak{p}} \twoheadrightarrow \Gamma \cup \infty$ is a Manis valuation with $A_{\tilde{v}} = A_{\mathfrak{p}}$, $\mathfrak{p}_{\tilde{v}} = \mathfrak{p}_{\mathfrak{p}}$. On the other hand, if $w: B_{\mathfrak{p}} \twoheadrightarrow \Gamma \cup \infty$ is a Manis valuation on $B_{\mathfrak{p}}$ with $A_w = A_{\mathfrak{p}}$, $\mathfrak{p}_w = \mathfrak{p}_{\mathfrak{p}}$ then the composition $v := w \circ \varphi$ with the natural map $\varphi: B \to B_{\mathfrak{p}}$ is a valuation with $v(B) = \Gamma \cup \infty$, hence v is Manis. We have $A_v = A^B_{[\mathfrak{p}]} = A$, $\mathfrak{p}_v = \mathfrak{p}^B_{[\mathfrak{p}]} = \mathfrak{p}$, since $(A, \mathfrak{p})$ is assumed to be saturated in B. $\square$

Theorem 5.14. Assume that $(A, \mathfrak{p})$ is saturated in R. Let C denote the preimage of the PM-hull $\mathrm{PM}(A_{\mathfrak{p}}, \mathfrak{p}_{\mathfrak{p}}, R_{\mathfrak{p}})$ in R under the natural homomorphism $R \to R_{\mathfrak{p}}$. If B is any R-overring of A then $(A, \mathfrak{p})$ is Manis in B iff $B \subset C$.

Proof. Let $D := \mathrm{PM}(A_{\mathfrak{p}}, \mathfrak{p}_{\mathfrak{p}}, R_{\mathfrak{p}})$. It is easily verified that $C_{\mathfrak{p}} = D$. By Lemma 13 we know that $(A, \mathfrak{p})$ is Manis in C. It follows that $(A, \mathfrak{p})$ is Manis in any R-overring B of A contained in C. On the other hand, if B is an R-overring of A with $(A, \mathfrak{p})$ Manis in B then $(A_{\mathfrak{p}}, \mathfrak{p}_{\mathfrak{p}})$ is Manis in $B_{\mathfrak{p}}$. Since this pair is local, it is even PM in $B_{\mathfrak{p}}$. Thus $B_{\mathfrak{p}} \subset D$, and we conclude that $B \subset C$. $\qquad\square$

Definition 5. If $(A, \mathfrak{p})$ is saturated in R then we call the maximal overring C of A in R with $(A, \mathfrak{p})$ Manis in C the *Manis valuation hull of* $(A, \mathfrak{p})$ *in* R and denote it by $\mathrm{MV}(A, \mathfrak{p}, R)$.

Contrary to the cases PM and TV we do not know how to find an "absolute" Manis valuation hull of a given pair $(A, \mathfrak{p})$, since we do not have some sort of "universal" extension $A \subset R'$ with $(A, \mathfrak{p})$ saturated in R' at our disposal.

We want to prove a result for MV-hulls similar to Proposition 12 above. For this we insert some elementary observations on Manis pairs, which may be also useful later on.

Proposition 5.15. Let B be an R-overring of A different from A. Let $\mathfrak{q}$ denote the conductor of A in B. Assume that $(A, \mathfrak{p})$ is Manis in B. Then $(A_{[\mathfrak{p}]}, \mathfrak{p}_{[\mathfrak{p}]})$ is Manis in $B_{[\mathfrak{q}]}$, and $B_{[\mathfrak{q}]} = B_{[\mathfrak{p}]} = A_{[\mathfrak{q}]}$, $\mathfrak{q}_{[\mathfrak{q}]} = \mathfrak{q}_{[\mathfrak{p}]}$. {Here $B_{[\mathfrak{q}]}$ and $\mathfrak{q}_{[\mathfrak{q}]}$ mean the saturations of B and $\mathfrak{q}$ in R with respect to the multiplicative set $B \setminus \mathfrak{q}$, while $A_{[\mathfrak{q}]}$ means the saturation of A in R with respect to $A \setminus \mathfrak{q}$.}

Proof. Let $v : B \twoheadrightarrow \Gamma \cup \infty$ be a surjective Manis valuation with $A_v = A$, $\mathfrak{p}_v = \mathfrak{p}$. This valuation extends uniquely to a valuation $v' : B_{[\mathfrak{q}]} \twoheadrightarrow \Gamma \cup \infty$ (cf. Lemma 4.9) which, of course, is again Manis. We have $A_{[\mathfrak{p}]} \subset A_{v'}$ and $\mathfrak{p}_{[\mathfrak{p}]} \subset \mathfrak{p}_{v'}$. Let $x \in A_{v'}$ be given. We choose $s \in B \setminus \mathfrak{q}$ with $sx \in B$. Then we choose $t \in B$ with $v(t) = -v(s)$. We have $st \in A \setminus \mathfrak{p}$ and $v(stx) = v'(x) \geq 0$, hence $stx \in A$. Thus $x \in A_{[\mathfrak{p}]}$. If $x \in \mathfrak{p}_{v'}$ then $v(stx) > 0$, hence $stx \in \mathfrak{p}$ and $x \in \mathfrak{p}_{[\mathfrak{p}]}$. This proves that $A_{v'} = A_{[\mathfrak{p}]}$ and $\mathfrak{p}_{v'} = \mathfrak{p}_{[\mathfrak{p}]}$. Let $x \in B_{[\mathfrak{q}]}$ be given. We choose $s \in B \setminus \mathfrak{q}$ with $sx \in B$. Then we choose $t \in B$ with $v(t) = -v(s)$. We have $st \in A \setminus \mathfrak{p}$ and $stx \in B$. Thus $x \in B_{[\mathfrak{p}]}$. This proves that $B_{[\mathfrak{q}]} = B_{[\mathfrak{p}]}$. If $stx \in A$ then $x \in A_{[\mathfrak{p}]}$. If $stx \notin A$ there exists $u \in A$ with $v(u) = -v(stx)$. We have $u \in A \setminus \mathfrak{q}$, hence $ust \in A \setminus \mathfrak{q}$, and $ustx \in A$. Thus $x \in A_{[\mathfrak{q}]}$ in both cases. We conclude that $B_{[\mathfrak{q}]} = A_{[\mathfrak{q}]}$. $\qquad\square$

This proposition gives us, among other things, the following additional information on $MV(A, \mathfrak{p}, R)$.

Corollary 5.16. Assume that $(A, \mathfrak{p})$ is saturated in R. Let $C := MV(A, \mathfrak{p}, R)$, and let $\mathfrak{q}$ denote the conductor of A in C if $A \neq C$. Otherwise put $\mathfrak{q} := \mathfrak{p}$. Then $(C, \mathfrak{q})$ is saturated in R, and $C = A_{[\mathfrak{q}]}$. Also $MV(C, \mathfrak{q}, R) = C$.

Proof. We know from the preceding Proposition that $(A, \mathfrak{p})$ is Manis in $C_{[\mathfrak{q}]}$. By the maximality property of C we have $C_{[\mathfrak{q}]} = C$. Let $D := MV(C, \mathfrak{q}, R)$. The pair $(A, \mathfrak{p})$ is Manis in D by Proposition 11.i. This implies $D \subset C$, i.e. $D = C$. $\square$

Proposition 5.17. Assume that $(A, \mathfrak{p})$ is Manis in R with associated surjective Manis valuation $v: R \twoheadrightarrow \Gamma \cup \infty$. Let B be an R-overring of A different from A, and let $\mathfrak{q}$ denote the conductor of A in B. Then there exists a convex subgroup H of Γ with $A_{[\mathfrak{q}]} = B_{[\mathfrak{q}]} = A_{v,H}$ and $\mathfrak{q} = \mathfrak{q}_{[\mathfrak{q}]} = \mathfrak{p}_{v,H}$ (cf. Notations 4.10). Thus $(B_{[\mathfrak{q}]}, \mathfrak{q})$ is Manis in R and v/H is the associated Manis valuation. {Here $\mathfrak{q}_{[\mathfrak{q}]}$ means the saturation of $\mathfrak{q}$ in R with respect to $B \setminus \mathfrak{q}$.}

Proof. Let Φ denote the set of values $v(x) \leq 0$ with x running through $B \setminus A$. There is a convex subgroup H of Γ such that H_- is the convex hull of Φ in Γ. Let $x \in A$ be given. Then $x \in \mathfrak{q}$ iff $Bx \subset \mathfrak{p}$ iff $v(bx) > 0$ for all $b \in B \setminus A$. This means that $v(x) > -\gamma$ for every $\gamma \in \Phi$ and implies $v(x) > h$ for every $h \in H$. Thus $\mathfrak{q} = \mathfrak{p}_{v,H}$. It is now immediate that $B_{[\mathfrak{q}]} \subset A_{v,H}$. On the other hand $A_{v,H} = A_{[\mathfrak{p}_{v,H}]} = A_{[\mathfrak{q}]}$ (cf. Prop.I.1.13). Thus also $A_{v,H} = B_{[\mathfrak{q}]}$. If $x \in \mathfrak{q}_{[\mathfrak{q}]}$ then there exists some $s \in B \setminus \mathfrak{q}$ with $sx \in \mathfrak{q}$. We have $v(sx) > H$, but $v(s) \leq h$ for some $h \in H$. Thus $v(x) > H$, i.e. $x \in \mathfrak{p}_{v,H} = \mathfrak{q}$. This proves $\mathfrak{q}_{[\mathfrak{q}]} = \mathfrak{q}$.

Proposition 5.18. Let B be an R-overring of A different from A, and let $\mathfrak{q}$ denote the conductor of A in B. Assume that $(A, \mathfrak{p})$ is Manis in B. Then $MV(A_{[\mathfrak{p}]}, \mathfrak{p}_{[\mathfrak{p}]}, R) = MV(B_{[\mathfrak{q}]}, \mathfrak{q}_{[\mathfrak{q}]}, R)$. Also $B_{[\mathfrak{q}]} = B_{[\mathfrak{p}]} = A_{[\mathfrak{q}]}$ and $\mathfrak{q}_{[\mathfrak{q}]} = \mathfrak{q}_{[\mathfrak{p}]}$.

Proof. We use the notations $A' := A_{[\mathfrak{p}]}$, $\mathfrak{p}' := \mathfrak{p}_{[\mathfrak{p}]}$, $B' := B_{[\mathfrak{q}]}$, and $\mathfrak{q}' := \mathfrak{q}_{[\mathfrak{q}]}$. Proposition 15 tells us that $(A', \mathfrak{p}')$ is Manis in $(B', \mathfrak{q}')$,

and that $B' = B_{[\mathfrak{p}]} = A_{[\mathfrak{q}]}$ and $\mathfrak{q}' = \mathfrak{q}_{[\mathfrak{p}]}$. Clearly $\mathfrak{q}'$ is the conductor of A' in B'. Let $C := \mathrm{MV}(A', \mathfrak{p}', R)$ and $D := \mathrm{MV}(B', \mathfrak{q}', R)$.
Since $(A', \mathfrak{p}')$ is Manis in B' and $(B', \mathfrak{q}')$ is Manis in D we know by
Proposition 11 above that $(A', \mathfrak{p}')$ is Manis in D. Thus $D \subset C$. On
the other hand $A' \subset B' \subset C$, $(A', \mathfrak{p}')$ is Manis in C, and $(B', \mathfrak{q}')$ is
saturated in C, hence $(B', \mathfrak{q}')$ is Manis in C by Proposition 17. Thus
$C \subset D$. We conclude that $C = D$.   □

§6 Inside and outside the Manis valuation hull

As before let $A \subset R$ be a ring extension and $\mathfrak{p}$ a prime ideal of
A. Having proved the existence of the valuation hulls $\mathrm{PM}(A, \mathfrak{p}, R)$,
$\mathrm{TV}(A, \mathfrak{p}, R)$ and – for $(A, \mathfrak{p})$ saturated – $\mathrm{MV}(A, \mathfrak{p}, R)$, we turn to
the question how to decide for a given element x of R whether it is
contained in any of these hulls.[*] This will occupy the present and
the next two sections. Of course, we may assume in advance that
$x \notin A$.

Theorem 6.1. Let x be an element of $R \setminus A$. The following conditions are equivalent.

(1) $(A, \mathfrak{p})$ is Manis in $A[x]$.
(2) For every $a \in A$ with $ax \notin A$ there exists some $y \in \mathfrak{p}$ with
    $yax \in A \setminus \mathfrak{p}$.
(3) $\mathfrak{p}x \not\subset \mathfrak{p}$. For every $a \in \mathfrak{p}$ with $ax \notin A$ there exists some $y \in \mathfrak{p}$
    with $yax \in A \setminus \mathfrak{p}$.

Proof. The implications $(1) \Rightarrow (2) \Rightarrow (3)$ are evident.
$(2) \Rightarrow (1)$: We have to verify the following (cf. Th.I.2.4): If $f(T)$ is
a polynomial in $A[T]$ with $f(x) \notin A$, there exists some $b \in \mathfrak{p}$ with
$bf(x) \in A \setminus \mathfrak{p}$. Then we will be done by Theorem I.2.4.

We proceed by induction on the degree d of $f(T)$. Let $d = 1$, $f(T) =$
$a_1 T + a_0$. We have $a_1 x + a_0 \notin A$, hence $a_1 x \notin A$. By hypothesis there

---

[*] Here we do not mean criteria which are useful in any practical sense,
but only criteria of theoretical interest. Questions of effectivity will be not
touched.

exists some $y \in \mathfrak{p}$ with $ya_1 x \in A \setminus \mathfrak{p}$, hence $yf(x) \in A \setminus \mathfrak{p}$. Assume now $d \geq 2$, $f(T) = a_d T^d + a_{d-1} T^{d-1} + \cdots + a_0$. We distinguish two cases.

Case 1. $a_d x \in A$. We introduce the polynomial

$$g(T) := (a_d x + a_{d-1}) T^{d-1} + a_{d-2} T^{d-2} + \cdots + a_0 \in A[T].$$

We have $g(x) = f(x) \notin A$. By induction hypothesis there exists some $y \in \mathfrak{p}$ with $yg(x) \in A \setminus \mathfrak{p}$.

Case 2. $a_d x \notin A$. There exists some $c \in \mathfrak{p}$ with $ca_d x \in A \setminus \mathfrak{p}$. Let $y := c^d a_d^{d-1} \in \mathfrak{p}$. The element

$$yf(x) = (ca_d x)^d + ca_{d-1}(ca_d x)^{d-1} + \cdots + c^d a_d^{d-1} a_0$$

lies in $A \setminus \mathfrak{p}$ since the first term on the right lies in $A \setminus \mathfrak{p}$ while the others lie in $\mathfrak{p}$.

$(3) \Rightarrow (2)$: Let $a \in A \setminus \mathfrak{p}$ be given with $ax \notin A$. We have to find an element y of $\mathfrak{p}$ with $yax \in A \setminus \mathfrak{p}$.

Case 1: $\mathfrak{p}x \not\subset A$. We choose $a_0 \in \mathfrak{p}$ with $a_0 x \notin A$. By our assumption there exists some $y_0 \in \mathfrak{p}$ with $y_0 a_0 x \in A \setminus \mathfrak{p}$. We have $(a_0 y_0)(ax) \in A \setminus \mathfrak{p}$ and $a_0 y_0 \in \mathfrak{p}$.

Case 2: $\mathfrak{p}x \subset A$. Since $\mathfrak{p}x \not\subset \mathfrak{p}$, there exists some $y \in \mathfrak{p}$ with $yx \in A \setminus \mathfrak{p}$. Then $yax = a(yx) \in A \setminus \mathfrak{p}$. $\qquad\square$

In the following we assume always that $(A, \mathfrak{p})$ is saturated in R. Let C denote the Manis valuation hull $\mathrm{MV}(A, \mathfrak{p}, R)$, and let $\mathfrak{q}$ denote the conductor of A in C if $A \neq C$. If $A = C$ then we put $\mathfrak{q} := \mathfrak{p}$ and, abusing language, we call this ideal still the conductor of A in C.

Corollary 6.2. An element x of $R \setminus A$ lies in the Manis valuation hull C iff it fulfills one of the following equivalent conditions:

(A) For every $a \in A$ with $ax \notin A$ there exists some $b \in A$ (hence $b \in \mathfrak{p}$) with $bax \in A \setminus \mathfrak{p}$.

(B) $\mathfrak{p}x \not\subset \mathfrak{p}$. For every $a \in \mathfrak{p}$ with $ax \notin A$ there exists some $b \in A$ (hence $b \in \mathfrak{p}$) with $bax \in A \setminus \mathfrak{p}$.

Proof. Notice that $x \in C$ iff $A[x] \subset C$ iff $(A, \mathfrak{p})$ is Manis in $A[x]$. $\square$

In Appendix C we will give a direct proof of the existence of Manis valuation hulls and of this corollary without using anything about Prüfer extensions.

Here is another description of the Manis valuation hull C.

Theorem 6.3. Let $x \in R \setminus A$ be given. The following are equivalent

(1) $x \in C$
(2) $(Ax) \cap A \not\subset \mathfrak{p}$. If x' is any element of A with $xx' \in A \setminus \mathfrak{p}$ and y is an element of $R \setminus A$ with $x'y \in A$ then $(Ay) \cap A \not\subset \mathfrak{p}$.
(3) There exists an element x' of A with the following properties: $xx' \in A$. If $y \in R \setminus A$ and $yx' \in A$ then $(Ay) \cap A \not\subset \mathfrak{p}$.

Proof. $(1) \Rightarrow (2)$: We have $C = A_{[\mathfrak{q}]}$ (cf. Cor.5.16). Let $x' \in A$ and $y \in R \setminus A$ be given with $xx' \in A \setminus \mathfrak{p}$ and $yx' \in A$. Since $C\mathfrak{q} \subset \mathfrak{p}$ we have $x' \in A \setminus \mathfrak{q}$. From $yx' \in A$ we conclude that $y \in A_{[\mathfrak{q}]} = C$. Since $(A, \mathfrak{p})$ is Manis in C it follows that $(Ay) \cap A \not\subset \mathfrak{p}$. Also $(Ax) \cap A \not\subset \mathfrak{p}$.
$(2) \Rightarrow (3)$: This implication is trivial.
$(3) \Rightarrow (1)$: Suppose that $x \notin C$. By Corollary 2 above there exists some $a \in A$ with $ax \notin A$ and $(Aax) \cap A \subset \mathfrak{p}$. Let x' be an element of A with the properties indicated in our hypothesis (3). We have $axx' \in A$, hence $(Aax) \cap A \not\subset \mathfrak{p}$ according to (3). This is a contradiction. We conclude that $x \in C$. $\square$

We now give a description of the conductor $\mathfrak{q}$ of A in C of similar style, but under a restriction on the extension $A \subset R$. Notice that $C = A_{[\mathfrak{q}]}$, as has been stated in Cor.5.16. Thus a good knowledge of $\mathfrak{q}$ implies a good knowledge of $C = \mathrm{MV}(A, \mathfrak{p}, R)$.

Theorem 6.4. Assume that $R \setminus A$ is closed under multiplication. Let $x \in A$ be given. The following are equivalent.

(1) $x \in \mathfrak{q}$.
(2) Either $(Rx) \cap A \subset \mathfrak{p}$, or there exist elements $x' \in R$, $y \in A$ with $xx' \in A \setminus \mathfrak{p}$, $yx' \notin \mathfrak{p}$, and $(Ry) \cap A \subset \mathfrak{p}$.
(3) Either $(Rx) \cap A \subset \mathfrak{p}$, or, for every $x' \in R$ with $xx' \in A \setminus \mathfrak{p}$, there exists an element y in A with $yx' \notin A$ and $(Ry) \cap A \subset \mathfrak{p}$.

Proof. The implication $(3) \Rightarrow (2)$ is trivial.
$(2) \Rightarrow (1)$: Let $u: C \twoheadrightarrow \Gamma \cup \infty$ denote the Manis valuation with $A_u = A$ and $\mathfrak{p}_u = \mathfrak{p}$. It has the support $\mathfrak{q}$. If $(Rx) \cap A \subset \mathfrak{p}$, then a fortiori

$(Cx) \cap A \subset \mathfrak{p}$. This implies $u(x) = \infty$, i.e. $x \in \mathfrak{q}$. Now assume that there exist elements $x' \in R$, $y \in A$ with $xx' \in A \setminus \mathfrak{p}$, $yx' \notin \mathfrak{p}$, $(Ry) \cap A \subset \mathfrak{p}$. We have $(Cy) \cap A \subset \mathfrak{p}$, hence $u(y) = \infty$. Suppose that $u(x) \neq \infty$. Then $x' \in A_{[\mathfrak{q}]} = C$ (cf.5.16) and $u(x') = -u(x)$. From $yx' \notin \mathfrak{p}$ we conclude that $u(y) \leq u(x)$. This contradicts the fact that $u(y) = \infty$. Thus $u(x) = \infty$, i.e. $x \in \mathfrak{q}$.

$(1) \Rightarrow (3)$: Assume that $(Rx) \cap A \not\subset \mathfrak{p}$. Let $x' \in R$ be given with $xx' \in A \setminus \mathfrak{p}$. Then $x' \notin C$, since $xC \subset \mathfrak{q} \subset \mathfrak{p}$. By Corollary 2 above there exists $y \in A$ with $yx' \notin A$ and $(Ayx') \cap A \subset \mathfrak{p}$. Suppose that $(Ry) \cap A \not\subset \mathfrak{p}$. We choose $y' \in R$ with $yy' \in A \setminus \mathfrak{p}$. Then $xx'yy' \in A \setminus \mathfrak{p}$. Since $(Ayx') \cap A \subset \mathfrak{p}$ we conclude that $xy' \notin A$. It follows that $xy'yx' \notin A$, since $R \setminus A$ is multiplicatively closed. But $xx'yy' \in A$, a contradiction. We conclude that $(Ry) \cap A \subset \mathfrak{p}$. $\square$

We recall some facts from I, §2.

Scholium 6.5. Let again u denote the Manis valuation on C with $A_u = A$, $\mathfrak{p}_u = \mathfrak{p}$. The prime ideals $\mathfrak{r}$ of A with $\mathfrak{q} \subset \mathfrak{r} \subset \mathfrak{p}$ form a chain, since they are precisely the u-convex prime ideals of A. The rings $A_{[\mathfrak{r}]}$ are the u-convex subrings of C, and A has in $A_{[\mathfrak{r}]}$ the conductor $\mathfrak{r}$.

All this is clear from Theorem I.2.6, if we replace $A_{[\mathfrak{r}]} = A_{[\mathfrak{r}]}^R$ by $A_{[\mathfrak{r}]}^C$. But $A_{[\mathfrak{r}]}^R \subset A_{[\mathfrak{q}]}^R = C$ (cf. Cor.5.16), and thus indeed $A_{[\mathfrak{r}]}^R = A_{[\mathfrak{r}]}^C$. $\square$

We now study the Manis valuation hull in the special case that a valuation $v: R \to G \cup \infty$ is given, and $A = A_v$, $\mathfrak{p} = \mathfrak{p}_v$. Notice that then $(A, \mathfrak{p})$ is saturated in R and $R \setminus A$ is closed under multiplication. How is the Manis valuation on $MV(A, \mathfrak{p}, R)$, which is given by the pair $(A, \mathfrak{p})$, related to v? We start with the following general observation.

Proposition 6.6. If B is an R-overring of $A = A_v$, such that $(A, \mathfrak{p}) = (A_v, \mathfrak{p}_v)$ is Manis in B, the associated Manis valuation on B is (up to equivalence) the special restriction $v|_B$.

Proof. We may assume that $A \neq B$. Let $u := v|_B$, and let w be a Manis valuation on B with $A_w = A$, $\mathfrak{p}_w = \mathfrak{p}$. Also $A_u = A$, $\mathfrak{p}_u = \mathfrak{p}$. By Proposition I.2.2 the valuations u and w have the same support, namely the conductor $\mathfrak{r}$ of A in B.

Let $x \in B \setminus \mathfrak{r}$ be given. Since w is Manis there exists some $y \in B$ with $w(x) + w(y) = 0$, i.e. $xy \in A \setminus \mathfrak{p}$. This implies that $u(x) + u(y) = 0$. We conclude that $u(B \setminus \mathfrak{r})$ is a group, i.e. u is Manis. □

Due to this proposition $C = \mathrm{MV}(A, \mathfrak{p}, R)$ is the unique maximal subring B of R containing A such that $v|_B$ is Manis.

Corollary 2 and also Theorem 3 tells us that, in the present situation, there exists a subset U of G such that an element x of $R \setminus A$ lies in C iff $v(x) \in U$. In other terms, the value $v(x)$ of a given element x of R decides already whether x is in C or not.

It seems to be worthwhile to make this observation more explicit. In the following proposition one should have in mind the example that G is as above and M is the set of values $v(R) \setminus \{\infty\}$.

Proposition 6.7. Let G be an ordered abelian group and M be a submonoid of G {i.e. $M \subset G$, $0 \in M$, $M + M \subset M$}. We use the notations $M_+ := M \cap G_+$, $M_- := M \cap G_-$, $G_0 := M \cap (-M)$. Notice that G_0 is the largest subgroup of G contained in M. Let U denote the subset of all $g \in M$ with $g < 0$ and the following property: For every $a \in M_+$ with $a + g < 0$ we have $-a - g \in M$, i.e. $a + g \in G_0$.

i) $\Gamma := U \cup (-U) \cup \{0\}$ is a subgroup of G_0.
ii) An element g of G_+ lies in Γ (i.e. $g \in (-U) \cup \{0\}$) iff $g \in M$ and $[0, g] \cap M \subset G_0$ [1]) . Thus Γ is the maximal subgroup H of G_0 such that H_+ is convex in M_+.
iii) An element g of G_- lies in Γ iff $g \in M$ and $[g, 0] \cap M \subset G_0$. Thus Γ is the maximal subgroup H of G_0 such that H_- is convex in M_-.
iv) Γ is the maximal subgroup of G_0 which is convex in M.

Proof. i): We have $U \subset G_0$ hence $\Gamma \subset G_0$. Let elements g_1 and g_2 of U be given. We prove the following:

a) $g_1 + g_2 \in U$,
b) If $g_1 < g_2$, then $g_1 - g_2 \in U$.

Once we know this it is immediate that $U \cup \{0\} \cup (-U)$ is a subgroup of G_0.

---

[1]) As usual $[0, g]$ denotes the interval $\{a \in G \mid 0 \leq a \leq g\}$.

a): We have $g_1 + g_2 \in M$. Let $h \in M_+$ be given with $g_1 + g_2 + h < 0$. We have to verify that $g_1 + g_2 + h \in G_0$.

Case 1: $g_2 + h < 0$. Now $g_2 + h \in G_0$, $g_1 \in G_0$, hence $g_1 + g_2 + h \in G_0$.

Case 2: $g_2 + h \geq 0$. Now $g_2 \in G_0 \subset M$, $h \in M$, thus $g_2 + h \in M_+$. Since $g_1 \in U$ we have again $g_1 + g_2 + h \in G_0$.

b): We have $g_1 - g_2 \in G_0 \subset M$. Let $h \in M_+$ be given with $g_1 - g_2 + h < 0$. Now $-g_2 \in G_0 \subset M$, and $-g_2 + h > 0$, thus $-g_2 + h \in M_+$. Since $g_1 \in U$ we conclude that $g_1 - g_2 + h \in G_0$.

ii): Let V denote the set of all $g \in M_+$ with $[0, g] \cap M \subset G_0$. We have to verify that $V = \Gamma_+$.

a) Let $g \in \Gamma_+$ be given. Then $g \in G_0 \subset M$. If $a \in M$ and $0 \leq a < g$, then $-g \in \Gamma_-$ and $-g + a < 0$. Thus $-g + a \in G_0$ by the definition of Γ. Since also $g \in G_0$, we have $a \in G_0$. We conclude that $g \in V$.

b) Let $g \in V$ be given. We verify that $-g \in \Gamma$, hence $g \in \Gamma$. We may assume that $g \neq 0$, hence $-g < 0$. Let $a \in M_+$ be given with $-g + a < 0$. Then $0 \leq a < g$. Since $g \in V$ we have $a \in G_0$. Also $g \in G_0$. Thus $-g + a \in G_0$. This proves that $-g \in \Gamma$.

iii): Let $g \in G_-$ be given. We may assume in advance that $g \in M_-$. Write $g = -\gamma$ with $\gamma \in M_+$.

Assume first that $g \in \Gamma$. Let $a \in M_+$ be given with $-\gamma \leq -a \leq 0$. We have to verify that $a \in G_0$. But this is evident from part ii) since $0 \leq a \leq \gamma \in \Gamma$.

Assume now that $[g, 0] \cap M \subset G_0$ and, without loss of generality, that $g \neq 0$. We have to verify that $g \in U$. Let $h \in M_+$ be given with $g + h < 0$. Then $g + h \in [g, 0] \cap M$. Thus $g + h \in G_0$. This proves that $g \in U$.

iv): Since Γ is the maximal subgroup H of G_0 such that H_+ is convex in M_+, and also such that H_- is convex in M_-, it is evident that Γ is the maximal subgroup H' of G_0 with H' convex in M.   □

Definition. We call this group $\Gamma = U \cup (-U) \cup \{0\}$ the *Manis subgroup of G with resepct to the submonoid M* and denote it by $\mathrm{MS}(G, M)$.

Up to the end of this section we stay with a valuation $v: R \to G \cup \infty$ and use the following notations, if nothing else is said.

Notations 6.8. $A := A_v$, $\mathfrak{p} := \mathfrak{p}_v$. $M := v(R) \setminus \{\infty\}$. $M_+ := M \cap G_+$, $M_- := M \cap G_-$. $\Gamma := \mathrm{MS}(G, M)$. $G_0 := M \cap (-M)$, i.e. G_0 is the largest subgroup of G contained in the submonoid M. $C := \mathrm{MV}(A, \mathfrak{p}, R)$, $u := v|_C$. $\mathfrak{q} := \mathfrak{q}_A^C =$ the conductor of A in C if $C \neq A$, while $\mathfrak{q} := \mathfrak{p}$ if $C = A$. By abuse of language we call $\mathfrak{q}$ the conductor of A in C also if $A = C$.

Theorem 6.9. i) u is a Manis valuation with value group Γ. An element x of $R \setminus A$ lies in C iff $v(x) \in \Gamma$. If B is an R-overring of A then $v|_B$ is Manis iff $B \subset C$.
ii) $\mathfrak{q}$ is the set of all $x \in R$ with $v(x) > \Gamma$ { i.e. $v(x) > \gamma$ for every $\gamma \in \Gamma$}.
iii) C is the set of all $x \in R$ with $v(x) \geq \gamma$ for some $\gamma \in \Gamma$. In particular, C is v-convex in R.
iv) Let $\hat{\Gamma}$ denote the convex hull of Γ in G. The coarsening $w := v/\hat{\Gamma}$ of v has the valuation ring $A_w = C$ and the center $\mathfrak{p}_w = \mathfrak{q}$.

Proof. Part i) is obvious from Corollary 2 (Condition A), Proposition 7.iii, and the meaning of the Manis valuation hull. The conductor of A in C is, of course, the support of u. Let an element x in R be given. Observe that $u(x) = v(x)$ if $x \in C$ and $u(x) \neq \infty$, since u is the special restriction of v to C. If $v(x) > \Gamma$ then $x \in A \subset C$, and we conclude that $u(x) = \infty$. On the other hand, if $x \in C$ and $u(x) = \infty$, then $x \in \mathfrak{q}$, hence $xC \subset A$, and this implies that $v(x) + \gamma \geq 0$ for every $\gamma \in \Gamma$, i.e. $v(x) > \Gamma$. Thus $\mathfrak{q} = \{x \in R \mid v(x) > \Gamma\}$. We have $v(C \setminus \mathfrak{q}) = \Gamma$. Assertions iii) and iv) now follow from Proposition I.1.13, since $C = A_{[\mathfrak{q}]}$ (cf. Cor.5.16). $\qquad\square$

Corollary 6.10. An element x of A lies in $A \setminus \mathfrak{q}$ iff $v(y) \in G_0$ for every $y \in A$ with $v(y) \leq v(x)$. $\qquad\square$

We also arrive at a pleasant description of the conductor $\mathfrak{q}$ of A in C.

Corollary 6.11. $\mathfrak{q}$ is the smallest v-convex ideal of A which contains all elements y of A with $v(y) \notin -M$.

Proof. We have $\mathfrak{q} = \{x \in A \mid v(x) > \Gamma\}$ (cf.Th.9). Thus $\mathfrak{q}$ is v-convex in A. Let $x \in \mathfrak{q}$ be given. By Corollary 10 there exists some $y \in A$ with $v(y) \leq v(x)$ and $v(y) \notin G_0$. Now $v(y) \notin G_0$ iff $v(y) \notin -M$. The claim follows.   □

A look at the previous Theorem 4 reveals that these corollaries are contained in that theorem. But we have gained some additional insight via the "abstract" Proposition 7.

Some part of Theorem 9 easily generalizes to convex subgroups of Γ instead of Γ itself. We leave the proof of the following proposition to the reader (cf. Prop.5.18 for the last statement there).

Proposition 6.12. Let H be a convex subgroup of Γ. Let $\mathfrak{p}_H$ denote the set of all $x \in R$ with $v(x) > H$, and let $A_{\Gamma,H}$ denote the set of all $x \in R$ with $v(x) \in \Gamma$ and $v(x) \geq h$ for some $h \in H$. Then $A_{\Gamma,H}$ is a subring of C containing A. The special restriction of v to $A_{v,H}$ is a Manis valuation with value group H. If $H \neq \{1\}$ then $A_{\Gamma,H} \neq A$ and $\mathfrak{p}_H$ is the conductor of A in $A_{\Gamma,H}$, while $A_{\Gamma,H} = A$ and $\mathfrak{p}_H = \mathfrak{p}$ if $H = \{1\}$. In both cases $(A_{\Gamma,H}, \mathfrak{p}_H)$ is saturated in R and $A_{\Gamma,H} = A_{[\mathfrak{p}_H]}$. Moreover $A_{\Gamma,H} \setminus \mathfrak{p}_H$ is the set of all $x \in R$ with $v(x) \in H$.[2] Finally, let $\hat{H}$ denote the convex hull of H in G, and $w := v/\hat{H}$. Then $A_w = A_{\Gamma,H}$, $\mathfrak{p}_w = \mathfrak{r}$, and $\mathrm{MV}(A_{\Gamma,H}, \mathfrak{r}) = C$.   □

Instead of starting with a convex subgroup H of Γ we may start with an C-overring B of A and look for a convex subgroup of Γ associated to B.

Proposition 6.13. Let B be an R-overring of A contained in the Manis valuation hull $C = \mathrm{MV}(A, \mathfrak{p}, R)$. Let $\mathfrak{r}$ denote the conductor of A in B if $A \neq B$. Otherwise let $\mathfrak{r} = \mathfrak{p}$. Then there exists a unique convex subgroup H of Γ with $\mathfrak{r} = \mathfrak{p}_H$ (cf. notations in Prop. 12). We have $H = v(B \setminus \mathfrak{r})$, and $A_{[\mathfrak{r}]} = B_{[\mathfrak{r}]} = B_{[\mathfrak{p}]} = A_{\Gamma,H}$. Also $\mathfrak{r} = \mathfrak{r}_{[\mathfrak{r}]}$. {Here $B_{[\mathfrak{r}]}$ and $\mathfrak{r}_{[\mathfrak{r}]}$ denote the saturations of B and $\mathfrak{r}$ in R with respect to $B \setminus \mathfrak{r}$, while $A_{[\mathfrak{r}]}$ denotes the saturation of A in R with respect $A \setminus \mathfrak{r}$.}

Proof. We have $\operatorname{supp} u = \mathfrak{q} \subset \mathfrak{r} \subset \mathfrak{p} = \mathfrak{p}_u$. By Theorem I.2.6.ii there exists a convex subgroup H of Γ such that $u(A \setminus \mathfrak{r}) = H_+$ and

---

[2] If necessary, we will write more systematically $A_{v,\Gamma,H}$ and $\mathfrak{p}_{v,H}$ instead of $A_{\Gamma,H}$ and $\mathfrak{p}_H$.

$\mathfrak{r} = \mathfrak{p}_{u,H}$. Since $u' := u|_B$ is Manis and supp $u' = \mathfrak{r}$, $A_{u'} = A$, we have $u(B \setminus \mathfrak{r}) = u'(B \setminus \mathfrak{r}) = H$. It follows that $v(A \setminus \mathfrak{r}) = H_+$, $\mathfrak{r} = \mathfrak{p}_{v,H}$, and $v(B \setminus \mathfrak{r}) = H$. Proposition 5.18 tells us that $B_{[\mathfrak{r}]} = B_{[\mathfrak{p}]} = A_{[\mathfrak{r}]}$. Clearly $B_{[\mathfrak{r}]} \subset A_{v,\Gamma,H}$. Let $x \in A_{v,\Gamma,H}$ be given. There exists some $h \in H$ with $v(x) \geq h$ and some $s \in B \setminus \mathfrak{r}$ with $v(s) = -h$. We have $v(sx) \geq 0$, hence $sx \in A \subset B$. Thus $x \in B_{[\mathfrak{r}]}$, and we conclude that $B_{[\mathfrak{r}]} = A_{v,\Gamma,H}$. Clearly $v(x) > H$ for every $x \in \mathfrak{r}_{[\mathfrak{r}]}$. Thus $\mathfrak{r}_{[\mathfrak{r}]} = \mathfrak{r}$. $\square$

§7 The TV-hull in a valuative extension

Definition 1. We call a ring extension $A \subset R$ *valuative* if there exists a valuation v on R with $A_v = R$. We then also say that A is *valuative in* R.

Remark. In Huckaba's book [Huc] and related literature another terminology is used: If A is valuative in R there A is called a *paravaluation ring* in R, and v is called a *paravaluation*. We have explained in the introduction why we do not follow this terminology.

Griffin has given an intrinsic characterization of valuative extensions, cf. [Huc, Th.5.5]. We will have no occasion to use his result.

Notice that the set $R \setminus A$ is closed under multiplication if A is valuative in R. If in addition $A \neq R$ then the set $\mathfrak{p}_A := \mathfrak{p}_A^R := \{x \in A \mid xy \in A$ for some $y \in R \setminus A\}$ is a prime ideal of A, and also the conductor $\mathfrak{q}_A = \mathfrak{q}_A^R$ of A in R is a prime ideal, as has been stated in Theorem I.2.1.

Remark 7.1. Assume that $A \subset R$ is a valuative extension, $A \neq R$, and $v: R \to G \cup \infty$ is a valuation on R with $A_v = A$. Let $v^* := v|_R$ denote the special restriction of v to R, i.e. the special valuation $v|_{c_v}(G)$ associated to v (cf. I, §1). Then $A_{v^*} = A$ and $\mathfrak{p}_v = \mathfrak{p}_{v^*} \supset \mathfrak{p}_A$. $\square$

For a valuative extension $A \subsetneq R$ there may be (up to equivalence) more than one special valuation v on R with $A_v = A$, cf. the example below. But, if v is a Manis valuation on R with $A_v = A$ and $A \neq$

R, then $\mathfrak{p}_v = \mathfrak{p}_A$, and v is the only special valuation on R (up to equivalence) with $A_v = A$, as is clear from I, §2 and Theorem 3.12.

Example 7.2. Let G be an ordered abelian group and H a convex subgroup of G with $H \neq G$. We choose an element ξ of G with $\xi > H$. Let k be a domain and $u : k \to H \cup \infty$ be a valuation with $\operatorname{supp} u = 0$. We extend u to a valuation $v : R \to G \cup \infty$ on the polynomial ring $R := k[x]$ in one indeterminate x over k by the formula

$$v\Big(\sum_{i=0}^{n} a_i x^i\Big) = \min_{0 \leq i \leq n} \big(u(a_i) - i\xi\big)$$

cf. [Bo, VI, §10, Lemma 1]. v is special. We have $A_v = A_u$, $\mathfrak{p}_v = \mathfrak{p}_u$, and $\operatorname{supp} v = 0$.

We now iterate this construction. Let G be an ordered abelian group containing two convex subgroups $H_1 \subsetneqq H_2 \subsetneqq G$. Choose $\xi \in H_2$ and $\eta \in G$ with $\xi > H_1$ and $\eta > H_2$. We start with a valuation $u : k \to H_1 \cup \infty$ which has support zero. We then extend u to a valuation $w_1 : k[x] \to H_2 \cup \infty$ in the same way as above with $w_1(x) = -\xi$ and then extend w_1 to a valuation $v_1 : k[x, y] \to G \cup \infty$ with $v_1(y) = -\eta$, again in the same way. On the other hand we extend u to $w_2 : k[y] \to H' \cup \infty$ in this way with $w_2(y) = -\xi$ and then extend w_2 in this way to a valuation $v_2 : k[x, y] \to G \cup \infty$ with $v_2(x) = -\eta$. Both valuations v_1 and v_2 are special and have $A_{v_1} = A_{v_2} = A_u$, $\mathfrak{p}_{v_1} = \mathfrak{p}_{v_2} = \mathfrak{p}_u$, $\operatorname{supp} v_1 = \operatorname{supp} v_2 = 0$. But v_1 and v_2 are not equivalent. □

Definition 2. Let $A \subset R$ be a ring extension and $\mathfrak{p}$ a prime ideal of A. We say that *the pair $(A, \mathfrak{p})$ is valuative in R* if there exists a valuation v on R with $A_v = A$ and $\mathfrak{p}_v = \mathfrak{p}$.

In the following $(A, \mathfrak{p})$ is a pair which is valuative in a given extension R of A. We will give somewhat explicit descriptions of the TV-hull $\mathrm{TV}(A, \mathfrak{p}, R)$. These descriptions can be used in principle if more generally $(A, \mathfrak{p})$ is saturated in R since then we may replace R in advance by the Manis valuation hull $\mathrm{MV}(A, \mathfrak{p}, R)$, which we have described in §6. The general case, where $(A, \mathfrak{p})$ is an arbitrary pair in a ring R, will remain open.

Notations 7.3. We choose a valuation $v: R \to G \cup \infty$ with $A_v = A$ and $\mathfrak{p}_v = \mathfrak{p}$. $M := v(R) \setminus \{\infty\}$, $M_+ := M \cap G_+$, $M_- := M \cap G_-$, $G_0 := M \cap (-M)$.

It follows from Proposition 6.6, that $\mathrm{TV}(A, \mathfrak{p}, R)$ (resp. $\mathrm{PM}(A, \mathfrak{p}, R)$) is the unique maximal R-overring B of A such that the special restriction $v|_B$ is tight (resp. PM).

Lemma 7.4. Let B be an R-overring of A, $u := v|_B$, $N := u(B) \setminus \{\infty\}$, and $N_- := N \cap G_-$. Then $N_- = v(B \setminus A)$. The following are equivalent.

(1) $I_{\gamma,v} = I_{\gamma,u}$ for every $\gamma \in N_-$.
(2) B is v-convex in R.
(3) $I_{\gamma,v} = I_{\gamma,u}$ for every $\gamma \in N$.

Proof. We may assume that $B \neq A$. It is evident from the definition of $u = v|_B$ that $v(B \setminus A) = N_-$.
(1) $\Rightarrow$ (2): B is the union of the sets $I_{\gamma,u}$ with γ running through N_-. Since $I_{\gamma,u} = I_{\gamma,v}$ for these γ the sets $I_{\gamma,u}$ are v-convex. Thus B is v-convex.
(2) $\Rightarrow$ (3): Let $\gamma \in N$ be given. We choose $x \in B$ with $v(x) = \gamma$. We pick some $y \in R$. If $v(y) \geq v(x)$ then $y \in B$, since B is v-convex. We conclude that $u(y) \geq v(y) \geq \gamma$. This proves that $I_{v,\gamma} \subset I_{u,\gamma}$. On the other hand, if $y \in B$ and $u(y) \geq \gamma$, then either $v(y) > H$ with H the characteristic group of $v|_B$ (cf. I, §1, Def.3) or $v(y) = u(y)$. In the first case certainly $v(y) > N$, since N is contained in this characteristic subgroup. Thus $v(y) \geq \gamma$ in both cases. This proves that $I_{u,\gamma} \subset I_{v,\gamma}$.
The implication (3) $\Rightarrow$ (1) is trivial. $\qquad\square$

Proposition 7.5. Assume that B is an R-overring of A such that the special restriction $u := v|_B$ is tight. Then B is v-convex in R.

Proof. Let $x \in B \setminus A$ be given and $\gamma := u(x)$, hence $\gamma < 0$. We verify that $I_{\gamma,u} = I_{\gamma,v}$, and then will be done by the preceding lemma. The A-module $I_{\gamma,u}$ is B-invertible. Proposition 4.2 tells us that $I_{\gamma,u}^{-1} = I_{-\gamma,u}$, and that there exists some $z \in A$ with $u(z) = -\gamma$. Then $v(z) = -\gamma$. We have $I_{-\gamma,u} = I_{-\gamma,v}|_A = I_{-\gamma,v}$. Thus $I_{-\gamma,v}$ is B-invertible and a fortiori R-invertible. We conclude, again by Proposition 4.2, that $I_{-\gamma,v}^{-1} = I_{\gamma,v}$. Thus $I_{\gamma,v} = I_{\gamma,u}$. $\qquad\square$

Lemma 7.6. Let $x \in R \setminus A$, $\gamma := v(x)$. Assume that $I_{\alpha,v}$ is R-invertible for every $\alpha \in [\gamma, 0[$ with $\alpha - \gamma \in M$. Then $x \in MV(A, \mathfrak{p}, R)$.

Proof. Let $a \in A$ be given with $ax \in R \setminus A$. According to Corollary 6.2 it suffices to verify that there exists some $y \in A$ with $yax \in A \setminus \mathfrak{p}$. Let $\alpha := v(ax)$. Then $\gamma \leq \alpha < 0$ and $\alpha - \gamma = v(a) \in M$. Thus $I_{\alpha,v}$ is R-invertible. By Proposition 4.2 there exists some $y \in A$ with $v(y) = -\alpha$. We have $yax \in A \setminus \mathfrak{p}$. $\square$

We are ready for a description of the tight valuation hull $TV(A, \mathfrak{p}, R)$. Since this hull is v-convex in R (Prop.5) we know that for an element $x \in R$ the value $v(x)$ alone decides whether x is in this hull or not.

Theorem 7.7. Let $x \in R \setminus A$ be given, and $\gamma := v(x)$. Then $x \in TV(A, \mathfrak{p}, R)$ iff for every $\alpha \in [\gamma, 0[\cap M$ the A-module $I_{\alpha,v}$ is R-invertible.[*]

Proof. Let U denote the set of all $\gamma \in M$ such that $\gamma < 0$ and $I_{\alpha,v}$ is R-invertible for every $\alpha \in M$ with $\gamma \leq \alpha < 0$. We define a subset D of R by $D := A \cup \{x \in R \mid v(x) \in U\}$. Notice that $[\gamma, 0[\cap M \subset U$ for every $\gamma \in U$, hence $D = A \cup \bigcup_{\gamma \in U} I_{\gamma,v}$. {If $U \neq \emptyset$ then $D = \bigcup_{\gamma \in U} I_{\gamma,v}$.}

Assume that $x \in D \setminus A$, hence $\gamma := v(x) \in U$. Let $\beta \in G$ be given with $\gamma \leq \beta < 0$ and $\beta - \gamma \in M$. Then $\gamma \in M$, $\beta \in M$, and $I_{\beta,v}$ is invertible by the definition of U. Lemma 6 tells us that $x \in C := MV(A, \mathfrak{p}, R)$. Thus $D \subset C$. We will now prove directly the following:

(1) D is a subring of R.
(2) The valuation $v|_D$ is tight.
(3) If B is any R-overring of A with $v|_B$ tight then $B \subset D$.

Then the claim of the theorem will be evident. The proof will not use in advance the existence of $TV(A, \mathfrak{p}, R)$, but will give this anew for valuative extensions.

(1): By definition the set D is v-convex in R. Let elements x_1 and x_2 of D be given, and $\gamma_1 := v(x_1)$, $\gamma_2 := v(x_2)$. We have to verify that $x_1 + x_2 \in D$ and $x_1 x_2 \in D$.

---

[*] $[\gamma, 0[$ denotes the half open interval $\{\alpha \in G \mid \gamma \leq \alpha < 0\}$.

We assume without loss of generality that $\gamma_1 \leq \gamma_2$. We have $v(x_1 + x_2) \geq \gamma_1$. Since D is v-convex it follows that $x_1 + x_2 \in D$. If $x_1 \in A$ or $x_2 \in A$ then $v(x_1 x_2) \geq \gamma_1$, and we conclude again that $x_1 x_2 \in D$.

In the following we assume that $\gamma_1 < 0$ and $\gamma_2 < 0$. Let $y \in R$ be given with $\gamma_1 + \gamma_2 = v(x_1 x_2) \leq v(y) < 0$. Let $\alpha := v(y)$. We verify that $I_{\alpha,v}$ is R-invertible. Then we will know that $\gamma_1 + \gamma_2 \in U$, hence $x_1 x_2 \in D$.

If $\alpha \geq \gamma_1$ then $I_{\alpha,v}$ is R-invertible since $\gamma_1 \in U$. Assume now that $\alpha < \gamma_1$. Then $\gamma_1 + \gamma_2 \leq \alpha < \gamma_1$, hence $\gamma_2 \leq \alpha - \gamma_1 < 0$. The element x_1 is contained in C. Since $v|_C$ is Manis we have $-\gamma_1 \in v|_C(C)$, hence $-\gamma_1 \in v(C) \subset v(R)$. Thus $\alpha - \gamma_1 \in v(R)$. Since $\gamma_2 \in U$ it follows that $I_{\alpha - \gamma_1,v}$ is invertible. Thus also $I_{\alpha,v} = I_{\alpha - \gamma_1,v} \cdot I_{\gamma_1,v}$ (cf. Prop.4.2) is R-invertible.

(2): We now know that D is a subring of R with $A \subset D \subset C$. Let $u := v|_D$. Since $v|_C$ is Manis, also $u = (v|_C)|_D$ is Manis. We have to prove that u is tight. For this it suffices to verify for a given element $x \in D \setminus A$ that the A-module $I_{\gamma,u}$ with $\gamma := u(x)$ is D-invertible. Since D is v-convex we know by Lemma 4 that $I_{\gamma,u} = I_{\gamma,v}$. Since $\gamma \in U$, it follows that $I_{\gamma,v}$ is R-invertible. From Proposition 4.2 we know that $I_{\gamma,v}^{-1} = I_{-\gamma,v}$. Since $-\gamma > 0$ the A-module $I_{-\gamma,v}$ is contained in A, hence in D, and we conclude that $I_{\gamma,u} = I_{\gamma,v}$ is D-invertible, as desired.

(3): Let finally B be an R-overring such that $u' := v|_B$ is tight. Then B is v-convex by Proposition 5. Let $x \in B \setminus A$ be given. We have to verify that $x \in D$, i.e. $\gamma := v(x) \in U$. Let $\alpha \in [\gamma, 0[\cap v(R)$ be given, $\alpha = v(y)$ for some $y \in R \setminus A$. Now $v(x) \leq v(y) < 0$. Since B is v-convex it follows that $y \in B$, and $\alpha = u'(y)$. Lemma 4 tells us that $I_{\alpha,u'} = I_{\alpha,v}$. Since u' is tight we know that the A-module $I_{\alpha,v}$ is B-invertible. A fortiori $I_{\alpha,v}$ is R-invertible. This proves that $\gamma \in U$, as desired. $\qquad\square$

Corollary 7.8. Assume that v is Manis. Let $x \in R \setminus A$ be given. Then x lies in $\mathrm{TV}(A, \mathfrak{p}, R)$ iff $(A : ax)$ is R-invertible for every $a \in A$ with $ax \notin A$.

Proof. We assume without loss of generality that $v(R) = G \cup \infty$. Let $\gamma := v(x) < 0$. We are done by the theorem if we prove that the following two assertions are equivalent.

i) The ideal $(A:ax)$ is R-invertible for every $a \in A$ with $ax \notin A$.

ii) The ideal $I_\alpha := I_{\alpha,v}$ is R-invertible for every $\alpha \in [\gamma, 0[$.

Assume that $a \in A$ and $ax \notin A$. Let $\beta := v(a) \geq 0$. Then $\gamma + \beta < 0$, and $(A:ax) = I_{-\gamma-\beta}$. Since $v(R) = G \cup \infty$, every $\alpha \in [\gamma, 0[$ is of the form $\gamma + \beta$ with such an element β, and Proposition 4.2 tells us that I_α is invertible iff $I_{-\alpha}$ is invertible. The equivalence of i) and ii) is evident. $\square$

We add a description of $TV(A, \mathfrak{p}, R)$ and its value group in the style of Proposition 6.7 and Theorem 6.9.

Proposition 7.9. Let $D := TV(A, \mathfrak{p}, R)$.

i) The set G_1 of all $g \in M$ with $I_{g,v}$ invertible in R is a subgroup of G_0.

ii) Let Δ denote the maximal subgroup H of G_1 such that H_- is v-convex in M_-. Then D is the set of all $x \in R$ with $v(x) \geq \gamma$ for some $\gamma \in \Delta$, and $v(D) = \Delta \cup \infty$.

iii) Δ_+ is the set of all $g \in M_+$ with $[0,g] \cap M \subset G_1$. Thus Δ is the maximal subgroup H of G_1 such that H_+ is v-convex in M_+.

iv) Δ is the maximal subgroup of G_1 which is v-convex in M, hence in G_0.

Proof. i): It follows from Proposition 4.2 that $G_1 + G_1 \subset G_1$, and that $-\gamma \in G_1$ for every $\gamma \in G_1$. Thus G_1 is a subgroup of G contained in M, hence in G_0.

ii): Theorem 7 now tells us that an element $x \in R \setminus A$ lies in D iff $v(x) \in \Delta$. Since $v|_D$ is Manis, the assertions in ii) follow.

iii): Let V denote the set of all $g \in M_+$ with $[0,g] \cap M \subset G_1$. We have to verify that $V = \Delta_+$.

a) Let $g \in \Delta_+$ be given, and assume without loss of generality that $g \neq 0$. If $a \in M$ and $0 \leq a < g$, then $-g \in \Delta_-$ and $-g \leq -g + a < 0$. Since $-g + a \in M$, it follows that $-g + a \in G_1$. Also $g \in G_1$. We conclude that $a \in G_1$. This proves that $g \in V$.

b) Let $g \in V$ be given and, $g \neq 0$. We verify that $-g \in \Delta$, hence $g \in \Delta$. If $a \in M$ and $-g \leq a \leq 0$ then $0 \leq g + a \leq g$. We have $g + a \in M$. It follows that $g + a \in G_0$. Also $g \in G_0$. (Take $a = 0$.) Thus $a \in G_0$. We conclude that $g \in \Delta_+$.

iv): Since Δ is the maximal subgroup H of G_1 with H_- v-convex in M and also the maximal subgroup H of G_1 with H_+ v-convex in M, it follows that H is the maximal subgroup of G_1 which is v-convex in M. □

§8 Principal valuations

In II, §10 we had introduced BM-valuations (BM = Bezout-Manis, cf. II, §10 Def.2) as a special class of PM-valuations. We now define "principal" valuations. They turn out to be a special class of tight valuations, related to BM-valuations in an analogous way as tight valuations are related to PM-valuations.

Definition 1. We call a valuation $v: R \to \Gamma \cup \infty$ on a ring R *principal*, if v is Manis and the A_v-module $I_\gamma := I_{\gamma,v}$ is principal for every $\gamma \in \Gamma_v$, i.e. $I_\gamma = A_v x$ with $x \in R$. {Recall that I_γ denotes the set of all $z \in R$ with $v(z) \geq \gamma$.}

We have briefly studied such valuations in II, §10. We read off from II, Lemma 10.6 and its proof:

Proposition 8.1. Let $v: R \to \Gamma \cup \infty$ be a valuation.
a) The following are equivalent:
(1) v is principal.
(2) v is Manis, and I_γ is a principal A_v-module for every $\gamma \in \Gamma_v$ with $\gamma < 0$.
(3) $v(R^*) = \Gamma_v$.

b) If v is principal, then for every $\gamma \in \Gamma_v$ there exists a *unit* x of R such that $I_\gamma = A_v x$. Moreover, if $\gamma \leq 0$, *every* generator of the A_v-module I_γ is a unit. □

Remarks 8.2. i) It follows that every principal valuation is tight, since an A_v-submodule $I = A_v x$ of R with $x \in R^*$ is clearly invertible, $I^{-1} = A_v x^{-1}$. More precisely, the principal valuations are just those tight valuations v, for which the modules $I_{\gamma,v}$ with γ running through Γ_v are principal.
ii) The BM-valuations, introduced in II, §10, Def.2, are the principal valuations which are also PM.

Examples 8.3. The valuation described in 4.1.c is principal but not BM, if the ideal pA there is not maximal. The valuation described in 4.1.d is tight but neither principal nor PM, if the k-module L there is not free and k is not a field.

Proposition 8.4. Assume that $v: R \twoheadrightarrow \Gamma \cup \infty$ is a principal surjective valuation.
a) If H is a convex subgroup of Γ, then $w := v/H$ is a principal valuation on R with $A_w = A_{v,H}$, $\mathfrak{p}_w = \mathfrak{p}_{v,H}$, and $u := v_H = v|_{A_{v,H}}$ is a principal valuation on $A_{v,H}$ with $A_u = A_v$, $\mathfrak{p}_u = \mathfrak{p}_v$.
b) The rings $A_{v,H}$ with H a convex subgroup of Γ (i.e. the v-convex R-overrings of A_v) are precisely the R-overrings B of A such that the extension $A \subset B$ is tight.

Proof. a): Of course $A_u = A_v$, $\mathfrak{p}_u = \mathfrak{p}_v$, $A_w = A_{v,H}$, $\mathfrak{p}_w = \mathfrak{p}_{v,H}$ (cf. I, § 1). It follows from $v(R^*) = \Gamma$ that $w(R^*) = \Gamma/H$. If $\gamma \in H$ is given, there exists some $x \in R^*$ with $v(x) = \gamma$. This implies that $x \in A_{v,H}$ and $x^{-1} \in A_{v,H}$. Thus x is a unit of B and $u(x) = v(x) = \gamma$. Thus $u(B^*) = H$, hence u is tight.
b): This is clear by Theorem 4.13, since principal valuations are tight. $\qquad\square$

Since a principal valuation $v: R \to \Gamma \cup \infty$ is Manis, it is – up to equivalence – uniquely determined by the pair $(A_v, \mathfrak{p}_v)$. This leads us to the following definition. Recall that in our terminoloy here a "*pair*" $(A, \mathfrak{p})$ consists of a ring A and a prime ideal $\mathfrak{p}$ of A. It is a "*pair in R*", if A is a subring of R.

Definition 2. Let R be a ring. We say that a pair $(A, \mathfrak{p})$ in R is *principally valuating* (abbreviated: pv) *in R* if there exists a principal valuation v on R with $A_v = A$, $\mathfrak{p}_v = \mathfrak{p}$. We say that $(A, \mathfrak{p})$ is *BM (= Bezout-Manis) in R* if in addition A is Prüfer in R, i.e. the valuation v is BM.

Scholium 8.5. Assume that $(A, \mathfrak{p})$ is pv in R and $A \neq R$. Then, since $(A, \mathfrak{p})$ is Manis in R, we have $\mathfrak{p} = \mathfrak{p}_A^R$. (Recall the notation in I, §2, Def.2). Let B be an R-overring of A with $B \neq R$. From Proposition 4 we read off the following: $(A, \mathfrak{p})$ is pv in B iff the extension $A \subset B$ is tight. In this case $(B, \mathfrak{p}_B^R)$ is pv in R, and $\mathfrak{p}_B^R$ coincides with the conductor $\mathfrak{q}_A^B$ of A in B. $\qquad\square$

Expanding on Proposition 5.11.i we present a transivity result for principal valuations. It is more satisfying than the result on tight valuations in 5.11.iii.

Proposition 8.6. Let $A \subsetneqq B \subsetneqq C$ be ring extensions and $\mathfrak{p}$ a prime ideal of A. Let $\mathfrak{q}$ denote the conductor of A in B. Assume that $(A, \mathfrak{p})$ is pv in B and $(B, \mathfrak{q})$ is pv in C. Then $(A, \mathfrak{p})$ is pv in C.

Proof. We know by Proposition 5.11.i that $(A, \mathfrak{p})$ is Manis in C. Let $v: C \twoheadrightarrow \Gamma \cup \infty$ be a surjective Manis valuation with $(A_v, \mathfrak{p}_v) = (A, \mathfrak{p})$. As explained in the proof of Prop.5.11.iii, there exists a convex subgroup H of Γ such that $w := v/H$ has the valuation ring $A_w = B$ and the center $\mathfrak{p}_w = \mathfrak{q}$. The special restriction $u := v|_B = v_H$ is a Manis valuation on B with $A_u = A$, $\mathfrak{p}_u = \mathfrak{p}$. By hypothesis both w and u are principal. We have to verify that v is principal.

Let $\gamma \in \Gamma$ be given. By Proposition 1 there exists some $s \in C^*$ with $w(s) = \gamma + H$, i.e. $v(s) = \gamma + h$ with some $h \in H$. Since u has the value group H, there also exists some $t \in B^*$ with $u(t) = -h$, hence $v(t) = -h$. It follows that $v(st) = \gamma$. This proves that v is principal (cf. Prop.1).   □

Corollary 8.7. In the situation of Proposition 8.6 assume that $(A, \mathfrak{p})$ is BM in B and $(B, \mathfrak{q})$ is BM in C. Then $(A, \mathfrak{p})$ is BM in C.

Proof. This follows from Proposition 6 and the fact that A is Prüfer in C (cf. Cor.I.5.3).   □

We mention that Corollary 7 is also a consequence of Prop.5.11.i and the transitivity result for Bezout extensions stated in Prop.II.10.11.

Given a pair $(A, \mathfrak{p})$ in a ring R, we want to establish and study a "PV-hull" $\mathrm{PV}(A, \mathfrak{p}, R)$ and a "BM-hull" $\mathrm{BM}(A, \mathfrak{p}, R)$ of A in R in full analogy to the TV-hull $\mathrm{TV}(A, \mathfrak{p}, R)$ and PM-hull $\mathrm{PM}(A, \mathfrak{p}, R)$ introduced in §5. In the case "BM" all the work is already done.

Definition 3. a) Given a pair $(A, \mathfrak{p})$ in R, we define the BM-hull $\mathrm{BM}(A, \mathfrak{p}, R)$ of $(A, \mathfrak{p})$ in R as the intersection of the PM-hull $\mathrm{PM}(A, \mathfrak{p}, R)$ and the Bezout-hull $\mathrm{Bez}(A, R)$ (cf. II, §10),

$$\mathrm{BM}(A, \mathfrak{p}, R) = \mathrm{Bez}(A, R) \cap \mathrm{PM}(A, \mathfrak{p}, R).$$

b) Given any pair $(A, \mathfrak{p})$, we define the *BM-hull of* $(A, \mathfrak{p})$ as

$$\mathrm{BM}(A, \mathfrak{p}) := \mathrm{BM}(A, \mathfrak{p}, Q(A)).$$

Proposition 8.8. Let $(A, \mathfrak{p})$ be a pair in a ring R.
i) $\mathrm{BM}(A, \mathfrak{p}, R)$ is the unique maximal overring B of A in R such that $(A, \mathfrak{p})$ is BM in R.
ii) If B is an R-overring of A and $(A, \mathfrak{p})$ is BM in B, there exists a unique homomorphism $\varphi \colon B \to \mathrm{BM}(A, \mathfrak{p})$ over A, and φ is injective.
iii) $\mathrm{BM}(A, \mathfrak{p}) = \mathrm{BM}(A, \mathfrak{p}, \mathrm{Quot}\, A) = \mathrm{BM}(A, \mathfrak{p}, P(A))$
iv) If the ideal $\mathfrak{p}$ is not maximal in A, then $\mathrm{BM}(A, \mathfrak{p}, R) = A$.

Proof. i) and ii) follow immediately from the analogous properties of PM-hulls and Bezout-hulls (cf. §5 and II, §10). Claim ii) follows from the facts that $\mathrm{Bez}(A) \subset \mathrm{Quot}\, A$ (cf. II, 10.15) and $\mathrm{Bez}(A) \subset P(A)$. {N.B. Also $\mathrm{PM}(A, \mathfrak{p}) \subset P(A)$.} Claim iv) is evident, since for a non trivial PM-valuation v on R the ideal $\mathfrak{p}_v$ is maximal in A_v (cf. Cor.1.4), hence $\mathrm{PM}(A, \mathfrak{p}, R) = A$. $\qquad\square$

Theorem 8.9. Let $(A, \mathfrak{p})$ be a pair in a ring R. Assume that $\mathfrak{B}$ is a set of R-overrings B of A such that $(A, \mathfrak{p})$ is pv in B for every $B \in \mathfrak{B}$. Then $\mathfrak{B}$ is totally ordered by inclusion, hence the union C of all $B \in \mathfrak{B}$ is again an R-overring. The pair $(A, \mathfrak{p})$ is pv in C.

Proof. Proposition 5.8 tells us that $\mathfrak{B}$ is totally ordered by inclusion, since $(A, \mathfrak{p})$ is tv in every $B \in \mathfrak{B}$. Now we run again through the proof of Lemma 5.9. In the setting explained there we have for every $\gamma \in \Gamma$ with $\gamma \leq 0$ some $i \in I$ with $I_{\gamma, v} = I_{\gamma, v_i}$. Since the valuation v_i is principal, the A-module $I_{\gamma, v}$ is principal. It follows by Proposition 1 that v is principal. Thus $(A, \mathfrak{p})$ is pv in C. $\qquad\square$

Choosing for $\mathfrak{B}$ the set of all R-overrings B of A with $(A, \mathfrak{p})$ pv in B we obtain

Corollary 8.10. Given a pair $(A, \mathfrak{p})$ in a ring R there exists a unique maximal R-overring C of A such that $(A, \mathfrak{p})$ is pv in C. It contains every other R-overring B of A with $(A, \mathfrak{p})$ pv in B. $\qquad\square$

Definition 3. a) We call this ring C the *principal valuative hull* (or *PV-hull* for short) *of* $(A, \mathfrak{p})$ *in* R, and we write $C = \mathrm{PV}(A, \mathfrak{p}, R)$.

b) If $(A, \mathfrak{p})$ is any pair we denote the ring $\mathrm{PV}(A, \mathfrak{p}, Q(A))$ more briefly by $\mathrm{PV}(A, \mathfrak{p})$, and we call $\mathrm{PV}(A, \mathfrak{p})$ be *principal valuative hull* (= *PV-hull*) *of* $(A, \mathfrak{p})$.

Scholium 8.11. a) If $(A, \mathfrak{p})$ is any pair, we have

$$\mathrm{PV}(A, \mathfrak{p}) = \mathrm{PV}(A, \mathfrak{p}, \mathrm{Quot}\, A) = \mathrm{PV}(A, \mathfrak{p}, T(A)).$$

b) Given a pair $(A, \mathfrak{p})$ in some ring R we have the following diagram of inclusions

$$\mathrm{PV}(A, \mathfrak{p}, R) \quad \subset \quad \mathrm{TV}(A, \mathfrak{p}, R)$$
$$\cup \qquad\qquad\qquad \cup$$
$$\mathrm{BM}(A, \mathfrak{p}, R) \quad \subset \quad \mathrm{PM}(A, \mathfrak{p}, R),$$

and $\mathrm{BM}(A, \mathfrak{p}, R) = \mathrm{PV}(A, \mathfrak{p}, R) \cap \mathrm{PM}(A, \mathfrak{p}, R)$. $\qquad\square$

Assume now – as in most of §7 – that there is a given valuation $R \to G \cup \infty$ and $A = A_v$, $\mathfrak{p} = \mathfrak{p}_v$. We are interested in a description of $D := \mathrm{PV}(A, \mathfrak{p}, R)$ in terms of the valuation v. This makes sense, since the special restriction $v|_D$ is the principal valuation u on D with $(A_u, \mathfrak{p}_u) = (A, \mathfrak{p})$, hence D is v-convex in R according to Proposition 7.5. Thus for any $x \in R$ the value $v(x) \in G$ decides whether x is an element of D or not. This ist just as in the case $D = \mathrm{TV}(A, \mathfrak{p}, R)$ studied in §7. Running again through the arguments in the second half of §7 one obtains a description of $\mathrm{PV}(A, \mathfrak{p}, R)$ completely analogous to the description of $\mathrm{TV}(A, \mathfrak{p}, R)$ there. We state:

Observation 8.12. Theorem 7.7, Corollary 7.8 and Theorem 7.9 remain true if we replace there everywhere $\mathrm{TV}(A, \mathfrak{p}, R)$ by $\mathrm{PV}(A, \mathfrak{p}, R)$ and the word "R-invertible" (or: "invertible in R") by "R-invertible and principal". $\qquad\square$

§9 Descriptions of the PM-hull

In the following $A \subset R$ is a ring extension and x is a given element of R. We look for a handy criterion that x is contained in the PM-hull $PM(A, R, \mathfrak{p})$ for a given prime ideal $\mathfrak{p}$ of A. We may always assume without loss of generality that $x \notin A$. We first need a new description of relative Prüfer hulls and an easy lemma about regularity of prime ideals.

Theorem 9.1. The following are equivalent.

(i) A is Prüfer in $A[x]$ {i.e. $x \in P(A, R)$}.

(ii) For every $a \in A$ the A-module $A + Aax$ is invertible.

(iii) $(A: ax) + ax(A: ax) = A$ for every $a \in A$.

Proof. If $y \in R$ and the A-module $A + Ay$ is invertible, then $A + Ay$ is R-invertible, since $(A + Ay)^{-1} \subset A \subset R$. Now the equivalence (ii) $\Leftrightarrow$ (iii) is evident, since $[A: A + Aax] = (A: ax)$. The implication (i) $\Rightarrow$ (iii) is covered by Theorem I.5.2. {(i) $\Rightarrow$ (ii) is also covered by Theorem II.1.13.}

(iii) $\Rightarrow$ (i): Let $B := A[x]$ and let $\mathfrak{p}$ be a prime ideal of A. We verify that the pair $(A^B_{[\mathfrak{p}]}, \mathfrak{p}^B_{[\mathfrak{p}]})$ is Manis in B. Replacing R by B we assume without loss of generality that $R = A[x]$.

Let $A' := A_{[\mathfrak{p}]}$, $\mathfrak{p}' := \mathfrak{p}_{[\mathfrak{p}]}$. We have $R = A'[x]$ and can apply Theorem 6.1. By that theorem it suffices to verify that, for a given $z \in A'$ with $zx \notin A'$, there exists some $y \in \mathfrak{p}'$ with $yzx \in A' \setminus \mathfrak{p}'$.

We choose $d \in A \setminus \mathfrak{p}$ with $dz \in A$. Since $zx \notin A'$ we have $dzx \notin A$. By our assumption (iii) we have $(A: dzx) + dzx(A: dzx) = A$. Now $(A: dzx) \subset \mathfrak{p}$, since $d \in A \setminus \mathfrak{p}$ and $zx \notin A_{[\mathfrak{p}]}$. We conclude that there exists some $y_0 \in (A: dzx) \subset \mathfrak{p}$ with $y_0 dzx \in A \setminus \mathfrak{p}$. Then $y := y_0 d \in \mathfrak{p} \subset \mathfrak{p}'$ and $yzx \in A \setminus \mathfrak{p} \subset A' \setminus \mathfrak{p}'$. $\qquad \square$

Remark 9.2. We mention that a proof of (ii) $\Rightarrow$ (i) is possible applying Theorem II.9.6 about Prüfer modules instead of Theorem 6.1. One argues in the same way as in the proof of Proposition II.10.29 on Bezout extensions, first observing that there exists an invertible ideal I of A with $Ix \subset A$. This implies that the extension $A \subset A[x]$ is tight, hence may be viewed as subextension of the tight hull $T(A)$

and then of $Q(A)$. For every $z \in A + Ax$ we have $A + Az = A + Aax$ with some $a \in A$. Thus $A + Az$ is invertible. We conclude that the $Q(A)$-overmodule $A + Ax$ is Prüfer. Now Theorem II.9.6 tells us that the extension $A \subset A[x]$ is Prüfer. $\qquad\square$

Lemma 9.3. Let $A \subset R$ be a weakly surjective ring extension and $\mathfrak{p}$ be a prime ideal of A.
i) $\mathfrak{p}$ is R-regular iff there exists some $x \in R$ with $(A\!:\!x) \subset \mathfrak{p}$.
ii) If $x \in R$ is given, then $\mathfrak{p}$ is $A[x]$-regular iff $(A\!:\!x) \subset \mathfrak{p}$.

Proof. i): We know by Lemma 1.1 that $\mathfrak{p}$ is R-regular iff $A_{[\mathfrak{p}]} \neq R$. This means that there exists some $x \in R$ with $(A\!:\!x) \subset \mathfrak{p}$.
ii): Replacing R by $A[x]$ we may assume that $R = A[x]$. Now $A_{[\mathfrak{p}]} \neq R$ iff $x \notin A_{[\mathfrak{p}]}$, and this means again that $(A\!:\!x) \subset \mathfrak{p}$. $\qquad\square$

Theorem 9.4. Assume that $x \in R \setminus A$ and $\mathfrak{p}$ is a prime ideal of A. The following are equivalent.

(1) $x \in \mathrm{PM}(A, \mathfrak{p}, R)$.
(2) $(A, \mathfrak{p})$ is saturated in $A[x]$, and the A-module $A + Aax$ is invertible for every $a \in A$.
(3) $(A, \mathfrak{p})$ is saturated in $A[x]$, and $(A\!:\!ax) + ax(A\!:\!ax) = A$ for every $a \in A$ (with $ax \notin A$).
(4) $\mathfrak{p}$ is the unique prime ideal of A containing $(A\!:\!x)$, and $A + Aax$ is invertible for every $a \in A$.

Proof. The implications $(1) \Rightarrow (2) \Leftrightarrow (3)$ are by now trivial. {Notice that (1) means that $(A, \mathfrak{p})$ is PM in $A[x]$.}
$(3) \Rightarrow (1)$: The extension $A \subset A[x]$ is Prüfer by Theorem 1. Thus $(A, \mathfrak{p}) = (A_{[\mathfrak{p}]}, \mathfrak{p}_{[\mathfrak{p}]})$ is Manis in $A[x]$. We conclude that $(A, \mathfrak{p})$ is PM in $A[x]$.
$(1) \Leftrightarrow (4)$: We know by Theorem 1.8 that A is PM in $A[x]$ iff A is Prüfer in $A[x]$ and $\mathfrak{p}$ is the unique R-regular maximal ideal of A. The claim now follows from Theorem 1 and Lemma 3.ii. $\qquad\square$

This theorem gives a criterion for membership in the PM-hull based on the modules $A + Aax$. We now look for a criterion based on the ideals $(A\!:\!ax)$ instead. Notice that for the TV-hull in a Manis extension such a criterion has been given in Corollary 7.8.

Lemma 9.5. Let $\mathfrak{p}$ be a prime ideal of A and x an element of $R \setminus A$. Assume that A is integrally closed in $B := A[x]$. Then $(A_{[\mathfrak{p}]}^B, \mathfrak{p}_{[\mathfrak{p}]}^B)$ is Manis in B iff for each $a \in A$ with $ax \notin A$ there exists a polynomial $F(T) \in A[T] \setminus \mathfrak{p}[T]$ with $F(ax) = 0$.

Proof. Replacing R by B we assume without loss of generality that $R = A[x]$. If $(A_{[\mathfrak{p}]}, \mathfrak{p}_{[\mathfrak{p}]})$ is Manis in R then such polynomials exist by Theorem I.2.12.

Assume now that conversely the polynomial condition is fulfilled. Let $a \in A$ be given with $ax \notin A_{[\mathfrak{p}]}$. By Theorem 6.1 it suffices to verify that there exists some $y \in \mathfrak{p}_{[\mathfrak{p}]}$ with $yax \in A_{[\mathfrak{p}]} \setminus \mathfrak{p}_{[\mathfrak{p}]}$. This can be done by arguing precisely as in the proof of the direction (ii) $\Rightarrow$ (i) in the proof of Theorem I.2.12, always working with ax instead of x. $\qquad \square$

Proposition 9.6. Assume that A is integrally closed in $A[x]$. The following are equivalent.

(1) A is Prüfer in $A[x]$.
(2) For every $a \in A$ (with $ax \notin A$) the ideal $(A : ax)$ of A is invertible in $A[ax]$.
(2') For every $a \in A$ (with $ax \notin A$) the ideal $(A : ax)$ is regular in $A[ax]$.
(3) For every $a \in A$ (with $ax \notin A$) the extension $A \subset A[ax]$ is weakly surjective.
(4) For every $\mathfrak{p} \in \operatorname{Spec} A$ and every $a \in A$ with $ax \notin A$ there exists some polynomial $F(T) \in A[T] \setminus \mathfrak{p}[T]$ with $F(ax) = 0$.

Proof. The implication (1) $\Rightarrow$ (2) is covered by Theorem 1 (or by I, §5), and (1) $\Leftrightarrow$ (4) is covered by Lemma 5. The implication (1) $\Rightarrow$ (3) is evident by the theory of Prüfer extensions in I, §5, and (3) $\Rightarrow$ (2') is evident by the theory in I, §3 (cf. Th.I.3.13). (2) $\Rightarrow$ (2') is trivial.

We prove (2') $\Rightarrow$ (4), and then will be done. Let $\mathfrak{p} \in \operatorname{Spec} A$ and $a \in A$ with $ax \notin A$ be given.

Case 1: $ax \in A_{[\mathfrak{p}]}$. We choose $b \in A \setminus \mathfrak{p}$ with $bax = c \in A$. Then $F(T) := bT - c$ does the job.

Case 2: $ax \notin A_{[\mathfrak{p}]}$, i.e. $(A : ax) \subset \mathfrak{p}$. Since $(A : ax)$ is $A[ax]$-regular, we have a relation

$$1 = \sum_{i=0}^{n} a_i (ax)^i$$

with $n \in \mathbb{N}_0$ and all coefficients $a_i \in (A : ax) \subset \mathfrak{p}$. Now the polynomial $F(T) := (a_0 - 1) + a_1 T + \cdots + a_n T^n$ does the job.   □

We turn to the special case that the pair $(A, \mathfrak{p})$ is valuative in R (§7, Def. 2). Thus there is given a valuation $v : R \to G \cup \infty$ with $A = A_v$, $\mathfrak{p} = \mathfrak{p}_v$. Notice that A is integrally closed in R and $(A, \mathfrak{p})$ is saturated in R.

If A is Prüfer in $A[x]$ then A is PM in $A[x]$, cf. the proof of $(3) \Rightarrow (1)$ in Theorem 4. Thus the PM-hull $\mathrm{PM}(A, \mathfrak{p}, R)$ coincides with the Prüfer hull $P(A, R)$.

Theorem 9.7. Let $x \in R \setminus A$ be given. The following are equivalent.

(1) A is Prüfer in $A[x]$ (i.e. $x \in \mathrm{PM}(A, \mathfrak{p}, R)$).
(2) $(A : ax) + ax(A : ax) = A$ for every $a \in A$.
(2′) The A-module $A + Aax$ is invertible for every $a \in A$ (hence invertible in $A[ax]$).
(3) The ideal $(A : ax)$ of A is invertible in $A[ax]$ for every $a \in A$.
(3′) The ideal $(A : ax)$ of A is regular in $A[ax]$ for every $a \in A$.
(4) A is weakly surjective in $A[ax]$ for every $a \in A$.
(5) $(A + Aax)(A \cap Aax) = Aax$ for every $a \in A$ (with $ax \notin A$).
(5′) For every $a \in A$ (with $ax \notin A$) there exists some $n \in \mathbb{N}$ such that $(ax)^n (A : ax) + (ax)^{n+1}(A : ax) = A(ax)^n$.
(6) The A-module $A + Ay$ is invertible for every $y \in A[x]$.
(7) The A-module $A + Ay$ is v-convex in R for every $y \in A[x]$.
(7′) The A-module $A + Aax^2$ is v-convex in $A[x]$ for every $a \in A$.
(7″) The A-module $A + A(ax)^2$ is v-convex in $A[ax]$ for every $a \in A$.
(8) $A[ax] = A[(ax)^2]$ for every $a \in A$.
(8′) For every $a \in A$ (with $ax \notin A$) there exists some $n \in \mathbb{N}$, $n \geq 2$, with $A[ax] = A[(ax)^n]$.

Proof. The equivalence of (1), (2), (2′) is covered by Theorem 1, and the equivalence of (1), (3), (3′), (4) is covered by Proposition 6. The implication $(1) \Rightarrow (5)$ is covered by Theorem II.1.4,(4). Since $A \cap Aax = ax(A : ax)$, the equation in (5) can also be read as $ax(A : ax) +$

$(ax)^2(A\colon ax) = Aax$. Thus the implication $(5) \Rightarrow (5')$ is trivial (and the implication $(2) \Rightarrow (5)$ as well).

$(5') \Rightarrow (1)$: Let $a \in A$ be given with $ax \notin A$. We have a relation

$$c_0(ax)^{n-1} + c_1(ax)^{n+1} = (ax)^n$$

with $c_0, c_1 \in A$. Thus condition (4) in Proposition 6 holds, and we conclude by that proposition that A is Prüfer in $A[x]$.

$(1) \Rightarrow (6)$: This is evident by Theorem II.1.13.

$(6) \Rightarrow (7)$: By Proposition 4.6 we have $A + Ay = I_{\gamma,v}$ with some $\gamma \in G$. Thus $A + Ay$ is v-convex.

The implications $(7) \Rightarrow (7')$, $(7) \Rightarrow (7'')$, $(8) \Rightarrow (8')$ are trivial.

$(7') \Rightarrow (8)$: Let $a \in A$ be given with $ax \notin A$. Then $v(ax^2) < v(x) < 0$. Since $A + Aax^2$ is assumed to be v-convex, we conclude that $x \in A + Aax^2$. This implies $ax \in A + A(ax)^2$ and $A[ax] = A[(ax)^2]$.

$(7'') \Rightarrow (8)$: The proof is similar.

$(8') \Rightarrow (1)$: We can apply $(4) \Rightarrow (1)$ in Proposition 6. $\qquad\square$

Remarks 9.8. i) In condition (2) of the theorem it is important to insist that $(A\colon ax)$ is invertible in $A[ax]$. We cannot replace here $A[ax]$ by R, since otherwise we would confuse our criteria for membership in $\mathrm{PM}(A, \mathfrak{p}, R)$ with the criterion 7.8 on membership in $\mathrm{TV}(A, \mathfrak{p}, R)$ (in the case that v is Manis). This can be illustrated well by our example 4.1.c. There we have $R = A[x^{-1}]$ and $(A\colon ax^{-1}) = Ax$ for every $a \in A$ with $ax^{-1} \notin A$. Thus $(A\colon ax^{-1})$ is R-invertible for every such a. But $(A\colon ax^{-1})$ is not invertible in $A[ax^{-1}]$. Thus x does not lie in the PM-hull of $(A, \mathfrak{p})$ in R. Indeed, v is tight, hence $\mathrm{TV}(A, \mathfrak{p}, R) = R$, but $\mathrm{PM}(A, \mathfrak{p}, R)$ is forced to be A, since the only proper subgroup of $\Gamma_v = \mathbb{Z}$ is the trivial group.

ii) It may be tempting to replace conditions (7) and $(7')$ in the theorem by the condition that $A + Aax$ is v-convex in $A[x]$ (or in R) for every $a \in A$ with $ax \notin A$. But this condition is too weak, as the following example shows.

Let k be a field and $v\colon R \to \mathbb{Z} \cup \infty$ be the valuation on the polynomial ring $R := k[X]$ with $v(k^*) = 0$ and $v(X) = -1$. Then $A := A_v = k$ and $A + AaX = k + kX$ for each $a \in A \setminus 0$. It is easy to see that $A + AaX$ is v-convex in R for every $a \in A$. But A is not Prüfer in R, not even weakly surjective in R. $\qquad\square$

We draw some profit from Proposition 6 and Theorem 7 for the theory of Bezout extensions started in II, §10.

Proposition 9.9. a) Assume that A is integrally closed in R. Let $x \in R \setminus A$ be given. The following are equivalent.

(1) A is Bezout in $A[x]$.

(2) For every $a \in A$ the ideal $(A : Aax)$ is generated by an element of $A \cap A[x]^*$.

b) Under the more special assumption, that there exists a valuation v on R with $A_v = A$, the conditions (1), (2) are also equivalent to

(3) For every $a \in A$ the A-module $A + Aax^2$ is principal.

N.B. Without any assumption on the extension $A \subset B$ we know by Theorem II.10.29 that (1) is equivalent to

(4) For every $a \in A$ the A-module $A + Aax$ is principal.

Proof. We assume without loss of generality that $R = A[x]$. It is clear by II, §10 that (1) implies (2) and (3).

(2) $\Rightarrow$ (1): We verify condition (4). We know by Proposition 6 that A is Prüfer in $A[x]$. Let $a \in A$. The A-module $A + Aax$ is invertible and $(A + Aax)^{-1} = (A : ax)$. By assumption, $(A : ax) = As$ with $s \in A \cap R^*$. It follows that $A + Aax = As^{-1}$.

(3) $\Rightarrow$ (1): Now $A = A_v$ for some valuation v on R. We again verify condition (4). Let $a \in A$ be given. Assume $ax \notin A$ without loss of generality. By assumption (3) there exists some $y \in R$ with $A + Ax^2 = Ay$. Clearly $y \in R^*$ and $v(x^2) = v(y)$. Now $v(\frac{x}{y}) > 0$, hence certainly $x = cy$ with $c \in A$. Also $y = a_1 + a_2 x^2$ with $a_1, a_2 \in A$. Thus

$$A + Aax = A + Aac(a_1 + a_2 x^2) = A + Aaca_2 x^2,$$

and this is principal by (3). □

We did not succeed to establish a criterion similar to $(7'')$ in Theorem 7 for A to be Bezout in $A[x]$.

The criteria in Theorem 7 leave something to be desired. We know by Proposition 7.5 that $\mathrm{PM}(A, \mathfrak{p}, R)$ is v-convex in R. In particular, for a given $x \in R$, the value $v(x)$ alone decides whether $x \in \mathrm{PM}(A, \mathfrak{p}, R)$ or not. But this is reflected by none of these criteria.

Notations 9.10. As before $v: R \to G \cup \infty$ is an arbitrary valuation, and $A := A_v$, $\mathfrak{p} := \mathfrak{p}_v$. Let $E := \mathrm{PM}(A, \mathfrak{p}, R) = P(A, R)$, and let $\mathfrak{q}$ denote the conductor of A in E. This is the support of the PM-valuation $u := v|_E$. Let Σ denote the subgroup $v(E \setminus \mathfrak{q})$ of G. If $g \in G$, and nothing else is said, then $I_g := I_{g,v}$. $\qquad \square$

Of course, $\Sigma = \Gamma_u$. Since E is v-convex we have

$$E = \{x \in R \mid v(x) \geq \gamma \text{ for some } \gamma \in \Sigma\} = A_{v,\widehat{\Sigma}} \quad ,$$

$$\mathfrak{q} = \{x \in R \mid v(x) > \gamma \text{ for all } \gamma \in \Sigma\} = \mathfrak{p}_{v,\widehat{\Sigma}} \quad ,$$

with $\widehat{\Sigma}$ the convex hull of Σ in G (cf. Notations 4.10). We start out to determine the group Σ. Here the criterion (5) in Theorem 3.4 will be of help for us.

Lemma 9.11. Let $s \in A \setminus \mathfrak{p}$ be given, and let Φ_s denote the set of all $g \in G_+$ with $I_g + As = A$. There exists a unique convex subgroup Ψ_s of G with $\Phi_s = (\Psi_s)_+$.

Proof. Clearly $0 \in \Phi_s$, and Φ_s is convex in G_+. We verify that $\Phi_s + \Phi_s \subset \Phi_s$, and then will be done.

Let $\gamma, \delta \in \Phi_s$ be given. We have $I_\gamma + As = A$ and $I_\delta + As = A$. Thus $I_\gamma I_\delta + I_\gamma s = I_\gamma$, and $I_\gamma I_\delta + I_\gamma s + As = A$. Since $I_\gamma I_\delta \subset I_{\gamma+\delta}$ and $I_\gamma \subset A$, it follows that $I_{\gamma+\delta} + As = A$, hence $\gamma + \delta \in \Phi_s$. $\qquad \square$

Theorem 9.12. The value group Σ of the PM-hull $E = \mathrm{PM}(A, \mathfrak{p}, R)$ is the intersection $\Psi \cap \Gamma$ of $\Psi := \bigcap_{s \in A \setminus \mathfrak{p}} \Psi_s$ and the value group Γ of the Manis valuation hull $C = \mathrm{MV}(A, \mathfrak{p}, R)$. {$\Gamma$ has been described in §6, cf. Th.6.9.}

Proof. $v' := v|_C$ is a Manis valuation with value group Γ. If $\gamma \in \Gamma$ then $I_{\gamma,v} = I_{\gamma,v'}$, since C is v-convex in R (Th.6.9). We obtain

$$(\Psi \cap \Gamma)_+ = \{\gamma \in \Gamma \mid I_{\gamma,v'} + As = A \quad \text{for every } s \in A \setminus \mathfrak{p}\}.$$

We have $E = \mathrm{PM}(A, \mathfrak{p}, C)$. Thus we may replace in our proof R by C and v by v'. From now on we assume without loss of generality that v is Manis with value group Γ.

The group Ψ is convex in Γ. Let $B := A_{v,\Psi} = \bigcup_{\gamma \in \Psi_+} I_{-\gamma}$, and $u :=$
$v|_B = v|_\Psi$ (cf. Notations 4.10). We prove directly:

(1) The valuation $u : B \twoheadrightarrow \Psi \cup \infty$ is PM.

(2) If B' is an R-overring of A with $u' := v|_{B'}$ PM , then $B' \subset B$.
Then the claim of the theorem will be obvious.

(1): u is Manis. Let $s \in A\setminus\mathfrak{p}$ and $z \in A$ be given with $\gamma := u(z) \neq \infty$.
By Theorem 3.4 (condition 5) it suffices to verify that $I_{\gamma,u} + As = A$.
Now $\gamma = v(z)$, $I_{\gamma,u} = I_{\gamma,v}$, and $\gamma \in \Psi_+ \subset \Phi_{s,+}$. Thus $I_{\gamma,u} + As = A$,
as desired.

(2): Consider $\gamma = u'(z) \neq \infty$ for some $z \in A$ not in the support of
u'. We have $\gamma = v(z) > 0$ and $I_{\gamma,v} = I_{\gamma,u'}$. Let $s \in A \setminus \mathfrak{p}$ be given.
Since u' is PM we know by Theorem 3.4 (or 2.7) that $I_{\gamma,u'} + As = A$,
hence $I_{\gamma,v} + As = A$. This proves that $\gamma \in \Psi_+$.

Let now $x \in B' \setminus A$ be given. Since u' is Manis there exists some
$z \in A$ with $\gamma := u'(z) = -u'(x)$. We have $\gamma \in \Psi_+$ and $x \in I_{-\gamma,u'} =$
$I_{-\gamma,v} \subset B$. Thus $B' \subset B$. $\square$

Corollary 9.13. An element x of A lies in the conductor $\mathfrak{q}_A^E$ of $E :=$
$\mathrm{PM}(A, \mathfrak{p}, R)$ iff either x is in the conductor $\mathfrak{q}_A^C$ of $C := \mathrm{MV}(A, \mathfrak{p}, R)$,
or $v(x) = \gamma \in \Gamma$, but I_γ is contained in some maximal ideal $\mathfrak{m} \neq \mathfrak{p}$ of
A. {The ideal $\mathfrak{q}_A^C$ had been described in §6, cf. Th.6.4 and Th.6.9.}.

Indeed, $I_\gamma \subset \mathfrak{m}$ for some maximal ideal $\mathfrak{m} \neq \mathfrak{p}$ of A iff there exists
some $s \in A \setminus \mathfrak{p}$ with $I_\gamma + As \neq A$. $\square$

§10 Composing valuations with ring homomorphisms

In this section we are given a surjective ring homomorphism
$\varphi : R \twoheadrightarrow \overline{R}$. Let N denote the kernel of φ.

If $w : \overline{R} \to \Gamma \cup \infty$ is a valuation on $\overline{R}$ then $w \circ \varphi : R \to \Gamma \cup \infty$ is a
valuation on R. Both valuations w and $v := w \circ \varphi$ have the same
value group $\Gamma_v = \Gamma_w$, and clearly $A_v = \varphi^{-1}(A_w)$, $\mathfrak{p}_v = \varphi^{-1}(\mathfrak{p}_w)$,
$\mathrm{supp}\, v = \varphi^{-1}(\mathrm{supp}\, w)$. In particular $\mathrm{supp}\, v \supset N$. Since $v(R) =$
$w(\overline{R})$, it is also evident that v is Manis iff w is Manis, and that v is
special iff w is special.

Conversely, if v is a valuation on R and $N \subset \operatorname{supp} v$, there exists a unique valuation w on $\overline{R}$ such that $v = w \circ \varphi$, as is easily seen.

Proposition 10.1. The nontrivial special valuations w on $\overline{R}$ correspond bijectively with the nontrivial special valuations v on R, such that $N \subset A_v$, via $v = w \circ \varphi$. The valuation v is Manis (resp. PM) iff w is Manis (resp. PM).

Proof. Assume that v is a special valuation on R with $N \subset A_v \neq R$. Proposition I.2.2 tells us that $\operatorname{supp} v$ is the conductor of A_v in R. Since N is an ideal of R, it follows that $N \subset \operatorname{supp} v$. As said above, this implies $v = w \circ \varphi$ with w a special valuation on $\overline{R}$, of course uniquely determined by v. We also observed that v is Manis iff w is Manis.

It remains to prove the claim concerning "PM". We have $\varphi(A_v) = A_w$ and $\varphi^{-1}(A_w) = A_v$. Proposition I.5.8 tells us that A_v is Prüfer in R iff A_w is Prüfer in $\overline{R}$. Thus v is PM in R iff w is PM in $\overline{R}$. □

We draw consequences from Proposition 1 for the PM-hulls and Manis valuation hulls introduced in §5. The easy proofs are left to the reader.

Corollary 10.2. (As before, $\varphi : R \to \overline{R}$ is a surjective ring homomorphism.) Assume that $\overline{A}$ is a subring of $\overline{R}$ and $\overline{\mathfrak{p}}$ a prime ideal of $\overline{A}$. Let $A := \varphi^{-1}(\overline{A})$ and $\mathfrak{p} := \varphi^{-1}(\overline{\mathfrak{p}})$.
a) $\operatorname{PM}(A, \mathfrak{p}, R) = \varphi^{-1}(\operatorname{PM}(\overline{A}, \overline{\mathfrak{p}}, \overline{R}))$.
b) $(A, \mathfrak{p})$ is saturated in R iff $(\overline{A}, \overline{\mathfrak{p}})$ is saturated in $\overline{R}$. In this case,

$$\operatorname{MV}(A, \mathfrak{p}, R) = \varphi^{-1}(\operatorname{MV}(\overline{A}, \overline{\mathfrak{p}}, \overline{R})).$$ □

In order to derive similar results for tight valuations and tight valuation hulls, we first study invertible ideals and tight extensions in general. {We could have done this study in Chapter II, §4.} As before, $\varphi : R \to \overline{R}$ is a surjective ring homomorphism.

Lemma 10.3. Let A be a subring of R and $\overline{A} := \varphi(A)$.
a) If I is an R-invertible A-submodule of R then $\varphi(I)$ is an $\overline{R}$-invertible $\overline{A}$-submodule of $\overline{R}$, and $\varphi(I)^{-1} = \varphi(I^{-1})$.

b) Assume that $N \subset A$, i.e. $A = \varphi^{-1}(\overline{A})$. Assume further that I is an A-submodule of R with $N \subset I$, and that the $\overline{A}$-submodule $\varphi(I)$ of $\overline{R}$ is $\overline{R}$-invertible. Then I is R-invertible.

Proof. a): This claim is evident.

b): Let $\overline{I} := \varphi(I)$ and $J := \varphi^{-1}(\overline{I}^{-1})$. Then $I = \varphi^{-1}(\overline{I})$ and $\varphi(IJ) = \overline{I}\,\overline{I}^{-1} = \overline{A}$. We prove that $N \subset IJ$. Then it will follow that $IJ = A$.

We have $\varphi(IR) = \overline{I}\,\overline{R} = \overline{R}$ and $N = NR \subset IR$, hence $IR = R$. Further $IN = IRN = RN = N$, and $N \subset J$. Thus indeed $IJ \supset N$. $\qquad\square$

Proposition 10.4. Let again A be a subring of R and $\overline{A} := \varphi(A)$.

a) If A is tight in R, then $\overline{A}$ is tight in $\overline{R}$.

b) Assume that $A = \varphi^{-1}(\overline{A})$, and that $\overline{A}$ is tight in $\overline{R}$. Then A is tight in R. $\qquad\square$

Proof. a): This claim is evident from part a) of Lemma 3.

b): Let $x \in R \setminus A$ be given. We have to find an R-invertible ideal I of A such that $Ix \subset A$.

There exists an $\overline{R}$-invertible ideal $\overline{I}$ of $\overline{A}$ such that $\overline{I}\varphi(x) \subset \overline{A}$. Let $I := \varphi^{-1}(\overline{I})$. Since $A = \varphi^{-1}(\overline{A})$ we have $Ix \subset A$. Lemma 3b tells us that I is R-invertible. $\qquad\square$

Lemma 10.5. (As before, $\varphi: R \to \overline{R}$ is a surjective ring homomorphism.) Let w be a valuation on $\overline{R}$ and $v := w \circ \varphi$. For any $\gamma \in \Gamma_v = \Gamma_w$ the A_v-module $I_{\gamma,v}$ is invertible in R iff the A_w-module $I_{\gamma,w}$ is invertible in $\overline{R}$.

Proof. Clearly $\varphi^{-1}(I_{\gamma,w}) = I_{\gamma,v}$. Thus $N \subset I_{\gamma,v}$ and $\varphi(I_{\gamma,v}) = I_{\gamma,w}$. Also $\varphi^{-1}(A_w) = A_v$. Lemma 3 gives the claim. $\qquad\square$

From this Lemma we obtain immediately

Proposition 10.6. Let w be a valuation on $\overline{R}$ and $v := w \circ \varphi$. Then v is tight iff w is tight. $\qquad\square$

As a consequence of this proposition one proves easily

Corollary 10.7. Assume that $\overline{A}$ is a subring of $\overline{R}$ and $\overline{\mathfrak{p}}$ a prime ideal of $\overline{A}$. Let $A := \varphi^{-1}(\overline{A})$ and $\mathfrak{p} := \varphi^{-1}(\overline{\mathfrak{p}})$. Then

$$\mathrm{TV}(A, \mathfrak{p}, R) = \varphi^{-1}(\mathrm{TV}(\overline{A}, \overline{\mathfrak{p}}, \overline{R})). \qquad \square$$

We return to PM-valuations.

Proposition 10.8. Assume that A is a subring of R which is PM in R. Then $\varphi(A)$ is PM in $\overline{R}$.

We give two proofs of this fact, the first one very short, the second one giving also information about the associated PM-valuations.

First proof. $\varphi(A) \subset \overline{R}$ is Prüfer by Proposition I.5.7. The subrings of $\overline{R}$ containing $\varphi(A)$ form a chain, since this holds for the subrings of R containing A. This gives the claim by Theorem 3.1. $\qquad \square$

Second proof. We may assume that $A \neq R$. Let v denote the PM-valuation on R with $A_v = A$. Since $A_v \subset \varphi^{-1}(\overline{A}) = A + N$, there exists a unique PM-valuation v' on R such that $A_{v'} = A + N$. It is a coarsening of v. Proposition 1 now tells us that there exists a PM-valuation w on $\overline{R}$, unique of course, with $v' = w \circ \varphi$. We have $\varphi^{-1}(A_w) = A_{v'}$, hence $A_w = \varphi(A_{v'}) = \varphi(A + N) = \overline{A}$. $\qquad \square$

These proofs may serve as an illustration of the good natured behaviour of PM-valuations. We are not able to deduce analogous results for Manis or tightly valuating extensions.

§11 Transfer of Valuations

We return to the situation in the main part of II, §6: Let $A \subset R$ be a Prüfer extension and $A \subset C$ a ws extension. We regard both R and C as subrings of $R \otimes_A C = RC$.

We know already from I, §5 that C is Prüfer in RC, and we have various transfer theorems at our disposal concerning overrings, regular ideals etc., cf.II, §6. These give us relations between valuations on R and on RC.

As a starting point for explaining this we mention a fact which could have been proved much earlier.

Proposition 11.1. (We assume that $A \subset R$ is Prüfer.) Let v be a special valuation on R with $A_v \supset A$, and $\mathfrak{p} := \mathfrak{p}_v \cap A$. Then v is PM and $(A_v, \mathfrak{p}_v) = (A_{[\mathfrak{p}]}, \mathfrak{p}_{[\mathfrak{p}]})$.

Proof. The set $R \setminus A_v$ is (empty or) multiplicatively closed. Proposition I.5.1.iii tells us that $A \subset R$ is Manis. Applying Proposition 6.6 (there with $B = R$) we see that v is Manis, hence PM since $A_v \subset R$ is Prüfer. We have $A_v \supset A_{[\mathfrak{p}]}$ and $\mathfrak{p}_v \supset \mathfrak{p}_{[\mathfrak{p}]}$. On the other hand, since $A \subset A_v$ is ws and $\mathfrak{p}A_v \neq A_v$, it follows from Theorem I.3.13 that $A_v \subset A_{[\mathfrak{p}]}$. Thus $A_v = A_{[\mathfrak{p}]}$. Now we conclude from $\mathfrak{p}_v \cap A = \mathfrak{p}_{[\mathfrak{p}]} \cap A = \mathfrak{p}$, say by Prop. I.4.6, that $\mathfrak{p}_v = \mathfrak{p}_{[\mathfrak{p}]}$. $\qquad\square$

Theorem 11.2. Let $\mathfrak{p}$ be a prime ideal of A with $\mathfrak{P} := \mathfrak{p}C \neq C$, hence $\mathfrak{P}$ a prime ideal of C. Let $C_{[\mathfrak{P}]} := C_{[\mathfrak{P}]}^{RC}$, $\mathfrak{P}_{[\mathfrak{P}]} := \mathfrak{P}_{[\mathfrak{P}]}^{RC}$, $A_{[\mathfrak{p}]} := A_{[\mathfrak{p}]}^{R}$, and $\mathfrak{p}_{[\mathfrak{p}]} := \mathfrak{p}_{[\mathfrak{p}]}^{R}$. Then $C_{[\mathfrak{P}]} = C \cdot A_{[\mathfrak{p}]}$ and $C_{[\mathfrak{P}]} \cap R = A_{[\mathfrak{p}]}$. Moreover $\mathfrak{P}_{[\mathfrak{P}]} = C \cdot \mathfrak{p}_{[\mathfrak{p}]}$ and $\mathfrak{P}_{[\mathfrak{P}]} \cap R = \mathfrak{p}_{[\mathfrak{p}]}$. If w is a PM-valuation of RC with $A_w = C_{[\mathfrak{P}]}$, $\mathfrak{p}_w = \mathfrak{P}_{[\mathfrak{P}]}$, and $R \cap C = A$, the restriction $v := w|R$ is a PM-valuation on R with $A_v = A_{[\mathfrak{p}]}$, $\mathfrak{p}_v = \mathfrak{p}_{[\mathfrak{p}]}$. The ideal $\mathfrak{P}$ is RC-regular iff $\mathfrak{p}$ is R-regular.

Proof. a) Let $A' := R \cap C$. The extension $A \subset A'$ is ws and the ideal $\mathfrak{p}' := \mathfrak{p}A'$ is different from A', hence a prime ideal of A' with $\mathfrak{p}' \cap A = \mathfrak{p}$, and $\mathfrak{p}'$ is R-regular iff $\mathfrak{p}$ is R-regular. Also $\mathfrak{p}'C = \mathfrak{p}C = \mathfrak{P}$, and $A'^{R}_{[\mathfrak{p}']} = A_{[\mathfrak{p}]}$ by Theorem I.3.13. Thus we may replace the pair $(A, \mathfrak{p})$ by $(A', \mathfrak{p}')$. We assume henceforth that $R \cap C = A$.

b) Since the extension $A \subset RC$ is ws we have $A_{[\mathfrak{p}]}^{RC} = C_{[\mathfrak{P}]}^{RC}$, again by Theorem I.3.13. Intersecting with R we see that $R \cap C_{[\mathfrak{P}]} = R \cap A_{[\mathfrak{p}]}^{RC} = A_{[\mathfrak{p}]}$. We introduce the notations $\tilde{A} := A_{[\mathfrak{p}]}$, $\tilde{C} := C_{[\mathfrak{P}]}$, $\tilde{\mathfrak{p}} := \mathfrak{p}_{[\mathfrak{p}]}$, $\tilde{\mathfrak{P}} := \mathfrak{P}_{[\mathfrak{P}]}$. Thus $A \subset \tilde{A} \subset R$, $C \subset \tilde{C} \subset RC$, and $\tilde{C} \cap R = \tilde{A}$. The transfer theorem for overrings (Cor.II.6.6) tells us that $\tilde{A}C = \tilde{C}$.

We have a 2-step ladder of ring extensions

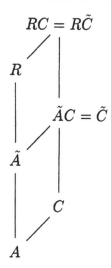

$$RC = R\tilde{C}$$

Here the verticals represent Prüfer extensions and the oblique lines represent ws extensions. The ideal $\mathfrak{p}$ is R-regular iff $\tilde{A} \neq R$ and $\mathfrak{P}$ is RC-regular iff $\tilde{C} \neq RC$, cf. Lemma 1.1. Since $\tilde{C} \cap R = \tilde{A}$ these conditions are equivalent (cf.II.6.6). If $\tilde{A} = R$ then $\tilde{C} = RC$, $\tilde{\mathfrak{p}} = R\mathfrak{p}$, $\tilde{\mathfrak{P}} = RC\mathfrak{P} = C\tilde{\mathfrak{p}}$, $\tilde{\mathfrak{P}} \cap R = \tilde{\mathfrak{p}}$ (cf. Prop.I.4.6 or Th.I.4.8), and the assertions in the theorem about w and v are trivially true. Starting from now we assume that $\tilde{A} \neq R$, hence $\tilde{C} \neq RC$.

c) The pair $(\tilde{A}, \tilde{\mathfrak{p}})$ is PM in R, and $(\tilde{C}, \tilde{\mathfrak{P}})$ is PM in RC. Thus $\tilde{\mathfrak{p}}$ is the unique R-regular maximal ideal of $\tilde{A}$ and $\tilde{\mathfrak{P}}$ is the unique RC-regular maximal ideal of $\tilde{C}$ (cf. Cor.1.4). It is clear by the transfer theorem for regular modules (Th.II.6.5), or by Corollary II.6.7, that $\tilde{\mathfrak{P}} = \tilde{C}\tilde{\mathfrak{p}} = C\tilde{A}\tilde{\mathfrak{p}} = C\tilde{\mathfrak{p}}$ and $\tilde{\mathfrak{P}} \cap R = \tilde{\mathfrak{p}}$. Let w denote "the" PM-valuation on RC with $A_w = \tilde{C}$, (hence) $\mathfrak{p}_w = \tilde{\mathfrak{P}}$, and let $v := w|R$. We have $A_v = R \cap A_w = \tilde{A}$ and $\mathfrak{p}_v = R \cap \tilde{\mathfrak{P}} = \tilde{\mathfrak{p}}$. We want to verify that v is special. Then we will know by Proposition 1 that v is Manis, and everything will be proved. That v is special means that $\operatorname{supp} v$ coincides with the conductor $\mathfrak{q}$ of $\tilde{A}$ in R (Prop.I.2.2). Let $\mathfrak{Q}$ denote the conductor of $\tilde{C}$ in RC. We have $\mathfrak{Q} = \operatorname{supp} w$, hence $R \cap \mathfrak{Q} = \operatorname{supp} v$. Now $\mathfrak{Q}$ is the set of all $x \in \tilde{C}$ with $R\tilde{C}x \subset \tilde{C}$, i.e. $Rx \subset \tilde{C}$. If $x \in R$ then $Rx \subset \tilde{C}$ means that $Rx \subset R \cap \tilde{C} = \tilde{A}$. Thus $\mathfrak{Q} \cap R$ is the conductor $\mathfrak{q}$ of $\tilde{A}$ in R, i.e. $\operatorname{supp} v = \mathfrak{q}$. The theorem is proved.   □

Corollary 11.3. In the situation of Theorem 2 we have $\operatorname{supp} w = (\operatorname{supp} v) \cdot C$.

Proof. Let $\mathfrak{q} := \operatorname{supp} v$, $\Omega := \operatorname{supp} w$. We have $\Omega \cap R = \mathfrak{q}$. Applying Theorem I.4.8 (or Prop. I.4.6) to the ws extension $R \subset RC$, we obtain $\Omega = C\mathfrak{q}$. □

Theorem 11.4. The R-overrings B of $R \cap C$, such that B is PM in R (resp. $R \cap C$ is PM in B), correspond bijectively with the RC-overrings D of C such that D is PM in RC (resp. C is PM in D) via $D = BC$, $B = R \cap D$. If w is a PM-valuation on RC with $A_w = BC$, then the restriction $w|R$ is a PM-valuation v on R with $A_v = B$. If u is a PM-valuation on BC with $A_u = C$, then the restriction $u|B$ is a PM-valuation v on B with $A_v = R \cap C$.

Proof. We may assume in advance that $R \cap C = A$. Now the R-overrings B of A correspond bijectively with the RC-overrings D of C via $D = BC$, $B = R \cap D$ (Cor. II.6.6). The ring A is PM in B iff the rings between A and B form a chain, and the ring C is PM in D iff the rings between C and D form a chain (Th. 3.1). Since the rings between A and B correspond with the rings between C and D by the above bijective correspondence, it is clear that A is PM in B iff C is PM in D. For the same reason B is PM in R iff D is PM in RC. (This can also be read off from Theorem 2.)

Let some B of A in R be given, and let $D := BC$. If there exists a PM-valuation w on RC with $A_w = D$, Theorem 2 tells us that $w|R$ is a PM-valuation on R and $A_{w|R} = B$. If there exists a PM-valuation u on BC, then, applying this to the extensions $A \subset B$, $A \subset C$, and the B-overring A of A, we learn that $u|B$ is a PM-valuation on B and $A_{u|B} = A$. □

Theorem 4 implies a result on PM-hulls (cf. §5 for the definiton of such hulls). Let $A \subset R$ and $A \subset C$ be subextensions of a ring extension $A \subset T$. We still assume that $A \subset C$ is ws but now do not assume that $A \subset R$ is Prüfer.

Corollary 11.5. Let $\mathfrak{p}$ be a prime ideal of A with $\operatorname{PM}(A, \mathfrak{p}, R) \neq A$. Assume that $R \cap C = A$. Then $\mathfrak{p}C \neq C$ and $\operatorname{PM}(A, \mathfrak{p}, R) \cdot C \subset \operatorname{PM}(C, \mathfrak{p}C, RC)$.

Proof. Let $B := \operatorname{PM}(A, \mathfrak{p}, R)$ and $D := BC$. We have $B \cap C = A$. By Theorem 4 the extension $C \subset D$ is PM. Moreover, by that theorem, if w is the Manis valuation on D with $A_w = C$, then $v := w|B$ is the Manis valuation on B with $A_v = A$. We have $\mathfrak{p}_w \cap A = \mathfrak{p}_v = \mathfrak{p}$. Since $A \subset C$ is ws, this implies $\mathfrak{p}_w = \mathfrak{p}C$. Thus $\mathfrak{p}C$ is a prime ideal of C, and $(C, \mathfrak{p}C)$ is PM in D, hence $D \subset \mathrm{PM}(C, \mathfrak{p}C, RC)$. $\square$

We hasten to give an example of a PM-valuation which does not have maximal support, as promised in §3.

Example 11.6. We choose a Prüfer domain A with quotient field K and a valuation w on K with $A_w \supset A$ and of rank one. This means that $\Gamma_w \neq 1$, but Γ_w has no nontrivial proper convex subgroups. We assume that there is given a subring B of K with $A \subset B$ but $A_w \not\subset B$, $B \not\subset A_w$. It is easy to create such a situation. For example, let u and w be valuations of a field K with w of rank 1, u not trivial, and $A_u \not\subset A_w$, i.e. w is not a coarsening of u. Then take $A = A_u \cap A_w$ and $B = A_u$. It is well known that A is Prüfer in K (cf. Th.I.6.10).

Replacing A by $A_w \cap B$ we assume without loss of generality that $A = A_w \cap B$. We have $A \neq B$ and $BA_w = K$, since BA_w is an overring of A_w in K different from A_w. Now Theorem 4 tells us, that the restriction $v := w|B$ of w is a PM valuation on B with $A_v = A$. We have $\operatorname{supp} w = \{0\}$, hence $\operatorname{supp} v = \{0\}$. Suppose that $\operatorname{supp} v$ is a maximal ideal of B. Then B would be a field. Since $B \subset K$ is Prüfer, this forces $B = K$, contradicting our assumption $A_w \not\subset B$. Thus v does not have maximal support. It is clear that $\Gamma_v = \Gamma_w$, since $\operatorname{Quot} B = K$ and $\operatorname{supp} v = \{0\}$. $\square$

Is it possible to obtain a transfer theorem near to Theorem 4 for valuations of weaker type than PM-valuations, say tight valuations? Here we run into difficulties. Anyway, we have the following theorem, where instead of A, C, R only a ws extension $R \subset T$ is given.

Theorem 11.7. Assume that $R \subset T$ is a ws ring extension and $v: R \to \Gamma \cup \infty$ is a valuation with $(\operatorname{supp} v) \cdot T \neq T$.

  i) v extends uniquely to a valuation $w: T \to \Gamma \cup \infty$, and $\Gamma_v = \Gamma_w$, $(\operatorname{supp} v) \cdot T = \operatorname{supp} w$. If v is Manis, then w is Manis.

  ii) Assume that v is tight. Then w is tight, and $I_{\gamma, w} = A_w \cdot I_{\gamma, v}$ for every $\gamma \in \Gamma_v = \Gamma_w$. Also $\mathfrak{p}_w = A_w \cdot \mathfrak{p}_v$. If v is principal then w is principal.

Proof. a) Let $\mathfrak{q} := \operatorname{supp} v$. We have $\mathfrak{q}T \neq T$. Since $R \subset T$ is ws, this implies that $T = R_{[\mathfrak{q}]}^T$, cf. Th.I.3.13. Now Lemma 4.9 tells us that v extends to a valuation $w: T \to \Gamma \cup \infty$ in a unique way, and, looking at the proof of this easy lemma, we see that $\Gamma_v = \Gamma_w$. We have $R \cap \operatorname{supp} w = \mathfrak{q}$. Since $R \subset T$ is ws, this implies $\operatorname{supp} w = T \cdot \mathfrak{q}$ (cf. Prop.I.4.6). Of course, if v is Manis, i.e. $v(R \setminus \mathfrak{q}) = \Gamma_v$, then w is Manis.

b) Let v be tight. We assume without loss of generality that $\Gamma_v = \Gamma_w = \Gamma$. Let $\gamma \in \Gamma$ be given. The A_v-module $I_{\gamma,v}$ is invertible in R. This implies that the A_w-module $A_w \cdot I_{\gamma,v}$ is invertible in $A_w \cdot R$, hence in T. Proposition 4.6 tells us that $A_w \cdot I_{\gamma,v} = I_{\delta,w}$ for some $\delta \in \Gamma$. Of course, $A_w \cdot I_{\gamma,v} \subset I_{\gamma,w}$. Thus $\delta \geq \gamma$. On the other hand, there exists some $x \in A_w \cdot I_{\gamma,v}$, namely $x \in I_{\gamma,v}$, with $w(x) = \gamma$. This forces $\delta = \gamma$. We have proved for every $\gamma \in \Gamma$ that $I_{\gamma,w} = A_w \cdot I_{\gamma,v}$. Since the ideal $\mathfrak{p}_v$ (resp. $\mathfrak{p}_w$) is the union of the ideals $I_{\gamma,v}$ (resp. $I_{\gamma,w}$) with γ running through the positive elements of Γ, it follows that $\mathfrak{p}_w = A_w \cdot \mathfrak{p}_v$. If $I_{\gamma,v}$ is principal for some $\gamma \in \Gamma$, then the same holds for $I_{\gamma,w}$. This gives the last claim. $\qquad\square$

Discussion 11.8. a) In the situation of Theorem 7 let us assume in addition that v is PM. We would like to conclude that w is PM. But this seems to be impossible. We only know (from Theorem 7) that w is tight.

More can be said about the restriction $u := w|RA_w$. This valuation is Manis, since $u(R) = \Gamma_v \cup \infty = \Gamma_w \cup \infty$, hence $u(A_w R) = \Gamma_w \cup \infty$. We have $A_u = A_w$, and the extension $A_w \subset RA_w$ is Prüfer since $A_v \subset R$ is Prüfer (cf. Th.I.5.10). Thus u is PM.

b) We return to a setting similar to the one used in the main body of this section: We are given two subextensions $A \subset R$ and $A \subset C$ of a ring extension $A \subset T$. We assume that $A \subset C$ is ws, but we no longer assume that $A \subset R$ is Prüfer. Let $v: R \to \Gamma \cup \infty$ be a Manis valuation on R and $\mathfrak{q} := \operatorname{supp} v$. We assume that $\mathfrak{q}RC = \mathfrak{q}C \neq RC$. Since $R \subset RC$ is ws (cf. Prop. I.3.10), we know by Theorem 7 that v extends uniquely to a Manis valuation $w: RC \to \Gamma \cup \infty$, and $\Gamma_w = \Gamma_v$, $\operatorname{supp} w = \mathfrak{q}C$.

Now assume in addition that $A_v \supset A$. Let $\mathfrak{p} := \mathfrak{p}_v \cap A$. Assume also that $\mathfrak{p}C \neq C$. Then $C = A_{[\mathfrak{p}]}^C$ (cf. Th. I.3.13), hence $C \subset A_w$. We have $(\mathfrak{p}_w \cap C) \cap A = \mathfrak{p}_w \cap A = \mathfrak{p}_v \cap A = \mathfrak{p}$, and we conclude that $\mathfrak{p}_w \cap C = \mathfrak{p}C$ (cf. Prop. I.4.6).

We also have $A_v C \subset A_w$, and $A_v \subset A_v C$ is again ws. Further $\mathfrak{p}_w \cap A_v = \mathfrak{p}_v$, hence $\mathfrak{p}_v(A_v C) \subset \mathfrak{p}_w \cap (A_v C)$, which is a prime ideal of $A_v C$, hence $\mathfrak{p}_v(A_v C) \neq A_v C$. Replacing A and C by A_v and $A_v C$, we see that $\mathfrak{p}_w \cap (A_v C) = \mathfrak{p}_v(A_v C) = \mathfrak{p}_v C$.

This is all very fine, but one major question remains open: Is $A_v C = A_w$? We have $A_v C \subset A_w$ and $A_w \cap R = A_v$. In the case, that $A_v \subset R$ is Prüfer, i.e. v is PM, we can conclude by our transfer theory in II, §6 from this, that indeed $A_v C = A_w$, but now we do not have such a sharp tool at our disposal. If v is tight, then w is tight by Theorem 7, but even then there seems to be no way to prove that $A_v C = A_w$ without further assumptions. □

We return to PM-valuations. We now prove some sort of generalization of Theorem 4, relaxing there the overall assumption, that the extension $A \subset R$ is Prüfer, to the weaker assumption, that $A \subset R$ is convenient (cf.I, §6 for the definition and examples of convenience). The following theorem also adds details to Theorem 4 in the case that $A \subset R$ Prüfer.

Theorem 11.9. Let $A \subset R$ and $A \subset C$ be subextensions of a ring extension $A \subset T$. Assume that $A \subset R$ is convenient and $A \subset C$ is ws. Then the extension $C \subset RC$ is convenient. The PM-valuations w on RC with $A_w \supset C$ correspond bijectively (up to equivalence) with the PM-valuations v on R with $A_v \supset A$ and $(\mathfrak{p}_v \cap A)C \neq C$, via $v = w|R$ and $A_w = A_v C$. The valuations v and w have the same value group $\Gamma_v = \Gamma_w$, and $\operatorname{supp} w = (\operatorname{supp} v)C$, $\mathfrak{p}_w = \mathfrak{p}_v C$. Also $I_{\gamma,w} = I_{\gamma,v}C$ for every $\gamma \in \Gamma_w$.

Proof. Let D be an overring of C in RC with $D \neq RC$ and $(RC) \setminus D$ closed under multiplication. We want to verify that RC is PM over D. The ring $B := R \cap D$ is an R-overring of A, and $R \setminus B$ is closed under multiplication. Since R is convenient over A the extension $B \subset R$ is PM. We have $BC \subset D$. The extension $B \subset BC$ is ws by Proposition I.3.10, and $B \subset R \cap (BC) \subset R \cap D = B$, hence $B = R \cap (BC)$. Now we know by Theorem 4, applied to the Prüfer extension $B \subset R$ and the ws extension $B \subset BC$, that $BC \subset RC$ is PM, and moreover $BC = D$, since BC and D have the same intersection B with R. Let w be a PM-valuation on RC with $A_w = BC$, and $v := w|R$. Theorem 4 also tells us that v is PM and $A_v = B$. We thus have proved that the extension $C \subset RC$ is convenient, and also, that every PM-valuation w on RC with $C \subset A_w$ restricts to a PM-valuation v on R. Of course $A_v \supset A$. Also $\mathfrak{p}_w \cap A_v = \mathfrak{p}_v$. This implies $\mathfrak{p}_w = \mathfrak{p}_v A_v C = \mathfrak{p}_v C$, since A_v is ws in $A_v C$. In particular, $\mathfrak{p}_v C \neq A_v C$, hence $(\mathfrak{p}_v \cap A)C \neq C$.

Let now a PM-valuation v on R be given with $A \subset A_v$. The extension $A_v \subset A_v C$ is ws. Applying Theorem 4 to the extensions $A_v \subset R$ and $A_v \subset A_v C$, we see that $A_v C$ is PM in RC. We have $A_v C \neq RC$. Let w' be a PM-valuation on RC with $A_{w'} = A_v C$. Further let $B := R \cap A_v C$. As already proved, $B = A_{v'}$ with $v' := w'|R$. We have $A_v \subset A_{v'}$, hence $\operatorname{supp} v \subset \operatorname{supp} v'$. Since $(\operatorname{supp} v')C = \operatorname{supp} w' \neq RC$, it follows that $(\operatorname{supp} v)C \neq RC$.

By Theorem 7.i the valuation v extends uniquely to a Manis valuation w on RC. Assume now in addition that $(\mathfrak{p}_v \cap A)C \neq C$. Running through the discussion 10.9 above, part b, we learn that $A_v C = A_w$ and $\mathfrak{p}_w = \mathfrak{p}_v C$. In particular $A_w \supset C$. Since RC is convenient over C, the valuation w is PM. {We could also argue that $A_v C$ is Prüfer in RC, since A_v is Prüfer in R.} Theorem 7.ii tells us that $I_{\gamma,w} = I_{\gamma,v}C$ for every $\gamma \in \Gamma_v$. {This could also be deduced from Th.II.6.4, since clearly $I_{\gamma,w} \cap R = I_{\gamma,v}$.} $\qquad\square$

Appendix

Appendix A (to I, §4 and I, §5): Flat epimorphisms

In this appendix all morphisms are meant in the category of rings (commutative, with 1, as always). In order to prove Theorem I.4.4 we need some basic facts about epimorphisms.

Let $\varphi\colon A \to B$ be a ring homomorphism. Related to φ we set up the following morphisms.

$$
\begin{array}{llll}
i_1\colon & B \to B \otimes_A B & , & x \mapsto x \otimes 1; \\
i_2\colon & B \to B \otimes_A B & , & x \mapsto 1 \otimes x; \\
m\colon & B \otimes_A B \to B & , & b_1 \otimes b_2 \mapsto b_1 b_2.
\end{array}
$$

Clearly, $m \circ i_1 = m \circ i_2 = id_B$. Notice also that $i_1 = \varphi \otimes_A B$ and $i_2 = B \otimes_A \varphi$.

Lemma A.1 (cf. [L, Lemma 1.0]). The following are equivalent:
(i) φ is an epimorphism.
(ii) $i_1 = i_2$.
(iii) i_1 is an isomorphism.
(iv) m is an isomorphism.

Proof. (i) $\Rightarrow$ (ii): $i_1 \circ \varphi = i_1 \circ \varphi$ implies $i_1 = i_2$.
(ii) $\Rightarrow$ (i): Let $g_1, g_2\colon B \to C$ be two morphisms with $g_1 \circ \varphi = g_2 \circ \varphi$. There is a unique morphism $g_1 \otimes_A g_2\colon B \otimes_A B \to C$ such that $g_k = (g_1 \otimes_A g_2) \circ i_k$ for $k = 1, 2$. Now $i_1 = i_2$ implies $g_1 = g_2$.
(ii) $\Rightarrow$ (iii): From $m \circ i_1 = id_B$ it follows that i_1 is injective. Since $i_1 = i_2$ and $B \otimes_A B$ is generated as a ring by the images of i_1 and i_2, we see that i_1 is surjective. Hence i_1 is an isomorphism.
(iii) $\Rightarrow$ (iv): Since $m \circ i_1 = id_B$ and i_1 is an isomorphism, m is an isomorphism.
(iv) $\Rightarrow$ (ii): Since $m \circ i_1 = id_B = m \circ i_2$ and m is an isomorphism, we have $i_1 = i_2$. $\qquad\square$

Lemma A.2 (cf. [L, Lemma 1.2]). A faithly flat epimorphism is an isomorphism.

Proof. Let $\varphi\colon A \to B$ be a faithfully flat epimorphism, and $K := \operatorname{Ker}\varphi$, $L := \operatorname{Coker}\varphi$. Since φ is flat, we have a natural exact sequence

$$
0 \longrightarrow K \otimes_A B \longrightarrow A \otimes_A B \overset{\varphi \otimes_A B}{\longrightarrow} B \otimes_A B \longrightarrow L \otimes_A B \longrightarrow 0.
$$

Lemma 1, (i) $\Rightarrow$ (iii), tells us that $\varphi \otimes_A B$ is an isomorphism. Thus $K \otimes_A B = 0$, $L \otimes_A B = 0$. Since φ is faithfully flat, we conclude that $K = 0$, $L = 0$. $\square$

Theorem I.4.4 (Lazard, Akiba). An injective homomorphism $\varphi: A \rightarrow B$ is weakly surjective iff φ is a flat epimorphism.

Proof. If φ is ws, we know by Prop.I.3.6, that φ is an epimorphism, and by Prop.I.4.1, that φ is flat.

Conversely, let φ be a flat epimorphism. Given a prime ideal $\mathfrak{p}$ of A with $\mathfrak{p}B \neq B$, we have to verify that $\varphi_\mathfrak{p}: A_\mathfrak{p} \rightarrow B_\mathfrak{p}$ is injective according to the definition of weak surjectivity in I, §3. It is easy to see that $\varphi_\mathfrak{p}: A_\mathfrak{p} \rightarrow B_\mathfrak{p}$ is a flat epimorphism. Since $\mathfrak{p}B \neq B$, it follows that $\varphi_\mathfrak{p}$ is faithfully flat [Bo, I §3, Prop.8]. Lemma 2 gives us that $\varphi_\mathfrak{p}$ is an isomorphism. $\square$

The following proposition is needed in the proof of Theorem I.5.2, (11) $\Rightarrow$ (4).

Proposition A.3 [L, Prop.1.7]. A finite epimorphism is surjective.

Proof. Let $\varphi: A \rightarrow B$ be a finite epimorphism. Let $\mathfrak{p}$ be a prime ideal of A and let $k(\mathfrak{p})$ denote the residue class field $\mathrm{Quot}(A/\mathfrak{p})$. One verifies in a straightforward way that $\varphi \otimes_A k(\mathfrak{p}): k(\mathfrak{p}) \rightarrow B \otimes_A k(\mathfrak{p})$ is an epimorphism. Since $k(\mathfrak{p})$ is a field, this epimorphism is faithfully flat [Bo, loc.cit.]. Lemma 2 tells us that $\varphi \otimes_A k(\mathfrak{p})$ is an isomorphism. Since φ is finite, it now follows by Nakayama's lemma that $\varphi_\mathfrak{p}: A_\mathfrak{p} \rightarrow B_\mathfrak{p}$ is surjective. Since this holds for every prime ideal $\mathfrak{p}$ of A, φ is surjective [Bo, Chap.II, §3]. $\square$

Appendix B (to II, §2): Arithmetical rings

In the following A is a ring (always commutative, with 1) and $\mathcal{J}(A)$ denotes the set of ideals of A, partially ordered by inclusion. $\mathcal{J}(A)$ is a lattice. The ring A is called *arithmetical*, if the lattice $\mathcal{J}(A)$ is distributive, i.e. $(\mathfrak{a} + \mathfrak{b}) \cap \mathfrak{c} = (\mathfrak{a} \cap \mathfrak{c}) + (\mathfrak{b} \cap \mathfrak{c})$, or equivalently,

$\mathfrak{a} + (\mathfrak{b} \cap \mathfrak{c}) = (\mathfrak{a} + \mathfrak{b}) \cap (\mathfrak{a} + \mathfrak{c})$ for any three ideals $\mathfrak{a}, \mathfrak{b}, \mathfrak{c}$ of A. We reproduce some results of C.U. Jensen $[J_1]$.

Theorem B.1 $[J_1$, Th.1$]$. A is arithmetical iff, for every maximal ideal $\mathfrak{p}$ of A, the set $\mathcal{J}(A_\mathfrak{p})$ is totally ordered.

Proof. If $\mathcal{J}(A_\mathfrak{p})$ is totally ordered then, of course, the lattice $\mathcal{J}(A_\mathfrak{p})$ is distributive. If this holds for every $\mathfrak{p} \in \operatorname{Max} A$ then clearly $\mathcal{J}(A)$ is distributive.

Assume now that the lattice $\mathcal{J}(A)$ is distributive. Let $\mathfrak{p} \in \operatorname{Max} A$ be given. Also $\mathcal{J}(A_\mathfrak{p})$ is distributive. Thus we may assume without loss of generality that A is a local ring. We verify in this case that for any two elements a, b of A either $a|b$ or $b|a$. Then it will follow that $\mathcal{J}(A)$ is totally ordered. {Indeed, if $I, J \in \mathcal{J}(A)$ and $I \not\subset J$, then choosing $x \in I \setminus J$, we have $y \nmid x$ for every $y \in J$, hence $Ay \subset Ax$ for every $y \in J$, and $J \subset Ax \subset I$.} We have

$$Aa = Aa \cap (Ab + A(a - b)) = (Aa \cap Ab) + (Aa \cap A(a - b)).$$

Thus $a = t + (a-b)c$ with $t \in Aa \cap Ab$ and $(a-b)c \in Aa$, i.e. $bc \in Aa$. If $c \in A^*$ then it follows that $b \in Aa$. Otherwise $1 - c \in A^*$, and we conclude from the equation $a(1 - c) = t - bc$ that $a \in Ab$. $\qquad\square$

Corollary B.2. If A is arithmetical and $\mathfrak{p}$ is a prime ideal of A then the set of generalizations of $\mathfrak{p}$ in $\operatorname{Spec} A$ is totally ordered. $\qquad\square$

Corollary B.3. If A is arithmetical, and $\mathfrak{p}_1, \mathfrak{p}_2$ are prime ideals of A with $\mathfrak{p}_1 \not\subset \mathfrak{p}_2$ and $\mathfrak{p}_2 \not\subset \mathfrak{p}_1$, then $A = \mathfrak{p}_1 + \mathfrak{p}_2$. $\qquad\square$

Theorem B.4 $[J_1$, Th.2$]$. The following are equivalent:
(1) A is arithmetical.
(2) Given ideals $\mathfrak{a} \subset \mathfrak{b}$ of A with $\mathfrak{b}$ finitely generated, there exists an ideal $\mathfrak{c}$ of A such that $\mathfrak{a} = \mathfrak{b}\mathfrak{c}$.[*]
(3) Ditto for ideals $\mathfrak{a} \subset \mathfrak{b}$, with $\mathfrak{a}$ principal and $\mathfrak{b}$ generated by two elements.

Proof. $(1) \Rightarrow (2)$: We verify that $\mathfrak{a} = \mathfrak{b}(\mathfrak{a}: \mathfrak{b})$. Since $\mathfrak{b}$ is finitely generated, it suffices to verify, for any $\mathfrak{p} \in \operatorname{Max} A$, that $\mathfrak{a}_\mathfrak{p} = \mathfrak{b}_\mathfrak{p}(\mathfrak{a}_\mathfrak{p}: \mathfrak{b}_\mathfrak{p})$, cf.

[*] i.e. $\mathfrak{b}$ is a multiplication ideal (II, §2, Def.2).

Lemma II.1.1. Since $\mathcal{J}(A_{\mathfrak{p}})$ is totally ordered (Th.1 above) and $\mathfrak{b}_{\mathfrak{p}}$ is finitely generated, the ideal $\mathfrak{b}_{\mathfrak{p}}$ is principal, $\mathfrak{b}_{\mathfrak{p}} = A_{\mathfrak{p}}x$. From $\mathfrak{a}_{\mathfrak{p}} \subset \mathfrak{b}_{\mathfrak{p}}$ we conclude that $\mathfrak{a}_{\mathfrak{p}} = Ix = I\mathfrak{b}_{\mathfrak{p}}$ with $I = (\mathfrak{a}_{\mathfrak{p}}: A_{\mathfrak{p}}x) = (\mathfrak{a}_{\mathfrak{p}}: \mathfrak{b}_{\mathfrak{p}})$.
$(2) \Rightarrow (3)$: trivial.
$(3) \Rightarrow (1)$: For every $\mathfrak{p} \in \mathrm{Max}A$ the property (3) is inherited by the ring $A_{\mathfrak{p}}$. Thus we may assume that A is local with maximal ideal $\mathfrak{m}$, and we have to prove that $\mathcal{J}(A)$ is totally ordered. It suffices to very for any two elements a, b' of A that $a|b'$ or $b'|a$. Let $\mathfrak{a} = Aa$ and $\mathfrak{b} = Aa + Ab'$. By assumption (3) we have $\mathfrak{a} = \mathfrak{b}\mathfrak{c}$, i.e. $Aa = \mathfrak{a}\mathfrak{c} + b'\mathfrak{c}$, with $\mathfrak{c} := (\mathfrak{a}: \mathfrak{b})$. This means, there exist elements x, y in A with $a = ax + b'y$, $b'x \in Aa$, $b'y \in Aa$. If $x \in A^{*}$ then $Ab' \subset Aa$. Otherwise $x \in \mathfrak{m}$, and we conclude from $a(1-x) = b'y$ that $Aa \subset Ab'$.
□

Theorem B.5 [J_1, Th.3]. The following are equivalent:
(1) A is arithmetical.
(2) If $\mathfrak{a}, \mathfrak{b}, \mathfrak{c}$ are ideals of A, and $\mathfrak{c}$ is finitely generated, then
$\quad (\mathfrak{a} + \mathfrak{b}): \mathfrak{c} = (\mathfrak{a}: \mathfrak{c}) + (\mathfrak{b}: \mathfrak{c})$.
$(2')$ Ditto for $\mathfrak{a}, \mathfrak{b}$ principal ideals and $\mathfrak{c} = \mathfrak{a} + \mathfrak{b}$.
(3) If $\mathfrak{a}, \mathfrak{b}, \mathfrak{c}$ are ideals of A with $\mathfrak{a}$ and $\mathfrak{b}$ finitely generated, then

$$\mathfrak{c}: (\mathfrak{a} \cap \mathfrak{b}) = (\mathfrak{c}: \mathfrak{a}) + (\mathfrak{c}: \mathfrak{b}).$$

$(3')$ Ditto for $\mathfrak{a}, \mathfrak{b}$ principal ideals and $\mathfrak{c} = \mathfrak{a} \cap \mathfrak{b}$.

Proof. We constantly use Lemma II.1.1.
$(1) \Rightarrow (2)$: We may assume that the ring A is local, replacing A by $A_{\mathfrak{p}}$ for any prime (or maximal) ideal $\mathfrak{p}$ of A. Now the set $\mathcal{J}(A)$ is totally ordered (cf. Th.B.1). Thus $\mathfrak{a} \subset \mathfrak{b}$ or $\mathfrak{b} \subset \mathfrak{a}$, and (2) is obvious.
$(1) \Rightarrow (3)$: We proceed exactly as at the end of the proof of Theorem II.1.4. Let $\mathfrak{p} \in \mathrm{Spec}\,A$ be given. Then $(\mathfrak{c}: \mathfrak{a} \cap \mathfrak{b})_{\mathfrak{p}} \subset (\mathfrak{c}_{\mathfrak{p}}: \mathfrak{a}_{\mathfrak{p}} \cap \mathfrak{b}_{\mathfrak{p}})$. We have $\mathfrak{a}_{\mathfrak{p}} \subset \mathfrak{b}_{\mathfrak{p}}$ or $\mathfrak{b}_{\mathfrak{p}} \subset \mathfrak{a}_{\mathfrak{p}}$. Thus certainly $(\mathfrak{c}_{\mathfrak{p}}: \mathfrak{a}_{\mathfrak{p}} \cap \mathfrak{b}_{\mathfrak{p}}) = (\mathfrak{c}_{\mathfrak{p}}: \mathfrak{a}_{\mathfrak{p}}) + (\mathfrak{c}_{\mathfrak{p}}: \mathfrak{b}_{\mathfrak{p}}) = [(\mathfrak{c}: \mathfrak{a}) + (\mathfrak{c}: \mathfrak{b})]_{\mathfrak{p}}$. This proves $(\mathfrak{c}: \mathfrak{a} \cap \mathfrak{b}) \subset (\mathfrak{c}: \mathfrak{a}) + (\mathfrak{c}: \mathfrak{b})$. The reverse inclusion is trivial.
$(2') \Rightarrow (1)$: Again we may assume that A is local. Let a and $b \in A$ be given. We verify that $Aa \subset Ab$ or $Ab \subset Aa$. Then we will know that A is arithmetical. We have $(Aa: Aa+Ab) = (Aa: Ab)$, $(Ab: Aa+Ab) = (Ab: Aa)$. Thus, by assumption $(2')$, $A = (Aa: Ab) + (Ab: Aa)$. We have elements $x, y \in A$ with $bx \in Aa$, $ay \in Ab$ and $x + y = 1$. If $x \in A^{*}$ then $Ab \subset Aa$. Otherwise $y \in A^{*}$ and $Aa \subset Ab$.

$(3') \Rightarrow (1)$: Let $a, b \in A$ be given and $\mathfrak{c} := (Aa) \cap (Ab)$. By assumption $A = (\mathfrak{c}: Aa) + (\mathfrak{c}: Ab) = (Ab: Aa) + (Aa: Ab)$. Let $\mathfrak{p} \in \operatorname{Spec} A$ be given. Then we conclude, as before, that in the ring $A_{\mathfrak{p}}$ either $A_{\mathfrak{p}} \cdot \frac{b}{1} \subset A_{\mathfrak{p}} \cdot \frac{a}{1}$ or $A_{\mathfrak{p}} \cdot \frac{a}{1} \subset A_{\mathfrak{p}} \cdot \frac{b}{1}$. This implies that A is arithmetical (cf. Th.B.1). $\qquad\square$

Appendix C (to III, §6): A direct proof of the existence of Manis valuation hulls

Let $A \subset R$ be a ring extension with $A \neq R$ and $\mathfrak{p}$ a prime ideal of A. Assume that the pair $(A, \mathfrak{p})$ is saturated in R (cf. §5, Def. 4). We prove the following proposition in a pedantic way but directly, i.e. without using anything from the theory of Prüfer extensions.

Proposition C.1. Let U denote the set of all $x \in R \setminus A$ such that for every $a \in A$ with $ax \notin A$ there exists some $a' \in A$ with $a'ax \in A \setminus \mathfrak{p}$. The set $C := A \cup U$ is a subring of R, and the pair $(A, \mathfrak{p})$ is Manis in C.

Proof. Let elements x_1 and x_2 of C be given.

a) We verify that $x_1 x_2 \in C$. We may assume in advance that $x_1 x_2 \notin A$. Let $a \in A$ be given with $ax_1 x_2 \notin A$. We have to find an element $a' \in A$ with $a'ax_1 x_2 \in A \setminus \mathfrak{p}$.

Case 1. $x_1 \notin A$, $x_2 \notin A$, $ax_1 \notin A$. We choose $b \in A$ with $bax_1 \in A \setminus \mathfrak{p}$. Then $(bax_1)x_2 \notin A$, since $A_{[\mathfrak{p}]} = A$. We choose $c \in A$ with $c(bax_1)x_2 \in A \setminus \mathfrak{p}$. The element $a' := bc$ lies in A and $a'ax_1 x_2 \in A \setminus \mathfrak{p}$.

Case 2. $x_1 \notin A$, $x_2 \notin A$, but $ax_1 \in A$. Since $x_2 \in U$ there exists $a' \in A$ with $a'(ax_1)x_2 \in A \setminus \mathfrak{p}$.

Case 3. $x_1 \in A$, $x_2 \notin A$. Since $x_2 \in U$ there exists $a' \in A$ with $a'(ax_1)x_2 \in A \setminus \mathfrak{p}$.

Consideration of these cases suffices to see that $x_1 x_2 \in C$.

b) We verify that $x_1 + x_2 \in C$. We may assume in advance that $x_1 + x_2 \notin A$. Let $a \in A$ be given with $a(x_1 + x_2) \notin A$. We have to find an element a' of A with $a'a(x_1 + x_2) \in A \setminus \mathfrak{p}$.

Case 1. $ax_1 \notin A$, $ax_2 \in A$. We choose $a' \in A$ with $a'ax_1 \in A \setminus \mathfrak{p}$. Since $A_{[\mathfrak{p}]} = A$ we have $a' \in \mathfrak{p}$, and $a'a(x_1 + x_2) = a'ax_1 + a'ax_2 \in A \setminus \mathfrak{p}$.

Case 2. $ax_1 \notin A$, $ax_2 \notin A$. We choose $b \in A$ with $bax_1 \in A \setminus \mathfrak{p}$.

Subcase 2a. $bax_2 \notin A$. We are in Case 1 for the elements x_2, x_1, ba instead of x_1, x_2, a. Thus there exists $c \in A$ with $c(ba)(x_2 + x_1) \in A \setminus \mathfrak{p}$. The element $a' := bc$ does the job.

Subcase 2b. $bax_2 \in A$. We have $(bax_1)(ax_1 + ax_2) = cax_1$ with $c := bax_1 + bax_2 \in A$. Now $bax_1 \in A \setminus \mathfrak{p}$ and $ax_1 + ax_2 \notin A$. Thus $cax_1 \notin A$. We choose $d \in A$ with $dcax_1 \in A \setminus \mathfrak{p}$, i.e. $(dbax_1)(ax_1 + ax_2) \in A \setminus \mathfrak{p}$. The element $a' := dbax_1$ does the job.

Consideration of these cases suffices to see that $x_1 + x_2 \in C$.

c) We now know that C is a subring of R. Let $x \in U = C \setminus A$ be given. There exists $x' \in A$ with $x'x \in A \setminus \mathfrak{p}$. Since $A_{[\mathfrak{p}]} = A$ we have $x' \in \mathfrak{p}$. This proves that $(A, \mathfrak{p})$ is Manis in C (cf. Th.I.2.4). $\qquad \square$

It is evident (again by Th.I.2.4) that C contains every R-overring B of A with $(A, \mathfrak{p})$ Manis in B. Thus we have constructed the Manis valuation hull of $(A, \mathfrak{p})$ and also obtained anew the description (A) of this hull in Corollary 6.2. {Recall that the equivalence (A) $\Leftrightarrow$ (B) there is a straightforward matter.}

References

[A]   T. Akiba, *Remarks on generalized rings of quotients, III.* J. Math. Kyoto Univ. 9-2 (1969), pp. 205–212.

[Al-M]   J. Alajbegović, J. Močkoř, Approximation theorems in commutative algebra. Kluwer Acad. Publ. Dordrecht (1992).

[AM]   M.F. Atiyah, I.G. Macdonald, *Introduction to commutative algebra.* Addison-Wesley 1969.

[An]   D. D. Anderson, *Multiplication ideals, multiplication rings and the ring $R(X)$.* Can. J. Math 28 (1976), 760–768.

[AP]   D. D. Anderson, J. Pascal, *Characterizing Prüfer rings via their regular ideals.* Comm. in Algebra 15(6) (1987), 1287–1295.

[AB]   J. T. Arnold, J. W. Brewer, *On flat overrings, ideal transforms and generalized transforms of a commutative ring.* J. Algebra 18 (1971), 254–263.

[Ba_1] A.D. Barnard, *Multiplication modules.* J. Algebra 71 (1981), 174-178.

[Ba_2] A.D. Barnard, *Distributive extensions of modules.* J. Algebra 70 (1981), 303-315.

[B]   E. Becker, *Valuations and real places in the theory of formally real fields.* Lecture Notes in Math. 959, pp. 1–40, Springer-Verlag, 1982.

[B_1]   E. Becker, *On the real spectrum of a ring and its applications to semialgebraic geometry.* Bull. Amer. Math. Soc. 15 (1986), pp. 19–60.

[B_2]   E. Becker, *Partial orders on a field and valuation rings.* Comm. Algebra 7(18) (1979), pp. 1933–1976.

[B_3]   E. Becker, *Theory of real fields and sums of powers.* forthcoming book.

[B_4]   E. Becker, *The real holomorphy ring and sums of 2n-th powers.* Lecture Notes in Math. 959, pp. 139–181, Springer-Verlag, New York, 1982.

[BP]   E. Becker, V. Powers, *Sums of powers in rings and the real holomorphy ring.* J. reine angew. Math. 480 (1996), 71–103.

[Be]   R. Berr, *Basic principles for a morphological theory of semialgebraic morphisms.* Habilitationsschrift, Univ. Dortmund 1994.

[BKW]   A. Bigard, K. Keimel, S. Wolfenstein, *Groupes et anneaux réticulés.* Lecture Notes Math. 608, Springer 1977.

[BCR]   J. Bochnak, M. Coste, M.-F. Roy, *Real algebraic geometry.* Ergeb. Math. Grenzgeb. 3. Folge, Band 36, Springer 1998.

[BL]   M.B. Boisen, M.D. Larsen, *Prüfer and valuation rings with zero divisors.* Pac. J. Math 49 (1972), 7–12.

[Bo]   N. Bourbaki, *Algèbre commutative.* Chap.1-7, Hermann Paris, 1961–1965.

[Br]   G. Brumfiel, *Partially ordered rings and semialgebraic geometry.* London Math. Soc. Lecture Notes 37, Cambridge Univ. Press 1979.

[BS]   L. Bröcker, J.-H. Schinke, *On the L-adic spectrum.* Schriften Math. Inst. Univ. Münster, 2. Serie, Heft 40 (1986).

[CE]   H. Cartan, S. Eilenberg, *Homological algebra.* Princeton Univ. Press, Princeton 1956.

[Da]   E.D. Davis, *Overrings of commutative rings III: Normal pairs.* Trans. Amer. Math. Soc. 182 (1973), pp. 175–185.

[Dvs]   T.M.K. Davison, *Distributive homomorphisms of rings and modules.* J. reine angew. Math. 271 (1974), 28-34.

[D]   A. Dress, *Lotschnittebenen mit halbierbarem rechten Winkel.* Arch. Math. 16 (1965), pp. 388–392.

[Eg]   N. Eggert, *Rings whose overrings are integrally closed in their complete quotient ring.* J. Reine Angew. Math. 282 (1976), pp. 88–95.

[E]   O. Endler, *Valuation theory.* Springer-Verlag, 1972.

[Er]   V. Erdogdu, *The distributive hull of a ring.* J. of Algebra 132 (1990), 263-269.

[FHP]   M. Fontana, J.A. Huckaba, I.J. Papick, *Prüfer domains.* Marcel Dekker, New York 1997.

[F]   P. A. Froeschl, *Chained rings.* Pacific J. Math. 65 (1976), 47–53.

[Fu]   L. Fuchs, *Über die Ideale arithmetischer Ringe.* Comment. Math. Helv. 23 (1949), 334–341.

[GJ]   L. Gillman, J. Jerison, *Rings of continuous functions*. Van Nostrand 1960. Reprint Springer 1976.

[GH]   R. Gilmer, J. Huckaba, Δ-*rings*. J. Algebra 28 (1974), 414–432.

[GH$_1$] R. Gilmer, J.A. Huckaba, *Maximal overrings of an integral domain not containing a given element*. Commun. Alg. 2(5) (1974), 377-401.

[Gi]   R. Gilmer, *Multiplicative ideal theory*. Marcel Dekker, New York, 1972.

[Gi$_1$] R. Gilmer, *Two constructions of Prüfer domains*. J. Reine Angew. Math. 239/240 (1970), 153–162.

[Gr]   J. Gräter, *Der allgemeine Approximationssatz für Manisbewertungen*. Mh. Math. 93 (1982), 277–288.

[Gr$_1$] J. Gräter, *Der Approximationssatz für Manisbewertungen*. Arch. Math. 37 (1981), 335–340.

[Gr$_2$] J. Gräter, *R-Prüferringe und Bewertungsfamilien*. Arch. Math. 41 (1983), 319–327.

[Gr$_3$] J. Gräter, *Über die Distributivität des Idealverbandes eines kommutativen Ringes*. Mh. Math 99 (1985), 267–278.

[G$_1$]   M. Griffin, *Prüfer rings with zero divisors*. J. Reine Angew. Math. 239/240 (1970), 55–67.

[G$_2$]   M. Griffin, *Valuations and Prüfer rings*. Canad. J. Math. 26 (1974), 412–429.

[HV]   D. K. Harrison, M. A. Vitulli, *V-valuations of a commutative ring*. J. Algebra 126 (1989), 264–292.

[Ho]   M. Hochster, *Prime ideal structure in commutative rings*. Trans. Amer. Math. Soc. 142 (1969), 43–60.

[Hu$_1$] R. Huber, *Bewertungsspektrum und rigide Geometrie*. Habilitationsschrift, Univ. Regensburg, 1990.

[Hu$_2$] R. Huber, *Continuous valuations*. Math. Z. 212 (1993), 455–477.

[Hu$_3$] R. Huber, *Semirigide Funktionen*. Preprint Univ. Regensburg 1990.

[HK]   R. Huber, M. Knebusch, *On valuation spectra*. Contemporary Mathematics 155 (1994), 167-206.

[Huc]   J. A. Huckaba, *Commutative rings with Zero Divisors*. Marcel Dekker, New York, 1988.

[J$_1$]   C. U. Jensen, *Arithmetical rings*. Acta Sci. Acad. Hungar. 17 (1966), 115–123.

[J$_2$]   C. U. Jensen, *A remark on the distributive law for an ideal in a commutative ring*. Proc. Glasgow Math. Assoc. 7 (1966), 193–198.

[K]   M. Knebusch, *Real closures of commutative ring I*. J. Reine & Angew. Math. 274/275, 61–89.

[K$_1$]   M. Knebusch, *Isoalgebraic geometry: First steps*. In: Seminaire de Theorie des Nombres Paris 1980-81 (M.-J. Bertin, ed.), Birkhäuser, Progress in Math. 22 (1982), pp.127-141.

[KS]   M. Knebusch, C. Scheiderer, *Einführung in die reelle Algebra*. Vieweg, 1989.

[KZ]   M. Knebusch, D. Zhang, *Manis valuations and Prüfer extensions* I, Doc. Math. J. DMV 1 (1996), 149-197.

[Ko]   S. Kochen, *Integer valued rational functions over the p-adic numbers: A p-adic analogue of the theory of real fields*. Proc. Symp. Pure Math., vol. XII, Number Theory (1976), pp. 57–73.

[La]   T. Y. Lam, *Orderings, valuations and quadratic forms*. CBMS Regional Conference Series in Math. vol. 52, Amer. Math. Soc. 1983.

[Lb]   J. Lambek, *Lectures on rings and modules*. Waltham-Toronto-London (1966).

[LM]   M. Larsen, P. McCarthy, *Multiplicative theory of ideals*. Academic Press, New York and London, 1971.

[L]   D. Lazard, *Autour de la platitude*. Bull. Soc. Math. France 97 (1969), pp. 81–128.

[Lo$_1$]   A. Loper, *On rings without a certain divisibility property*. J. Number Theory 28 (1988), 132–144.

[Lo$_2$]   A. Loper, *On Prüfer non-D-rings*. J. Pure and Applied Algebra 96 (1994), 271–278.

[LS]   G.M. Low, P.F. Smith, *Multiplication modules and ideals*. Commun. Alg. 18 (12) (1990), 4353–4375.

[M]   M. E. Manis, *Valuations on a commutative ring*. Proc. Amer. Math. Soc 20 (1969), pp. 193–198.

[Ma₁]   J. Marot, *Une généralisation de la notion d'anneau de valuation*. C.R. Acad. Sci. Paris 268 (1969), 1451-1454.

[Ma₂]   J. Marot, *Une extension de la notion d'anneau de valuation*. In: Thèse Univ. Paris-Sud, Centre d'Orsay N° 1859 (1977), Part B, pp. 1-22.

[Mar]   M. Marshall, *Orderings and real places on commutative rings*. J. Algebra 140 (1991), 484–501.

[Mat]   H. Matsumura, *Commutative algebra*. Benjamin publishing company, 1970.

[McA]   S. McAdam, *Two conductor theorems*. J. Algebra 23 (1972), 239–240.

[Pe]   R.L. Pendleton, *A characterization of Q-domains*. Bull. AMS 72 (1966), 499–500.

[P]   V. Powers, *Valuations and higher level orders in commutative rings*. J. Algebra 172 (1995), 255–272.

[Pt]   M. Prechtel, *Universelle Vervollständigungen in der Kategorie der reell abgeschlossen Räume*. Dissertation der Univ. Regensburg, 1992.

[PR]   A. Prestel, P. Roquette, *Formally p-adic fields*. Lecture Notes in Math. 1050, Springer-Verlag, 1984.

[Pu]   M.J. de la Puente Muñoz, *Riemann surfaces of a ring and compactifications of semi-algebraic sets*. Thesis, Stanford University, 1988.

[Re]   R. Remmert, *Funktionentheorie 2, 2*. Auflage. Springer-Verlag 1995.

[Rh]   C.P.L. Rhodes, *Relative Prüfer pairs of commutative rings*. Commun. in Algebra 19 (12), 3423–3445 (1991).

[Rib]   P. Ribenboim, *Théorie des valuations*. Univ. of Montreal Press, Montreal, 1964.

[Ri]   F. Richman, *Generalized quotient rings*. Proc. Amer. Math. Soc. 16 (1965), 794–799.

[R]   P. Roquette, *Principal ideal theorems for holomorphy rings in fields*. J. Reine & Angew. Math. 262/263 (1973), 361–374.

[R₁] P. Roquette, *Bemerkungen zur Theorie der formal p-adischen Körper*. Beitr. z. Algebra und Geometrie 1 (1971), 177–193.

[Sa] P. Samuel, *La notion de place dans un anneau*. Bull. Soc. Math. France 85 (1957), pp. 123–133.

[Sa₁] Séminaire d'algebre commutative dirigé par P. Samuel, *Les épimorphismes d'anneaux*. 1967/68.

[S] H.-W. Schülting, *Real holomorphy rings in real algebraic geometry*. Lecture Notes in Math. 959, pp. 433–442, Springer-Verlag, 1982.

[Sch] N. Schwartz, *The basic theory of real closed spaces*. Regensburger Math. Schriften 15, 1987.

[Sch₁] N. Schwartz, *The basic theory of real closed spaces*. Memoirs Amer. Math. Soc. 397 (1989).

[Sch₂] N. Schwartz, *Epimorphic extensions and Prüfer extensions of partially ordered rings*. manuscr. math. 102 (2000), 347-381.

[Sch₃] N. Schwartz, *Rings of continuous functions as real closed rings*. In: W.C. Holland, J. Martinez (eds), Ordered algebraic structures, Kluwer Acad. Publ. 1997, pp.277-313.

[Sch₄] N. Schwartz, *Letter to the authors*. 8.8.1995.

[St] H. H. Storrer, *Epimorphismen von kommutativen Ringen*. Comment. Math. Helv. 43 (1968), 378–401.

[St₁] H.H. Storrer, *A characterization of Prüfer domains*. Canad. Bull. Math. 12 (1969), 809–812.

Index

Subject Index

Symbol Index

Druck: Strauss Offsetdruck, Mörlenbach
Verarbeitung: Schäffer, Grünstadt

Recent Reprints and New Editions

4. Lecture Notes are printed by photo-offset from the master-copy delivered in camera-ready form by the authors. Springer-Verlag provides technical instructions for the preparation of manuscripts. Macro packages in T_EX, L^AT_EX2e, $L^AT_EX2.09$ are available from Springer's web-pages at

http://www.springer.de/math/authors/b-tex.html.

Careful preparation of the manuscripts will help keep production time short and ensure satisfactory appearance of the finished book.

The actual production of a Lecture Notes volume takes approximately 12 weeks.

5. Authors receive a total of 50 free copies of their volume, but no royalties. They are entitled to a discount of 33.3% on the price of Springer books purchase for their personal use, if ordering directly from Springer-Verlag.

Commitment to publish is made by letter of intent rather than by signing a formal contract. Springer-Verlag secures the copyright for each volume. Authors are free to reuse material contained in their LNM volumes in later publications: A brief written (or e-mail) request for formal permission is sufficient.

Addresses:

Professor Jean-Michel Morel
CMLA, École Normale Supérieure de Cachan
61 Avenue du Président Wilson
94235 Cachan Cedex France
e-mail: Jean-Michel.Morel@cmla.ens-cachan.fr

Professor Bernard Teissier
Institut de Mathématiques de Jussieu
Equipe "Géométrie et Dynamique"
175 rue du Chevaleret
75013 PARIS
e-mail: Teissier@ens.fr

Professor F. Takens, Mathematisch Instituut
Rijksuniversiteit Groningen, Postbus 800
9700 AV Groningen, The Netherlands
e-mail: F.Takens@math.rug.nl

Springer-Verlag, Mathematics Editorial, Tiergartenstr. 17
D-69121 Heidelberg, Germany
Tel.: +49 (6221) 487-701
Fax: +49 (6221) 487-355
e-mail: lnm@Springer.de

4. Lecture Notes are printed by photo-offset from the master-copy delivered in camera-ready form by the authors. Springer-Verlag provides technical instructions for the preparation of manuscripts. Macro packages in T_EX, L^AT_EX2e, $L^AT_EX2.09$ are available from Springer's web-pages at

http://www.springer.de/math/authors/b-tex.html.

Careful preparation of the manuscripts will help keep production time short and ensure satisfactory appearance of the finished book.

The actual production of a Lecture Notes volume takes approximately 12 weeks.

5. Authors receive a total of 50 free copies of their volume, but no royalties. They are entitled to a discount of 33.3 % on the price of Springer books purchase for their personal use, if ordering directly from Springer-Verlag.

Commitment to publish is made by letter of intent rather than by signing a formal contract. Springer-Verlag secures the copyright for each volume. Authors are free to reuse material contained in their LNM volumes in later publications: A brief written (or e-mail) request for formal permission is sufficient.

Addresses:

Professor Jean-Michel Morel
CMLA, École Normale Supérieure de Cachan
61 Avenue du Président Wilson
94235 Cachan Cedex France
e-mail: Jean-Michel.Morel@cmla.ens-cachan.fr

Professor Bernard Teissier
Institut de Mathématiques de Jussieu
Equipe "Géométrie et Dynamique"
175 rue du Chevaleret
75013 PARIS
e-mail: Teissier@ens.fr

Professor F. Takens, Mathematisch Instituut
Rijksuniversiteit Groningen, Postbus 800
9700 AV Groningen, The Netherlands
e-mail: F.Takens@math.rug.nl

Springer-Verlag, Mathematics Editorial, Tiergartenstr. 17
D-69121 Heidelberg, Germany
Tel.: +49 (6221) 487-701
Fax: +49 (6221) 487-355
e-mail: lnm@Springer.de